D0208010

Introduction to Scanning Tunneling Microscopy

OXFORD SERIES IN OPTICAL AND IMAGING SCIENCES

EDITORS
MARSHALL LAPP
JUN-ICHI NISHIZAWA
BENJAMIN B. SNAVELY
HENRY STARK
ANDREW C. TAM
TONY WILSON

Introduction
to Scanning Tunneling
Microscopy

C. JULIAN CHEN

IBM Research Division
Thomas J. Watson Research Center
Yorktown Heights, New York

New York   Oxford
OXFORD UNIVERSITY PRESS
1993

Oxford University Press

Oxford   New York   Toronto
Delhi   Bombay   Calcutta   Madras   Karachi
Kuala Lumpur   Singapore   Hong Kong   Tokyo
Nairobi   Dar es Salaam   Cape Town
Melbourne   Auckland   Madrid

and associated companies in
Berlin   Ibadan

Library of Congress Cataloging-in-Publication Data
Chen, C. Julian.
Introduction to scanning tunneling microscopy / C. Julian Chen.
p. cm.   (Oxford series in optical and imaging sciences ; 4)
Includes bibliographical references (p.   ) and index.
ISBN 0-19-507150-6
1. Scanning tunneling microscopy.
I. Title.   II. Series.
QH212.S35C44   1993
502'.8'2—dc20   92-40047

Printed in the United States of America
on acid-free paper

3   5   7   9   8   6   4   2

PREFACE

It has been more than 10 years since the scanning tunneling microscope (STM) made its debut by resolving the structure of Si(111)-7×7 in real space (Binnig, Rohrer, Gerber, and Weibel, 1983). This new instrument has proved to be an extremely powerful tool for many disciplines in condensed-matter physics, chemistry, and biology. The STM can resolve local electronic structure at an atomic scale on literally every kind of conducting solid surface, thus also allowing its local atomic structure to be revealed. An extension of scanning tunneling microscopy, atomic force microscopy (AFM), can image the local atomic structure even on insulating surfaces. The ability of STM and AFM to image in various ambiances with virtually no damage or interference to the sample made it possible to observe *processes* continuously. For example, the entire process of a living cell infected by viruses was investigated in situ using AFM (Häberle et al., 1992).

The field of scanning tunneling microscopy has enjoyed a rapid and sustained growth, phenomenal for a new branch of science. The growing number of papers presented at the six International STM Conferences documents the rising interest in this field:

Year	Date	Location	Papers
1986	July 14–18	Santiago de Compostela, Spain	59
1987	July 20–24	Oxnard, California	110
1988	July 4–8	Oxford, UK	157
1989	July 9–14	Oarai, Japan	213
1990	July 23–27	Baltimore, Maryland	357
1991	August 12–16	Interlaken, Switzerland	580

Similarly, STMs are being developed commercially at an astonishing speed. As of summer 1991, over 30 companies have manufactured and marketed STMs and parts. Many beginning companies dedicated to STMs and AFMs are seeing their business expand rapidly. The recent performance of Digital Instruments, Park Scientific Instruments, WA Technology, Angstrom

Technology, TopoMetrix, and RHK Technology are typical of such companies. Likewise, a number of major manufacturers of scientific instruments have expanded their wares to include STMs. Vacuum Generator, Newport Instrument, Omicron Vacuumphysik GmbH, Leica, JOEL, Nikon, and Seiko number among these manufacturers. Several other companies are supplying STM parts and accessories, such as piezoelectric elements, steppers, and probe tips. Because of their broad endorsement by the scientific and commercial spheres, STMs and AFMs will stand alongside optical and electron microscopes for the foreseeable future as important instruments in laboratories around the world.

Ten years in contemporary science is a long, long period. Myriad startling developments can take place in that amount of time. As Ziman says in *Principles of the Theory of Solids:* "Today's discovery will tomorrow be part of the mental furniture of every research worker. By the end of next week it will be in every course of graduate lectures. Within the month there will be a clamor to have it in the undergraduate curriculum. Next year, I do believe, it will seem so commonplace that it may be assumed to be known by every schoolboy." With the prolific activity in the field of scanning tunneling microscopy over the past 10 years, there remains a visible gap in the published material on that topic. Several edited collections of review articles on STM have already been published (for example, Behm, Garcia, and Rohrer, editors, *Scanning Tunneling Microscopy and Related Methods,* Kluwer, 1990, Güntherodt and Wiesendanger, editors, *Scanning Tunneling Microscopy,* Vols. I through III, Springer Verlag, 1991-1992, and a number of others.) A textbook on scanning force microscopy has already appeared (Sarid, *Scanning Force Microscopy,* Oxford University Press, 1991). Nevertheless, a coherent treatise or textbook on STM is still lacking. We have seen the use of STM and the information derived from it continue to expand rapidly. We have seen the potential of the STM in microelectronic and chemical industries for process control and diagnostics gradually becomes a reality. The need for a basic reference book and a textbook on STM is clearly evident. To satisfy this need is my goal in writing this work.

In the iterative process of planning and writing the chapters of this book, I have encountered a number of unexpected difficulties. First, a treatise or a textbook must be presented in a logical sequence starting from a common knowledge background, for example, the standard undergraduate physics courses. However, there are many inconsistencies and discontinuities in the existing STM literature. To make a logically coherent presentation, the materials have to be digested carefully, and the numerous gaps have to be filled. A second unexpected difficulty springs from the fact that the field of STM is inherently cross-disciplinary. The roots of STM run deeply into quantum mechanics, solid-state physics, chemical physics, electronic engineering, mechanical engineering, and control theory. To organize the necessary information in a comprehensible and coherent manner for a broad spectrum of readers is no trivial task. Finally, the field of STM is evolving so rapidly that a painstakingly written treatise or textbook might be obsolete just as it has been printed. Based on those considerations, in this book I have chosen, to the best of my knowledge and judgement, only the topics that are fundamental, indispensable, and having a lasting value.

The organization of the book is as follows: The first chapter, Overview, describes the basic facts, concepts, and a brief account of its history. This chapter is written at a general physics level and can be read as an independent unit. The last section, Historical Remarks, is an integrated part of the presentation of basic concepts in STM. As an introductory chapter of a textbook, it is not intended to be an authoritative and comprehensive treatment of the history of STM. However, serious efforts have been made to ensure the authenticity and accuracy of the historical facts. In addition to conducting an extensive literature search, I have consulted several key scientists in STM and related fields.

Part I of the book is a systematic presentation of much of the fundamental physics in STM and AFM from a single starting point, a time-dependent perturbation theory that is a modified version of the Bardeen approach. The Bardeen approach to tunneling phenomena is a perturbation theory for the understanding of the classical tunneling junction experiments. For atomic-scale phenomena as in STM and AFM, modifications to the original Bardeen approach are necessary. This approach can provide conceptual understanding and analytic predictions for both tunneling current and attractive atomic forces, as well as a number of tip–sample interaction effects. All essential derivations are given in full detail, to make it suitable as a textbook and a reference book. To understand Part I, some familiarity with elementary quantum mechanics and elementary solid-state physics is expected. For example, the reader is assumed to know elementary quantum mechanics equivalent to the first seven chapters of the popular Landau–Lifshitz *Quantum Mechanics (Nonrelativistic Theory)* and elementary solid-state physics equivalent to the first ten chapters of Ashcroft and Mermin's *Solid State Physics*.

Part II of the book deals with basic physical principles of STM instrumentation and applications, with many concrete working examples. The reader is expected to know relevant materials routinely taught in the standard undergraduate curricula in the fields of physics, chemistry, materials science, or related engineering disciplines. Effort is taken to make every chapter independent, such that each chapter can be understood with little reference to other chapters. Piezoelectricity and piezoelectric ceramics, which are not taught routinely in colleges, are presented from the basic concepts on. Elements of control theory, necessary for the understanding of STM operation, are presented for physicists and chemists in the main text and in an Appendix. As in Part I, all essential derivations are given in full detail. For specific applications, preliminary knowledge in the specific fields of the reader's interest (such as physics, chemistry, electrochemistry, biochemistry, or various engineering sciences) is assumed.

The Appendices cover a number of topics that are not standard parts of an average science or engineering undergraduate curriculum, and are relatively difficult to glean from popular textbooks. To understand those Appendices, the reader is assumed to have the average undergraduate background of a science or engineering major.

This introductory book, moderate in size and sophistication, is not intended to be the ultimate STM treatise. The first part of this book, especially, is not intended to be a comprehensive review of all published STM theories. More sophisticated theoretical approaches, such as those directly based on first-principle numerical calculations, are beyond the scope of this introductory book. With its moderate scope, this book is also not intended to cover all applications of STM. Rather, the applications presented are illustrative in nature. Several excellent collections of review articles on STM applications have already been published or are in preparation. An exhaustive presentation of STM applications to various fields of science and technology needs a book series, with at least one additional volume per year. Moreover, this book does not cover the numerous ramifications of STM, except a brief chapter on AFM. The references listed at the back of the book do not represent a catalog of existing STM literature. Rather, it is a list of references that would have lasting value for the understanding of the fundamental physics in STM and AFM. Many references from related fields, essential to the understanding of the fundamental processes in STM and AFM, are also included.

A preliminary camera-ready manuscript of this book was prepared in 1991. To ensure that its content was well-rounded, reasonably truthful and accurate, the book manuscript was sent to many fellow scientists for reviewing, mostly arranged by Oxford University Press. I am greatly indebted to those reviewers who have spent a substantial amount of time in scrutinizing it in detail and providing a large number of valuable comments and suggestions, both as reviewing reports and as marked on the manuscript. Those comments greatly helped me to correct omissions, inaccuracies, and inconsistencies, as well as improving the style of presentation. Among the reviewers are, in alphabetical order, Dr. A. Baratoff (IBM Zurich Laboratory), Dr. I. P. Batra (IBM Almaden Laboratory), Professor A. Briggs (University of Oxford), Dr. S. Chiang (IBM Almaden Laboratory), Dr. R. Feenstra (IBM Yorktown Laboratory), Professor N. Garcia (Universidad Atonoma de Madrid), Professor R. J. Hamers (University of Wisconsin, Madison), Professor J. B. Pethica (University of Oxford), Professor C. F. Quate (Stanford University), Dr. H. Rohrer (IBM Zurich Laboratory), Professor W. Sacks (Universiti Pierre et Marie Curie), Professor T. T. Tsong (Pennsylvania State University), and Dr. R. D. Young (National Institute of Standard and Technology). In addition, a number of fellow scientists reviewed specific parts of it. Also, in alphabetical order, Dr. N. Amer (IBM Yorktown Laboratory, Chapter 15), R. Borroff (Burleigh Instrument Inc., Chapter 12), Dr. F. J. Himpsel (IBM Yorktown Laboratory, Chapter 3), R. I. Kaufman (IBM Yorktown Laboratory, Chapter

11), Dr. V. Moruzzi (IBM Yorktown Laboratory, Chapter 3), C. Near (Morgan Matroc Inc., Vernitron Division, Chapter 9), Dr. E. O'Sullivan (IBM Yorktown Laboratory, sections on electrochemistry), R. Petrucci (Staveley Sensors Inc., EBL Division, Chapter 9), Dr. D. Rath (IBM Yorktown Laboratory, sections on electrochemistry), and Dr. C. Teague (National Institute of Standard and Technology, Chapter 1). After all the revisions were made, I felt that this book would serve the readers better than its preliminary version, and I had no regrets about delaying its printing.

I would also like to thank the senior editor in Physical Sciences, Mr. J. Robbins, and Miss A. Lekhwani of Oxford University Press in assisting in the publication, especially the painstaking reviewing process of the manuscript. Also, special thanks to Elizabeth McAuliffe of IBM Yorktown Laboratory for her development and assistance in the BookMaster software enabling me to prepare the camera-ready manuscript, which was essential for the preprinting reviewing process. Numerous colleagues provided original STM and AFM images, which enabled me to compile a gallery of 30 spectacular images in the high-quality printing section, which are acknowledged on each page.

Finally, I would like to admit that, even with such extensive reviews and revisions, the book is still neither perfect nor does it represent the last word. Especially in such an active field as STM and AFM, new concepts and new measurements come out every day. I expect that substantial progress will be made in the years to come. Naturally, I am looking forward to future editions. I am anxious to hear any comments and suggestions from readers, with whose help, the future editions of this book would be more useful, more truthful, and more accurate. Thus spake Johann Wolfgang von Goethe:[1]

Oft, wenn es erst durch Jahre durchgedrungen,
Erscheint es in vollendeter Gestalt.
Was glänzt, ist für den Augenblick geboren,
Das Echte bleibt der Nachwelt unverloren.

October 1992, at Yorktown Heights, New York   C. J. C.

---

1 Often, after years of perseverance, it emerges in a completed form. What glitters, is born for the moment. The Genuine lives on to the afterworld. *Faust, Vorspiel auf dem Theater.*

CONTENTS

LIST OF FIGURES

LIST OF PLATES

1. The IBM Zurich Laboratory soccer team.
2. A humble gadget that shocked the science community.
3. "Stairway to Heaven" to touch atoms and molecules.
4. Details of the Si(111)-7×7 structure.
5. The nascent Si(111) surface.
6. Voltage-dependent images of the Si(111)-2×1 surface.
7. Large-scale topographic image of the Si(100)-2×1 surface.
8. Voltage-dependent images of the defects on the Si(100)-2×1 surface.
9. Voltage-dependent images of the Ge(111)-2×1 surface.
10. Metastable three-dimensional Ge islands on Si(100).
11. Si(100) surface covered with hydrogen.
12. Individual π and π^* molecular orbitals observed by STM.
13. A large-scale STM image of the Ge(111) surface.
14. Details of the Ge(111)-c(2×8) surface.
15. Thermal conversion of Sb_4 clusters on Si(100).
16. Tip-induced conversion of the Sb_4 clusters on Si(100).
17. Three constant-current topographic images of the $1T$-TaS_2 surface.
18. Three constant-z current images of $1T$-TaS_2.
19. Domain structure in the charge density waves on $1T$-TaS_2.
20. Quasicrystal symmetry observed in real space.
21. Epitaxial growth of the Si(100) surface by chemical vapor deposition.
22. Initial reaction of oxygen with Si(100) surface.
23. Organic molecules observed by STM.
24. AFM images of insulating surfaces.
25. Progress of STM resolution on metal surfaces: Au(110).
26. Atomic-resolution STM images of Au(111).
27. Large-scale image of the Au(111)-22×$\sqrt{3}$ reconstruction.
28. Nucleation and growth of Fe on Au(111).
29. Tip induced modification of the herringbone structure on Au(111).
30. Manipulating individual atoms using the STM tip.
31. "It's a Small World": A miniature map of the Western Hemisphere.

Plate 1. The IBM Zurich Laboratory soccer team. On October 15, 1986, the soccer teams of IBM Zurich Laboratory and Dow Chemical played a game which had been arranged earlier. To everyone's surprise, a few hours before the game,the Swedish Academy announced the Nobel Prize for Gerd Binnig (right, holding flowers) and Heinrich Rohrer (left, holding flowers). Newspaper reporters rushed in for a press conference.

Towards the end of the conference, Binnig and Rohrer said that it had to end immediately because both were members of the laboratory soccer team. The reporters followed them to the soccer field. A photographer for the Swiss newspaper *Blick* took this photograph before the game started. The IBM Zurich Laboratory lost 2:4.

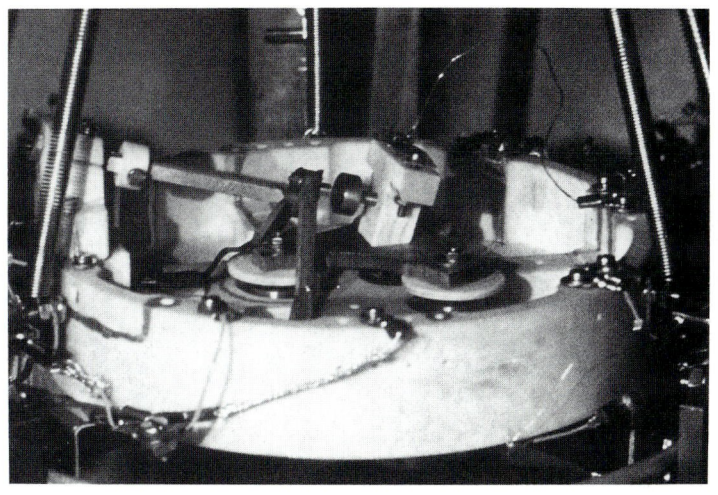

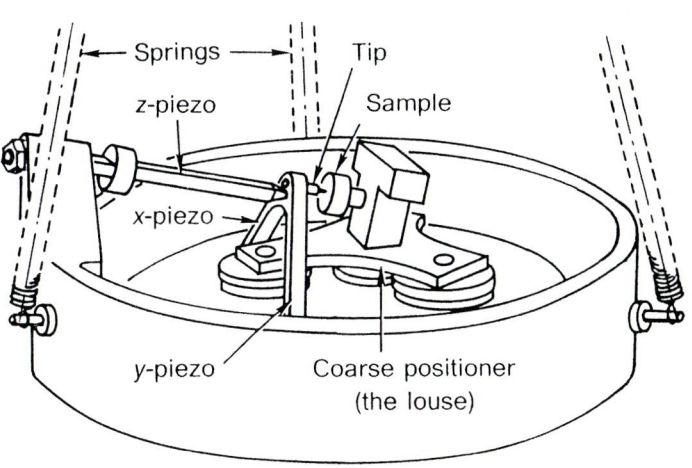

Plate 2. A humble gadget that shocked the science community. Upper, a photograph of an earlier version of the scanning tunneling microscope that enabled real-space imaging of the Si(111)-7×7 structure. Lower, a schematic of it. See Binnig and Rohrer (1982). Original photograph courtesy of Ch. Gerber.

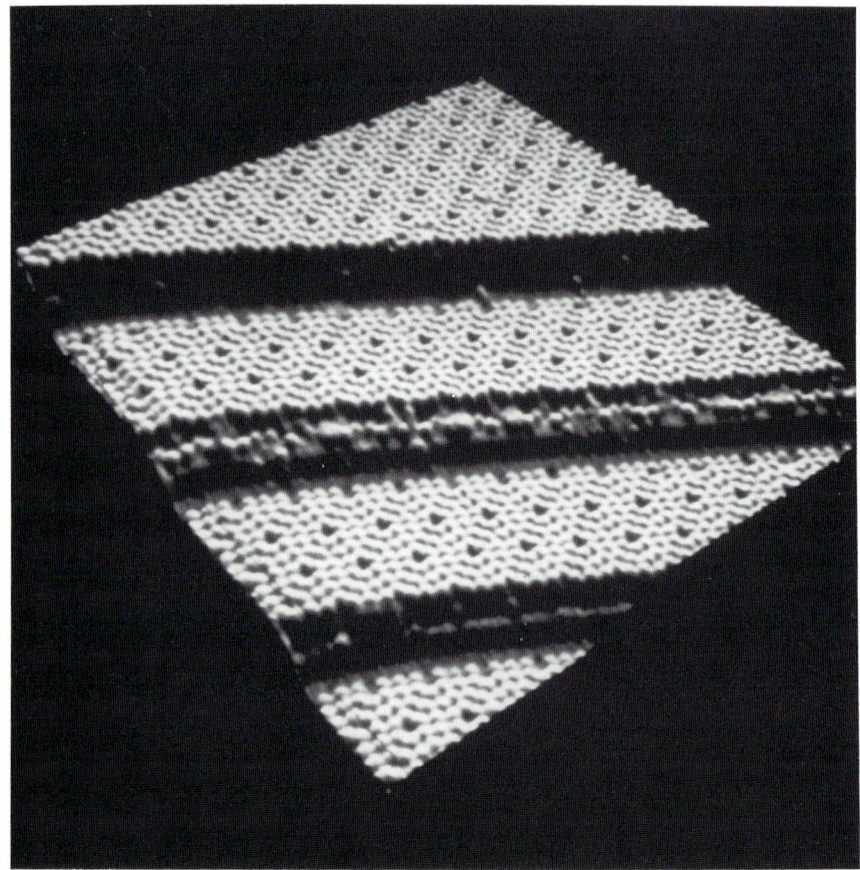

Plate 3. "Stairway to Heaven" to touch atoms and molecules. For almost two centuries, atoms and molecules were entities in the theoretical scientist's heaven, which *"cannot be perceived, by any one, ever"* (Robinson, 1984). The invention of the scanning tunneling microscope enables scientists to perceive directly the electronic structures of individual atoms and molecules. The first demonstration of the atomic-resolution capability of STM was made on the Si(111)-7×7 surface by Binnig, Rohrer, Gerber, and Weibel (1983). See Section 1.3.1 for details. Since then, the electronic structure of individual atoms and molecules in real space become a tangible reality, *indubitability.* [Image showing Si(111)-7×7 surface with steps. Size of image: 320×360 Å². The height of each step is ≈12 Å. For details, see Wiesendanger et al. (1990). Original image courtesy of R. Wiesendanger.]

0   50   100 Å

Plate 4. Details of the Si(111)-7×7 structure. (a) The dangling bonds at the top-layer adatoms of the large 7×7 reconstruction revealed by STM. The nearest-neighbor distance of these dangling bonds is 7.68 Å. Such resolution was achieved in the first year of STM experiments. The actual intrinsic resolution of STM far exceeds this value, as exemplified by the lower image. (b) After reacting with chlorine, the top-layer Si atoms of the Si(111)-7×7 surface are stripped off. The structure of its underlying rest-atom layer is exposed. The spacing between the rest atoms is 3.84 Å. Such resolution is now routinely obtained. See Boland (1991). Original images courtesy of J. Boland.

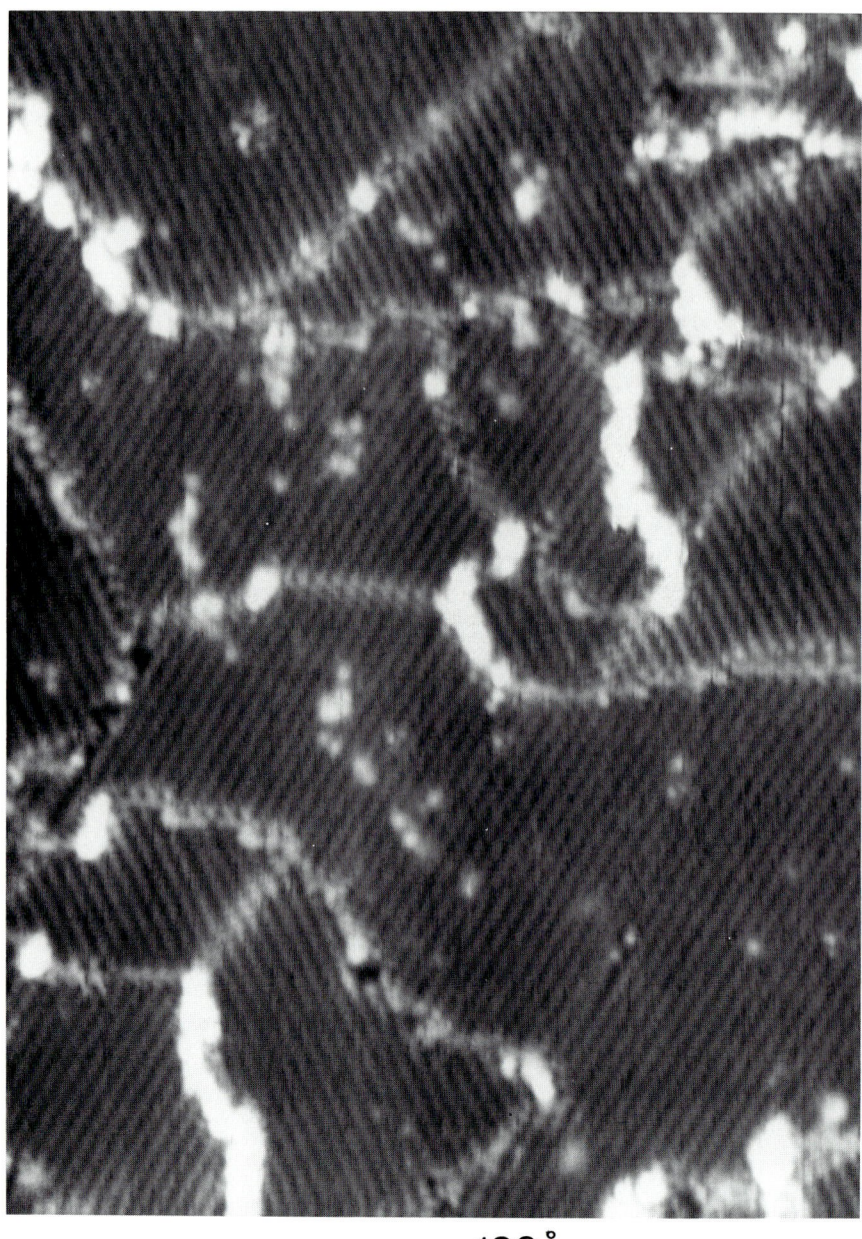

├────────────────────────┤ 100Å

Plate 5. The nascent Si(111) surface. The Si(111) plane is the natural cleavage plane of the Si crystal. After cleaving, the surface immediately reconstructs into a 2×1 structure. The threefold symmetry is locally degenerated to a twofold symmetry. There are three equivalent orientations of the Si(111)-2×1 structure. On a large scale, different orientations coexist to make a mosaic. See Feenstra and Lutz (1991) for details. Original image courtesy of R. M. Feenstra.

a)

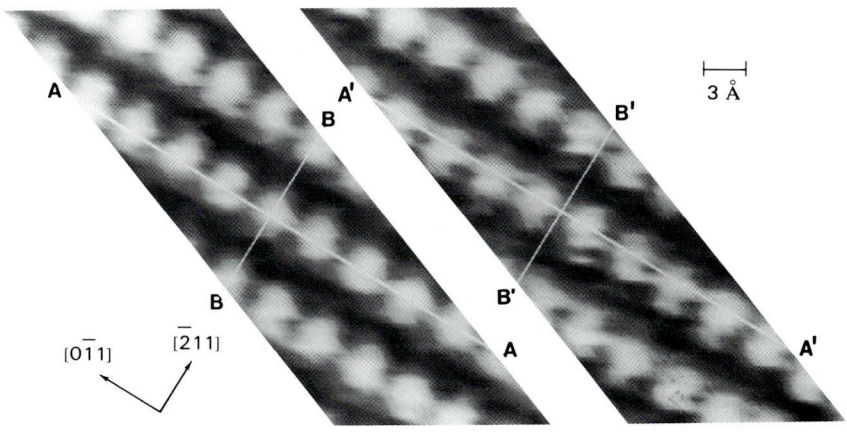

b)

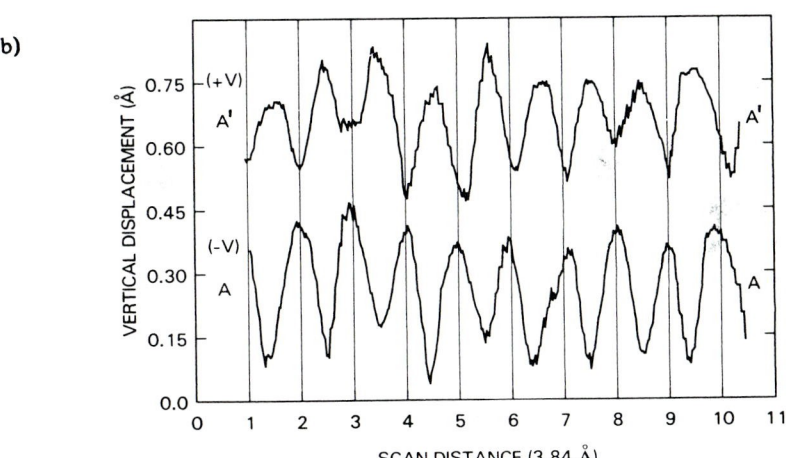

Plate 6. Voltage-dependent images of the Si(111)-2×1 surface. (a) The topographic image of Si(111)-2×1 at a positive bias looks very different from the topographic image of the same surface at a negative bias. In particular, in the $(0\bar{1}1)$ direction the positions of the peaks are reversed. This becomes even clearer in (b), where the topographic contours of the images at both biases are positioned together. Therefore, none of these topographic images is simply related to the positions of the nuclei of the surface atoms. The STM images reflect the electronic structure of the surface. In this particular case, the STM topographic images at different biases reflect the electronic structures of the surface either above the Fermi level, or below the Fermi level. See Feenstra and Stroscio (1987a) for details. Original images courtesy of R. M. Feenstra.

Plate 7. Large-scale topographic image of the Si(100)-2×1 surface. From a technical point of view, Si(100) is probably the most important surface because all the silicon integrated circuits are made on this surface. This large-scale STM image of the Si(100) surface reveals a rich plethora of defects and fine structures: two kinds of steps and various kinds of point defects. These local structures play an important role in semiconductor processing. See Hamers et al. (1987) for details. Size: 580×860 Å². Image taken at a bias -1.6 V at the sample, and with set current 1 nA. Original image courtesy of R. J. Hamers.

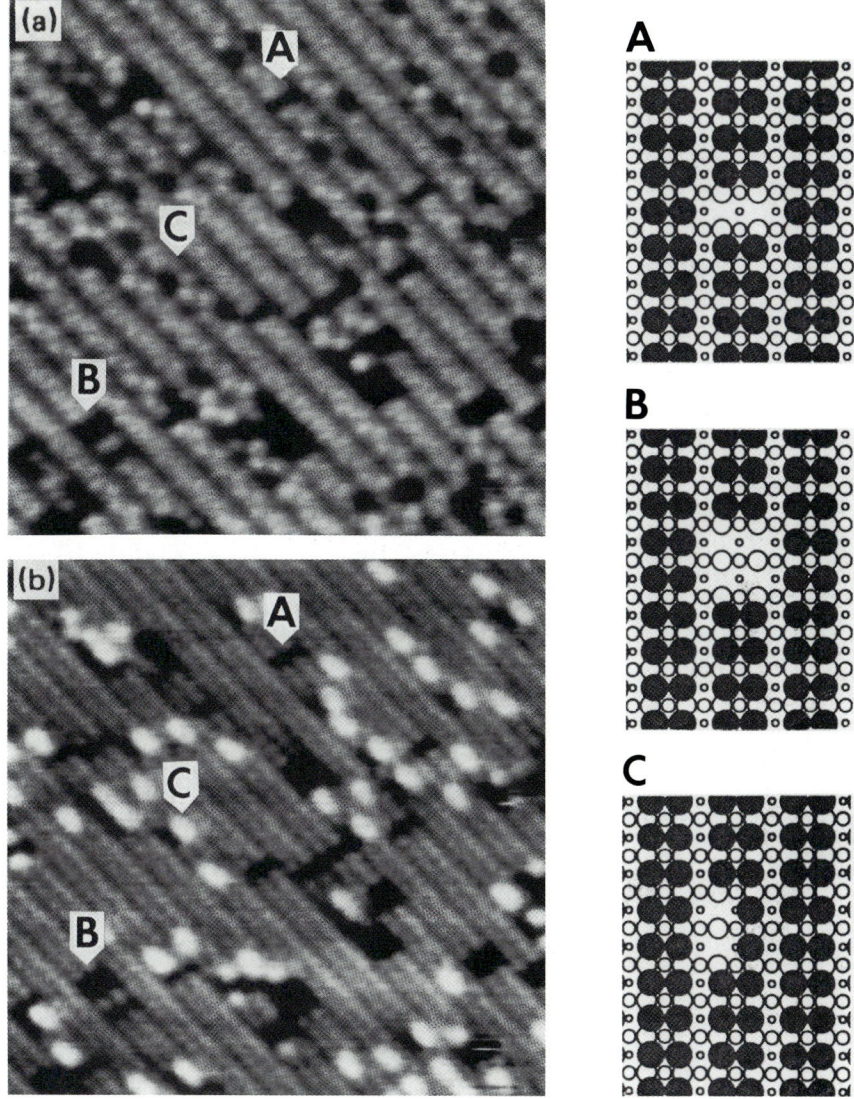

Plate 8. Voltage-dependent images of the defects on the Si(100)-2×1 surface. (a) At −1.6 V; (b) at +1.6 V. At different biases, the STM images of the Si(100) surface appear differently. For example, the type C defect appears as a pair of vacancies at negative bias, but appears as a strong protrusion at positive bias. The topographic STM images are related to the electronic structure of the surface. The geometric structure can be inferred from the STM images. The models for the geometric structures of the three types of defects, proposed by Hamers and Köhler, are shown on the right-hand side. The dark dots are top-layer Si atoms. Open circles are bulklike Si atoms. Circles with larger diameters represent atoms closer to the viewer. See Hamers and Köhler (1989) for details. Original images courtesy of R. J. Hamers.

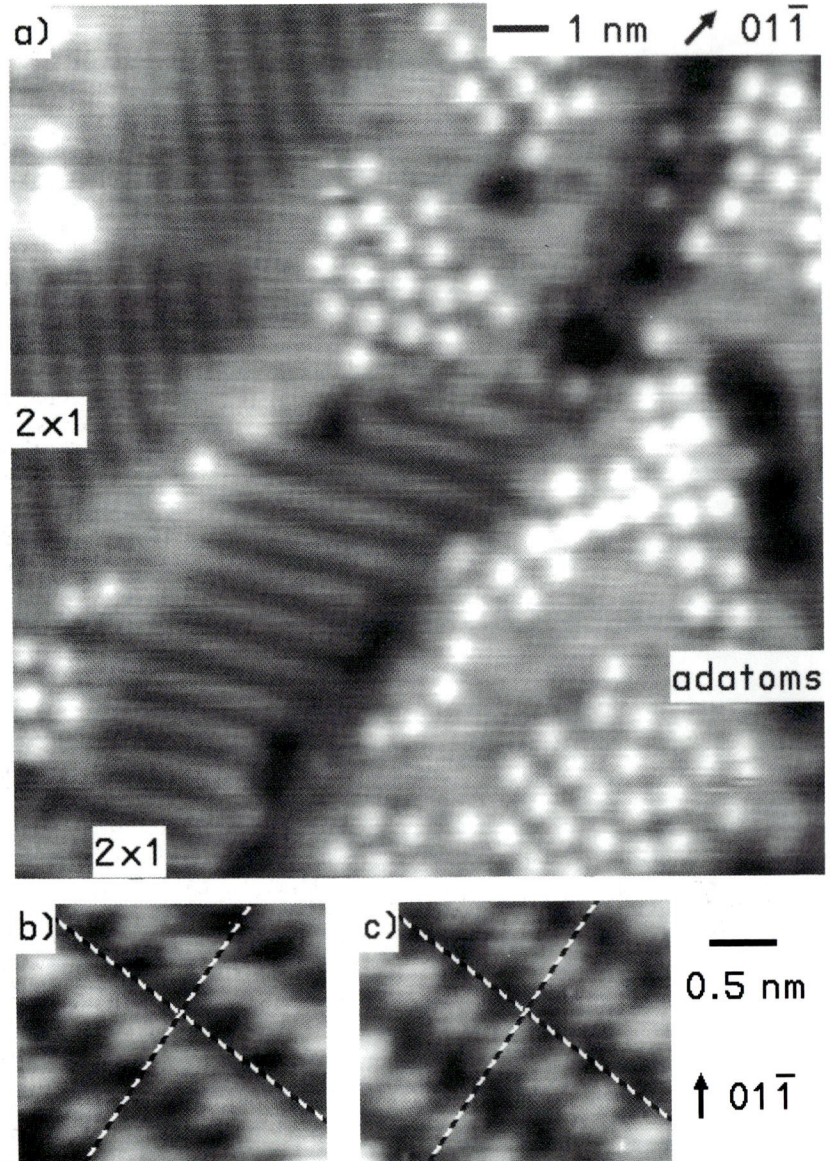

Plate 9. Voltage-dependent images of the Ge(111)-2×1 surface. The crystallographic structure of Ge is very similar to that of Si. The (111) plane is also the natural cleavage plane for Ge. (a) The STM image of a freshly cleaved Ge(111) surface is similar to that of the Si(111) surface (see Plate 5). Large areas of the Ge(111) surface are reconstructed to a 2×1 structure, which has three equivalent orientations. Roughly one-half of the cleaved Ge(111) surfaces are covered with adatoms which form local regions of 2×2 or $c2\times4$ symmetry. (b) and (c): Voltage dependence of the STM images in the 2×1 area. (b) Taken at a bias 0.8 V at the sample, and (c) at -0.8 V. The set current is 0.1 nA. Similar to the case of Si(111)-2×1, the corrugations in the (01$\bar{1}$) direction at opposite biases are reversed (see Plate 6). Again, these STM images reflect the electronic structures of the Ge(111) surface above or below the Fermi level, not at the Fermi level. See Feenstra (1991a) for details. Original image courtesy of R. M. Feenstra.

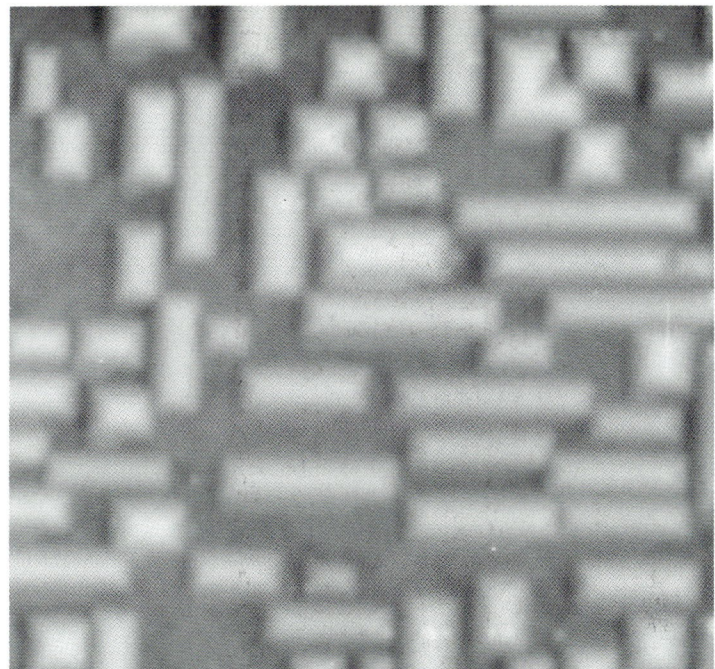

Plate 10. Metastable three-dimensional Ge islands on Si(100). The lattice constant of Ge is larger than that of Si by 4%. Therefore, the growth of Ge on the Si surface cannot be epitaxial. The details of the growth process follow a pattern which is called the Krastanlov-Stransky process. These hudlike islands serve as the kinetic passway for the 2-d to 3-d transition in the Krastanlov-Stransky growth process. (a) An 2500×2500 Å^2 gray-scale top view. (b) A 400×400 Å^2 perspective view of one hud. See Mo et al. (1990) for details. Original images courtesy of Y. W. Mo.

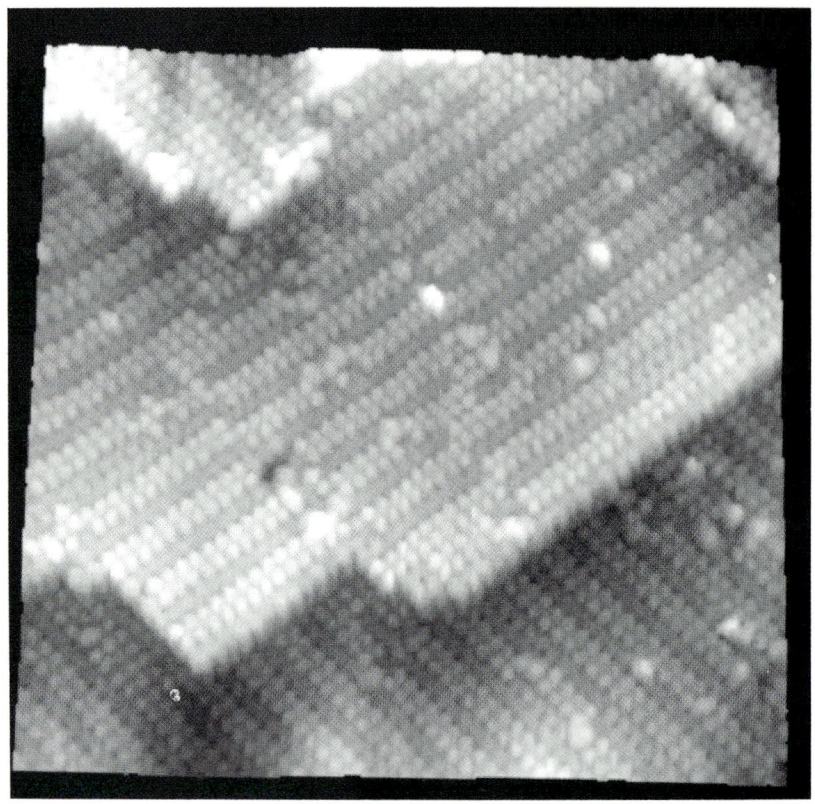

Plate 11. Si(100) covered with hydrogen. The existence of three kinds of structure of the Si(100) surface covered with hydrogen have been known for years. The details of their structure in the real space was revealed by STM. Upper, the Si(100)-3×1 surface is composed of alternating rows of bright monohydride dimer units and dark dihydride units. It is formed by exposing the Si(100) surface to atomic hydrogen at 100°C. Lower, the models for three kinds of structures of the hydrogen-covered Si(100) surface. See Boland (1992) for details. Original image courtesy of J. Boland.

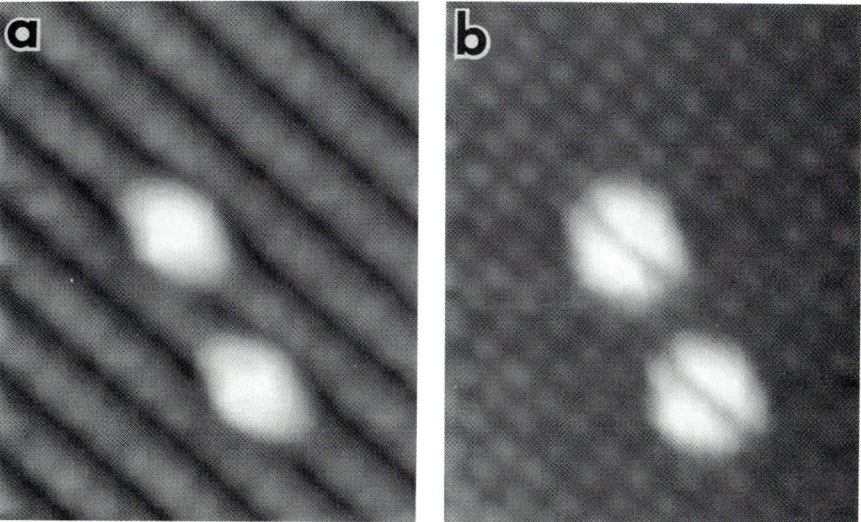

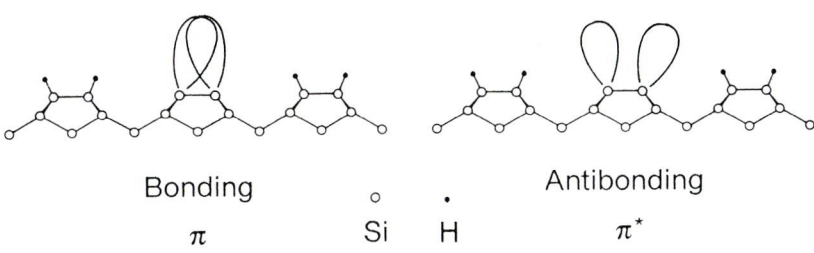

Bonding   ○   •   Antibonding

π Si   H π^*

Plate 12. Individual π and π^* molecular orbitals observed by STM. On the nascent Si(100) surface, every surface Si atom has a p_z-type dangling bond with it. By saturating the Si(100) surface with hydrogen, all the p_z-type dangling bonds are capped with a hydrogen atom. By heating the hydrogen-saturated Si(100) surface carefully, a small fraction of the hydrogen atoms are desorbed. The p_z-type dangling bonds on the Si atoms are paired to form bonding and antibonding orbitals. (a) The filled π orbital which has a symmetric structure. This is a typical example of the chemical bond which binds atoms together to form molecules and solids. (b) As expected, the empty π^* orbital exhibits a nodal plane between the dimer atoms. (c) Schematics of these orbitals. See Boland (1991b, 1991c) for details. Original images courtesy of J. Boland.

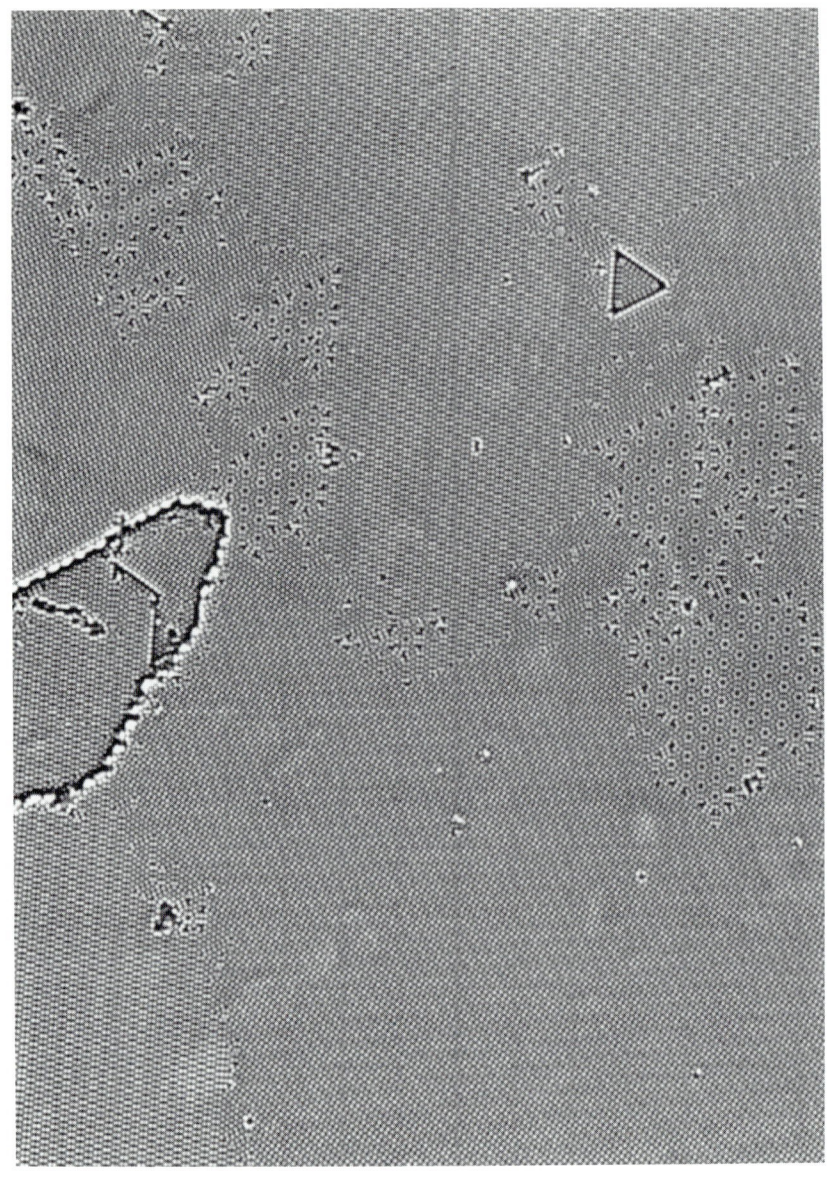

Plate 13. A large-scale STM image of the Ge(111) surface. The Ge sample was cleaved in an ultra-high vacuum, and annealed at about 400°C. On most areas of the surface, a $c(2 \times 8)$ superstructure is observed. Because the nascent Ge(111) surface has threefold symmetry and the $c(2 \times 8)$ reconstruction has twofold symmetry, there are three equivalent orientations. In the transition regions, 7×7 reconstruction appears. Image taken with a bias +1.8 V at the sample, and with set current 0.1 nA. Size: 1300×1700 Å^2. See Feenstra (1991) for details. Original image courtesy of R. M. Feenstra.

Plate 14. Details of the Ge(111)-c(2×8) surface. Upper, the top-layer dangling bonds on the top-layer Ge atoms of the Ge(111)-c(2×1) structure are revealed by STM. Lower, after reacting with hydrogen, the top-layer Ge atoms are removed. The underlying Ge atoms are arranged in a bulklike structure. The distance between the Ge atoms is 3.99 Å. See Boland (1992) for details. Original images courtesy of J. Boland.

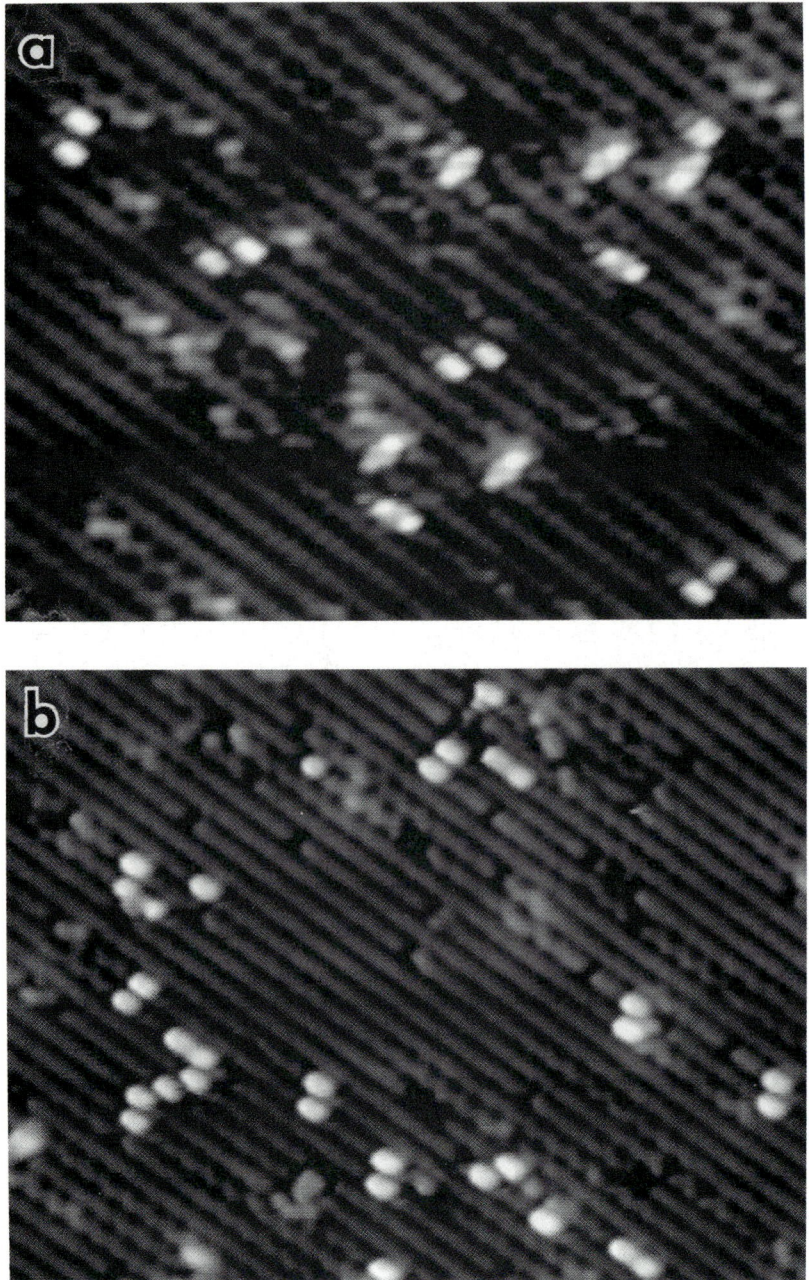

Plate 15. Thermal conversion of Sb$_4$ clusters on Si(100). (a) Sb$_4$ clusters deposited on Si(001) surface at room temperature show three distinct types of precursors as well as the final-state clusters of two dimers. (b) After annealing at 410 K for 20 minutes, almost all precursors are converted to the final-state clusters. See Mo (1992) for details. Original images courtesy of Y. W. Mo.

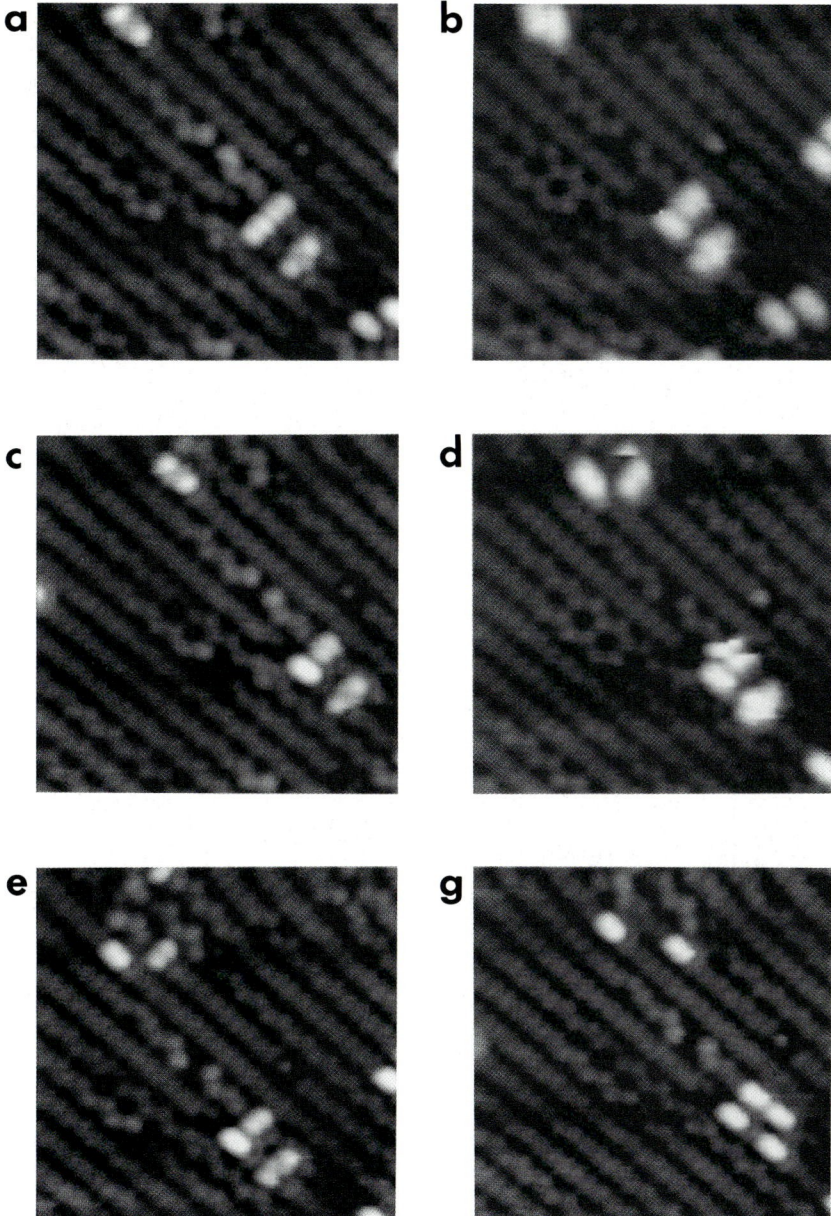

Plate 16. Tip-induced conversion of the Sb$_4$ clusters on Si(100). (a), (c), (e), and (g): "Observation" scans, with a bias of -1.0 V at the sample, and a set current of 0.4 nA. No changes are observed under such conditions. (b) and (d): "Conversion" scans, with a bias of -3.0 V at the sample and 0.4 nA set current. [The conversion scan between (e) and (g) is not shown here.] The different precursor clusters are converted step by step to the final-state clusters. Original images courtesy of Y. W. Mo.

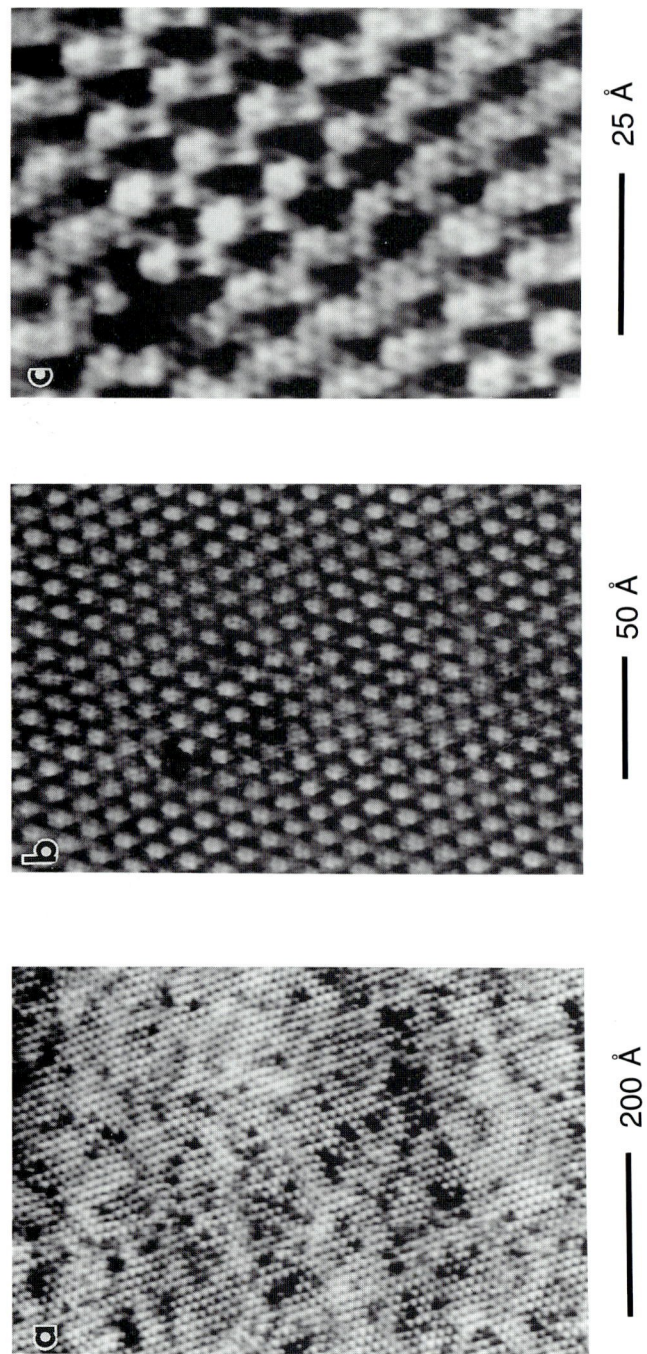

—— 200 Å   —— 50 Å   —— 25 Å

Plate 17. Three constant-current topographic images of the 1T-TaS$_2$ surface. (a) On a large scale, the charge density wave, with a periodicity of 11.8 Å, is the prominent feature. The observation of the charge density waves only weakly depends on the tip electronic structure. (b) When the tip is in good condition, the atomic structure (with nearest-neighbor distance 3.35 Å) can be clearly identified. (c) The corrugation amplitude of the atomic structure, depending on the tip con- dition and tunneling conditions, can be larger than 1 Å. Single atom defects are clearly resolved. The large corrugation is partly due to the deformation of the soft sample surface (see Section 8.2.1). Images obtained by the author with a home-made STM designed by the author, were taken (in 1988) in air using a tip mechanically cut from a Pt–Ir wire, with a bias 20 mV and a set current 2 nA.

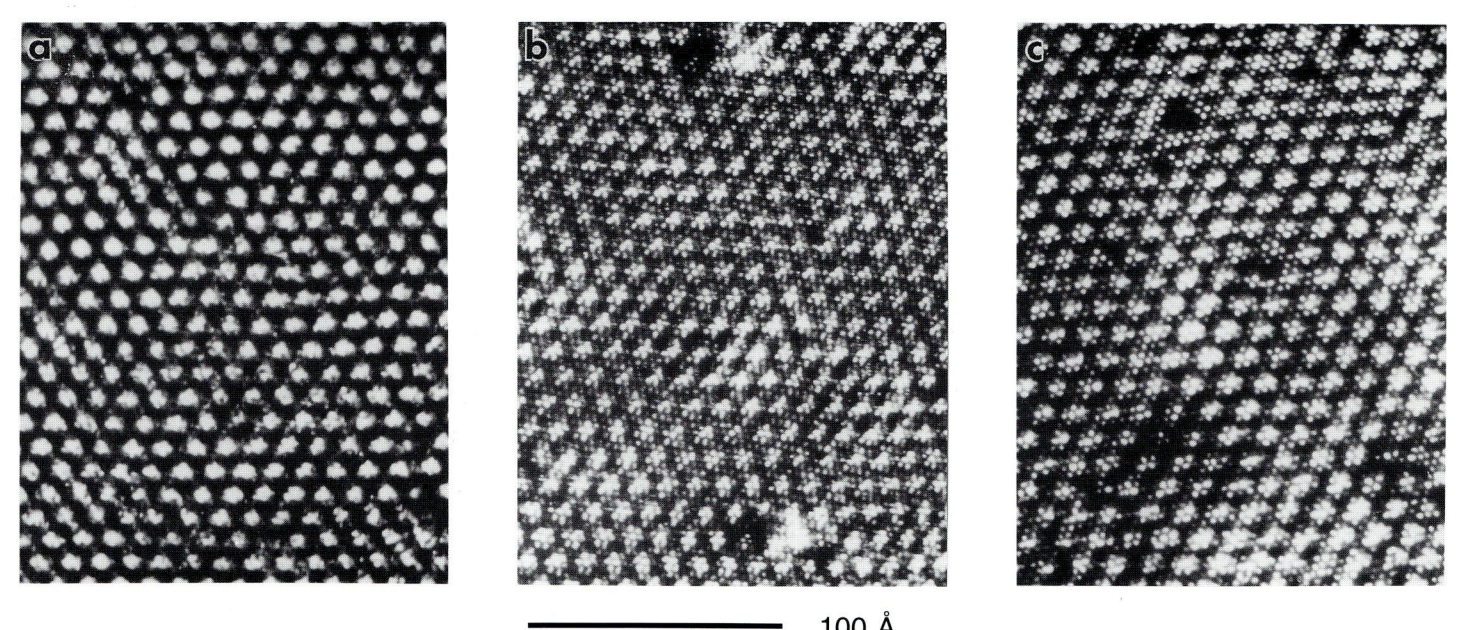

100 Å

Plate 18. Three constant-z current images of 1T-TaS$_2$. As the corrugations of the constant-current topographic images may be amplified by the deformation of the sample surface, the corrugation of the current images at a fixed z can only be reduced by the deformation of the sample surface. Yet, large corrugations in the tunneling current are observed. The appearance of the image depends on the materials and the conditions of the tip. (a) An image dominated by CDWs with weakly superimposed atomic modulation (current set point $I = 2$ nA, bias $V = 10$ mV, Pt–Ir tip). (b) A tunneling current image shows a prominent atomic corrugation ($I = 1$ nA, $V = 10$ mV, Ni tip). (c) The same as (b), with a Co tip. All images are 30×30 nm in size, obtained in air, with a Nanoscope II. By courtesy of Wenhai Han and E. R. Hunt.

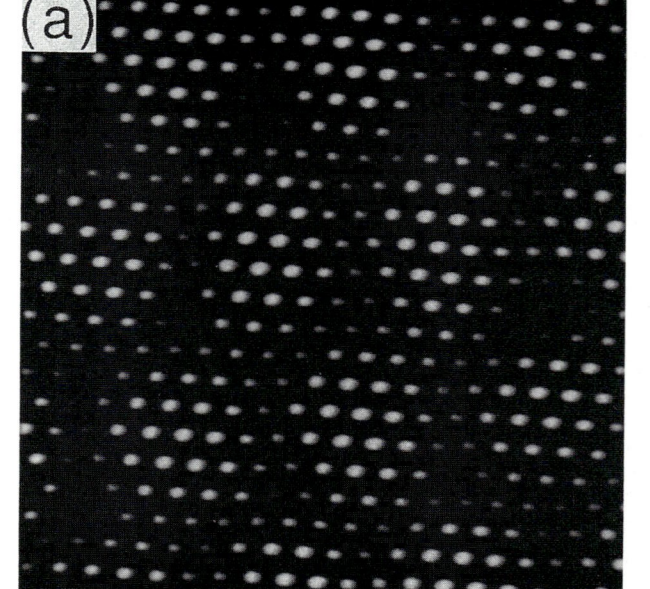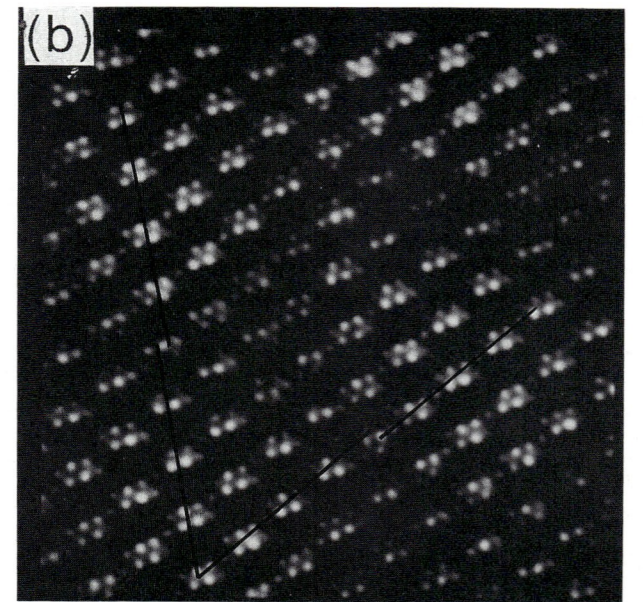

Plate 19. Domain structure of the charge density waves on 1*T*-TaS$_2$.
At room temperature, the charge density waves (CDW) on 1*T*-TaS$_2$ are partially commensurate with the underlying atomic structure. For decades there has been controversy about the microscopic nature of this structure. The STM images with atomic resolution resolved this controversy: The CDWs form domains. Within each domain, the CDW is commensurate with the underlying atomic structure. Near the boundaries, the charge density distributions are no longer commensurate with the underlying atomic structure. (a) A 250×250 Å^2 image with 10 mV bias and 2 nA set current. The CDW maxima exhibit a 72 Å periodic amplitude modulation that defines the hexagonal CDW domain structure. (b) High-resolution image exhibiting hexagonal atomic lattice (period 3.35 Å), hexagonal CDW superlattice (period 11.8 Å), and the hexagonal CDW domain structure (period 72 Å). Courtesy of Xianliang Wu and Charles M. Lieber.

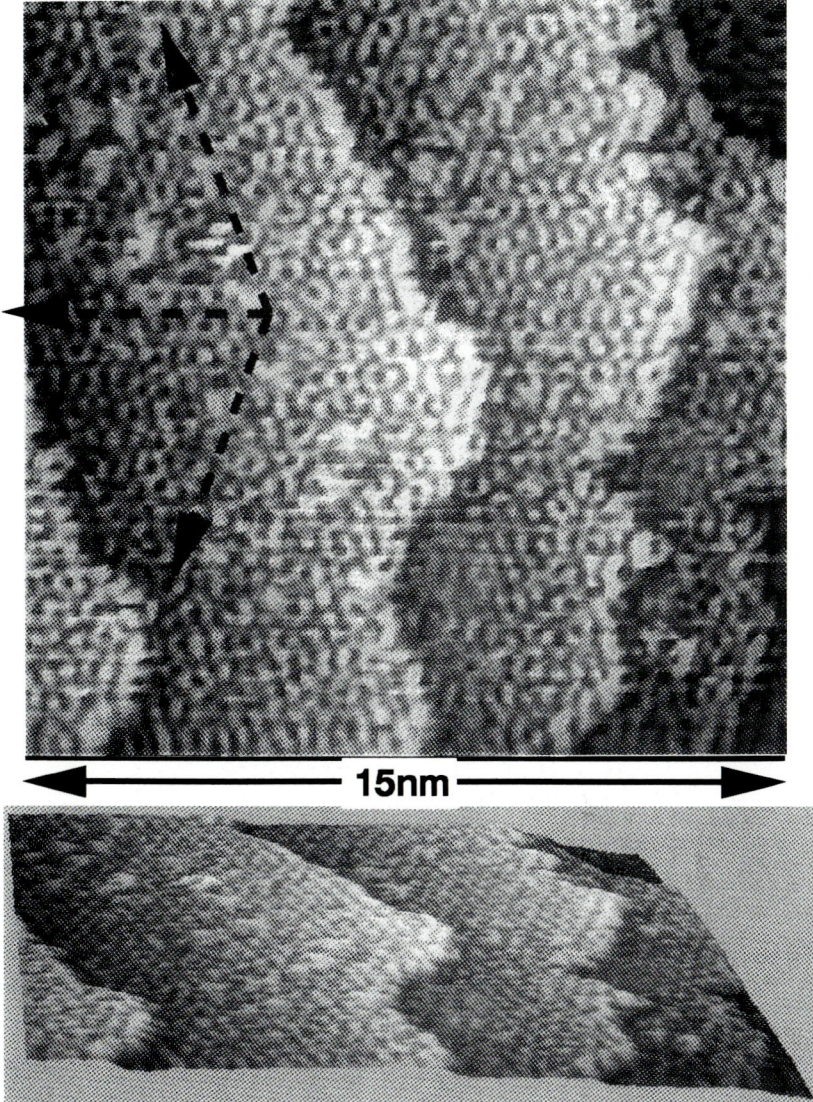

Plate 20. Quasicrystal symmetry observed in real space. Tunneling image of decagonal $Al_{65}Co_{20}Cu_{15}$ shown in both plane view (upper) and perspective (lower). The dashed arrows depict three of the five symmetry directions in the decagonal plane. Others are also visible. Four atomic steps, 2 Å in height, are prominent and represent slices through the complementary stacked quasilattices which have a periodicity of 8.2 Å normal to the surface plane. Inspection shows a number of centers of local decagonal symmetry as well as lines of mass density crisscrossing the surface along the fivefold symmetry directions. This image was taken with a polycrystalline W tip under conditions 100 meV and 1.0 nA while probing unoccupied sample states. [See Kortan, Becker, Thiel, and Chen (1990) for details. Original image courtesy of R. S. Becker.]

Plate 21. Epitaxial growth of the Si(100) surface by chemical vapor deposition. By exposing a Si(100) surface at 400°C to disilane (Si_2H_6), new Si layers are grown. During this process, disilane decomposes into SiH_2 groups which pair up, eliminate hydrogen, and form the dimer structure characteristic of the Si(100) surface. Upper, an STM image of a Si(100) surface after exposure to hydrogen and annealed at 650 K for 10 sec. Area shown is 250×250 Å. Lower, schematic of the decomposition of the SiH_2 fragments which yield the epitaxial monohydride surface. See Boland (1991a) for details. Original image courtesy of J. Boland.

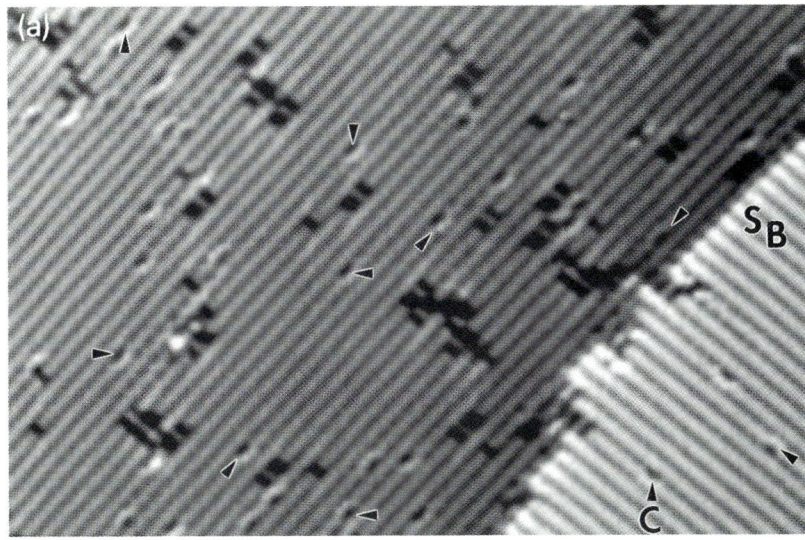

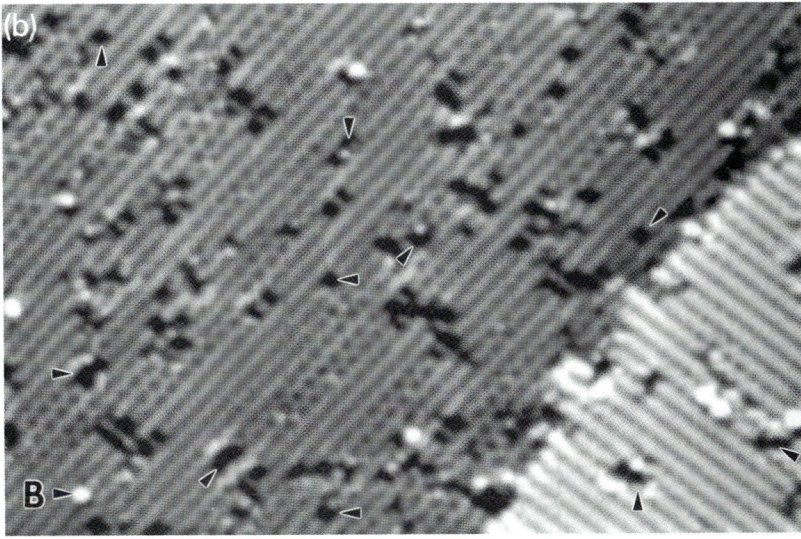

Plate 22. Initial reaction of oxygen with Si(100) surface. The oxidation of the Si(100) surface is one of the most important processing steps in microelectronics manufacturing. The STM provides a tool for studying this process at an atomic level. (a) A clean Si(100) surface prepared in ultra-high vacuum. A number of C-type defects are marked by arrows. The C-type defect was first identified by Hamers and Köhler (see Plate 8). (b) After exposure to about 1 langmuir of O_2. A lot of new defects, obviously due to oxidation, are formed. The initial reaction appears to be localized at defect sites, particularly C-type defects. See Avouris and Lyo (1992) for details. Original image courtesy of Ph. Avouris.

Plate 23. Organic molecules observed by STM. Upper, an STM image of n-$C_{32}H_{66}$ adsorbed on graphite. The length of these linear molecules is approximately 4 nm. Note the high degree of two-dimensional ordering; the long molecule axes are parallel, and troughs are formed where the ends of the molecules abut. The area is approximately 10×10 nm^2. Tip bias was 1 nA at 0.4 V (sample positive). Lower, a proposed model. (a) Graphite lattice with n-$C_{32}H_{66}$ molecules superimposed. (b) Careful observation of the angles and spacings in the STM image shows that its features correspond to those expected for graphite rather than n-$C_{32}H_{66}$ molecules. This implies an imaging mechanism whereby the tunneling sites inherent in the substrate are enhanced by the presence of the intermediate organic layer. See McGonigal, Bernhardt, and Thomson (1990) for details. Original image courtesy of D. J. Thomson.

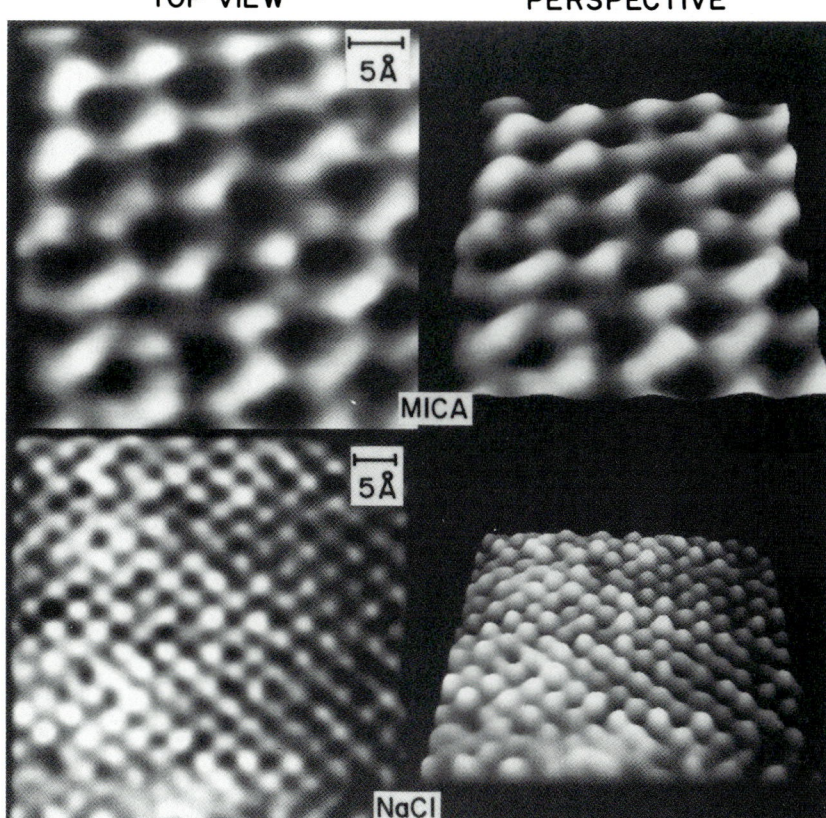

TOP VIEW   **PERSPECTIVE**

5Å

MICA

5Å

NaCl

Plate 24. AFM images of insulating surfaces. Atomic force microscopy (AFM) enables the direct investigation of electrically insulating surfaces. These two images of typical insulators, rock salt and mica, were obtained with an AFM using the optical-beam-deflection method, operating in the short-range repulsive-force regime. Atomic structures are clearly resolved. See Meyer and Amer (1990) for details. Original images courtesy of N. Amer.

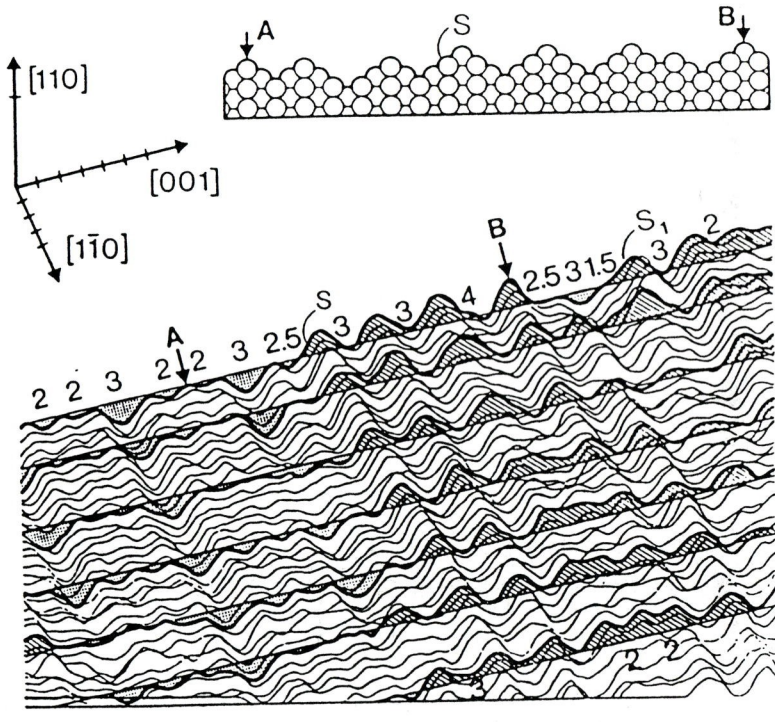

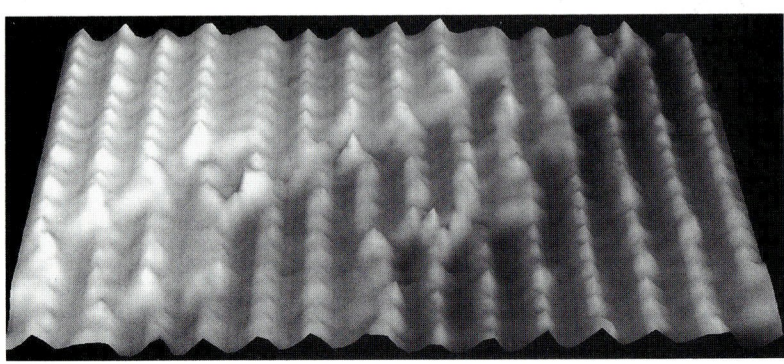

Plate 25. Progress of STM resolution on metal surfaces: Au(110). (a) In the spring of 1982, the IBM Zurich Laboratory obtained the STM images of the reconstruction profiles of Au(110)-2×1 and Au(110)-3×1. Overshadowed by the more interesting images of Si(111)-7×7 (obtained in Fall 1982), the gold image was not submitted for publication until April 1983 (see Binnig and Rohrer, 1987). (b) Later, the atomic structure of the Au(110) surface was resolved by STM. See, for example, Berndt et al. (1992) and Schuster et al. (1992). Today, atomic resolution on rigid surfaces has become a "must" in STM experiments (Rohrer, 1992). Original image courtesy of J. Gimzewski.

50 Å

Plate 26. Atomic-resolution STM images of Au(111). Atomic resolution on Au(111) was first observed by Chiang et al. (1988), as reported at the second International STM Conference (1987). Using a tip treatment procedure as described by Wintterlin et al. (1989), atomic resolution is routinely achieved on Au(111). Often, the corrugation is reversed, that is, the atomic sites appear as depressions instead of protrusions. In the lower image, the atomic arrangement near a step is resolved. See Barth et al. (1990). Original images courtesy of J. V. Barth.

500 Å

Plate 27. Large-scale image of the Au(111)-22×$\sqrt{3}$ reconstruction. The Au(111) surface reconstructs at room temperature to form a 22×$\sqrt{3}$ structure, which has a two-fold symmetry. On a large scale, three equivalent orientations for this reconstruction coexist on the surface. Furthermore, on an intermediate scale, a herring-bone pattern is formed. See Barth et al. (1990) for details. Original image be courtesy of J. V. Barth.

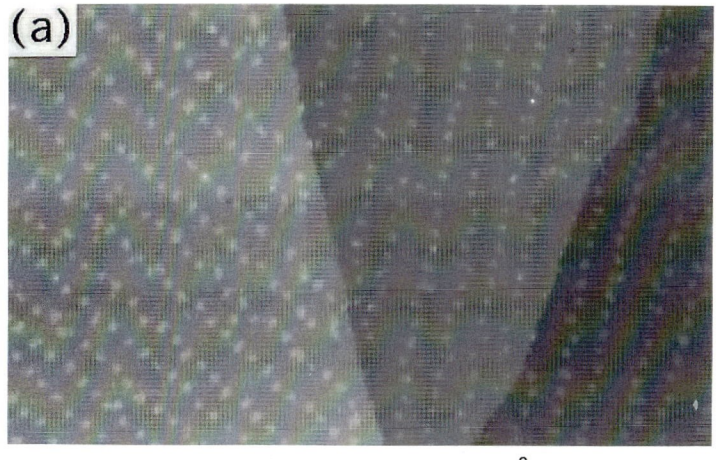

(a)

————————— 1000 Å

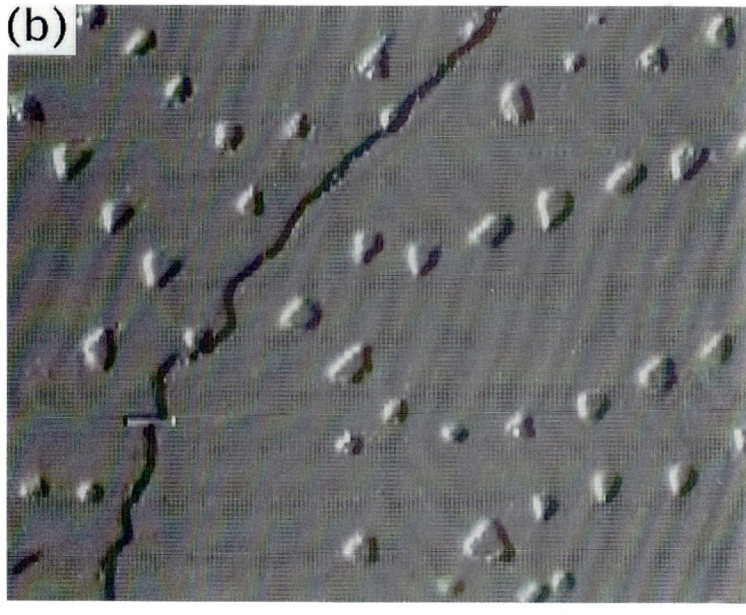

(b)

————————— 250 Å

Plate 28. Nucleation and growth of Ni and Fe on Au(111). By evaporating Ni or Fe on the Au(111) surface, the Ni or Fe atoms form islands with a regular pattern. (a) STM image of Au(111)+0.1 ML Fe. Each dot is a monolayer-high island containing 100-200 atoms. (b) Close-up STM image of the Fe islands. The Fe islands decorate domain boundaries on elbows of the herring-bone pattern to form a regular pattern. See Chambliss et al. (1991a,b) for details. Original images courtesy of D. D. Chambliss.

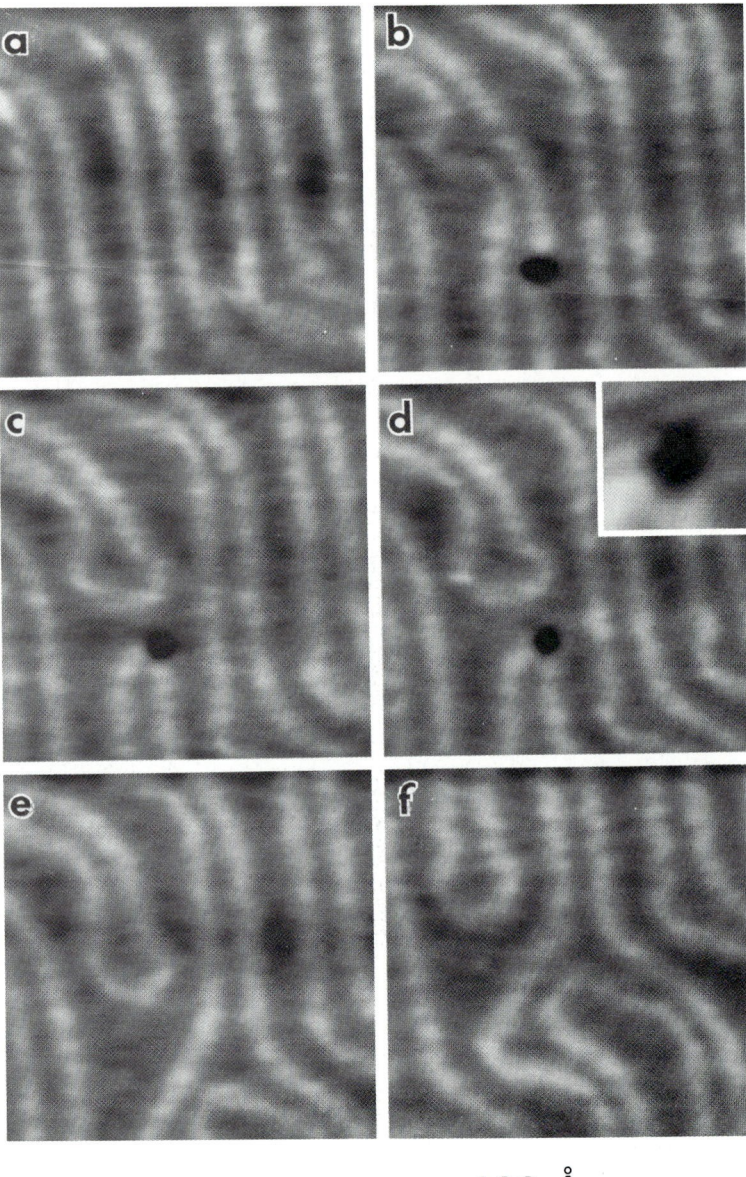

100 Å

Plate 29. Tip induced modification of the herringbone structure on Au(111). The herringbone structure is a result of a subtle balance of energy and stress on the Au(111) surface which involves a large number of atoms. Using a voltage pulse to create a hole on the Au(111) surface, the herringbone pattern deforms accordingly. The evolution of the pattern is observed by STM in real time. (a) Before making a hole. (b) After the hole was formed. (c) About 3 minutes after the hole was formed. (d) After about 9 minutes. (e) After about 15 minutes. (f) After 50 minutes. All images are taken with a bias of $+0.5 \approx +1$ V and a set current 0.1 nA. See Hasegawa and Avouris (1992) for details. Original images courtesy of H. Hasegawa and Ph. Avouris.

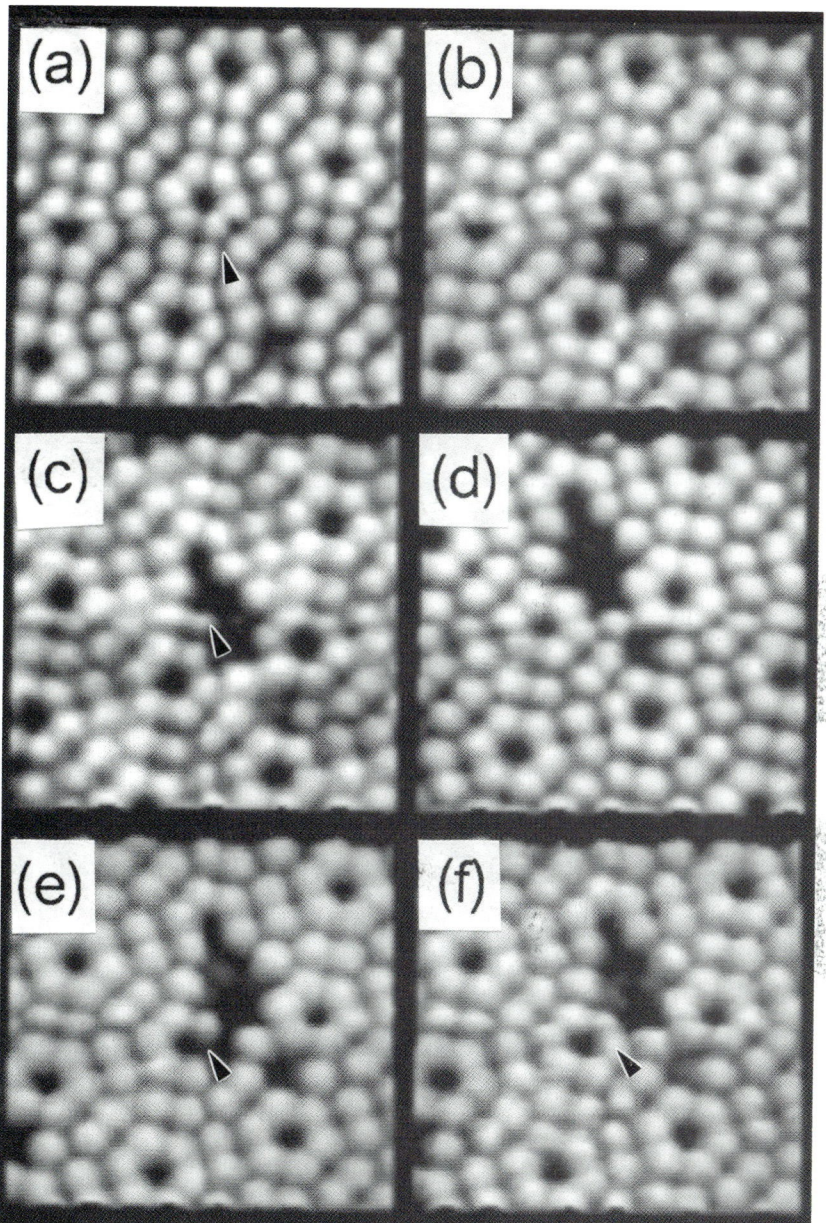

Plate 30. Manipulating individual atoms using the STM tip. The series of experiments was performed on Si(111)-7×7 surface at room temperature in an ultra high
vacuum. (a) The tip is placed at ≈ 1 Å from electronic contact over the site indicated
by the arrow. (b) A 1.0 V pulse removes 3 atoms leaving the fourth under the tip. (c)
The first attempt to remove this atom leads to its migration to the left (see arrow) and
bonding as a center adatom. (d) A second pulse removes this fourth atom. (e) A new
corner-atom is removed and in (f) it is returned to its original position. See Avouris
and Lyo (1992) for details. Original image courtesy of Ph. Avouris.

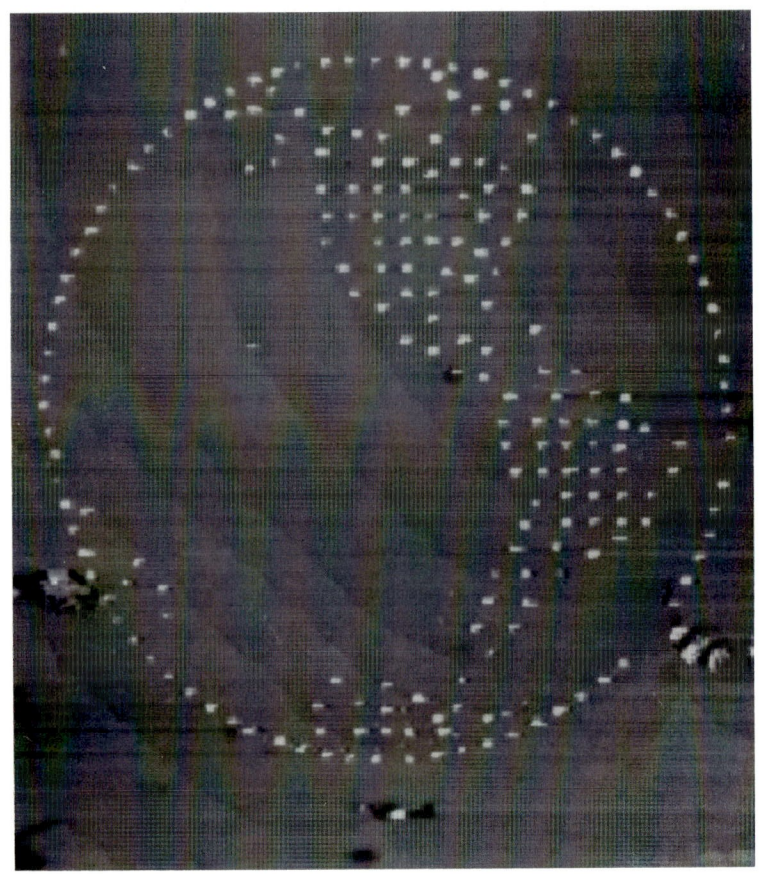

Plate 31. "It's a Small World": A miniature map of the Western Hemisphere. By applying a voltage pulse between a gold tip and a gold surface, a mound of 100–200 Å in diameter and 10–20 Å in height is formed. The location of the mound can be precisely controlled. By programming the positions of the mounds, a gold map is constructed. The diameter of the map is about 1 μm, giving the map a scale of about 10 trillion to 1. For the deposition process, see Mamin, Guenter, and Rugar (1990) for details. Original image courtesy of H. J. Mamin.

Introduction to Scanning Tunneling Microscopy

CHAPTER 1
OVERVIEW

1.1 The scanning tunneling microscope in a nutshell

The scanning tunneling microscope (STM) was invented by Binnig and Rohrer (1982, 1987) and implemented by Binnig, Rohrer, Gerber, and Weibel (1982a, 1982b). Figure 1.1 shows its essential elements. A probe tip, usually made of W or Pt–Ir alloy, is attached to a *piezodrive,* which consists of three mutually perpendicular piezoelectric transducers: x piezo, y piezo, and z piezo. Upon applying a voltage, a piezoelectric transducer expands or contracts. By applying a sawtooth voltage on the x piezo and a voltage ramp on the y piezo, the tip scans on the xy plane. Using the coarse positioner and the z piezo, the tip and the sample are brought to within a few ångströms of each other. The electron wavefunctions in the tip overlap electron wavefunctions in the sample surface. A bias voltage, applied between the tip and the sample, causes an electrical current to flow. Such a current is a quantum-mechanical phenomenon, *tunneling,* which is explained briefly in Section 1.2, and is discussed in detail in Part I of this book.

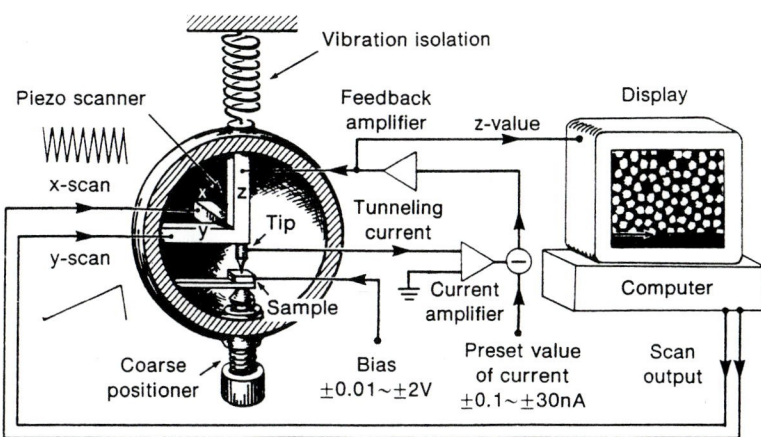

Fig. 1.1. Schematic diagram of the scanning tunneling microscope.

The tunneling current is amplified by the current amplifier to become a voltage, which is compared with a reference value. The difference is then amplified again to drive the z piezo. The phase of the amplifiers is chosen to provide negative feedback: If the tunneling current is larger than the reference value, then the voltage applied to the z piezo tends to withdraw the tip from the sample surface, and vice versa. Therefore, an equilibrium z position is established through the feedback loop. As the tip scans over the xy plane, a two-dimensional array of equilibrium z positions, representing a contour plot of the equal tunneling-current surface, is obtained and stored.

The contour plot is displayed on a computer screen, either as a line-scan image or as a gray-scale image (Fig. 1.2). The line-scan image is a sequence of curves, each of which represents a contour along the x direction with constant y. The curves with different ys are displaced vertically. The gray-scale image is similar to a black-and-white television picture. The bright spots represent high z values (protrusions), and the dark spots represent low z values (depressions).

To achieve atomic resolution, vibration isolation is essential. There are two ways to achieve a suitable solution. The first is to make the STM unit as rigid as possible. The second is to reduce the transmission of environmental vibration to the STM unit. A commonly used vibration isolation system consists of a set of suspension springs and a damping mechanism.

The STM experiments can be performed in a variety of ambiences: in air, in inert gas, in ultrahigh vacuum, or in liquids, including insulating and cryogenic liquids, and even electrolytes. The operating temperature ranges from absolute zero (-273.16°C) to a few hundred degrees centigrade.

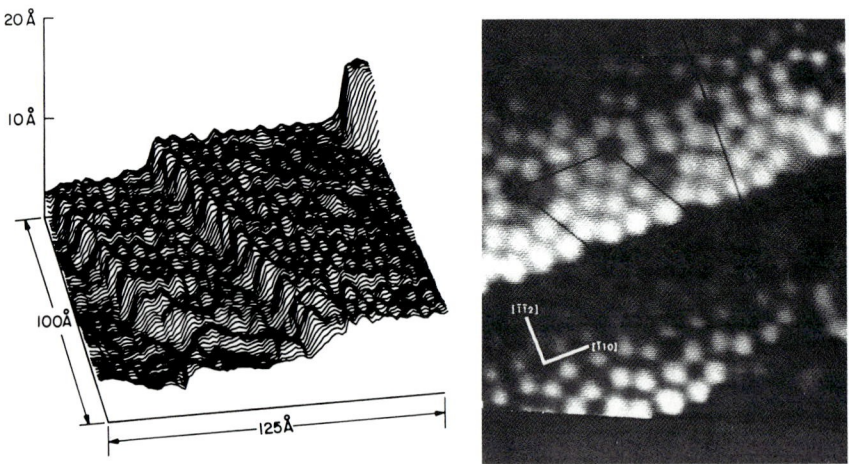

Fig. 1.2. Line-scan image and gray-scale image. The two images represent the same set of data collected by STM on a Si(111)-7×7 surface with steps. Dimensions: 100×125 Å². (Reproduced from Becker et al., 1985a, with permission.)

1.2 Tunneling: an elementary model

In this section, we discuss the concept of tunneling through an elementary one-dimensional model. In classical mechanics, an electron with energy E moving in a potential $U(z)$ is described by

$$\frac{p_z^2}{2m} + U(z) = E, \tag{1.1}$$

where m is the electron mass, 9.1×10^{-28} g. In regions where $E > U(z)$, the electron has a nonzero momentum p_z. On the other hand, the electron cannot penetrate into any region with $E < U(z)$, or a *potential barrier*. In quantum mechanics, the state of the same electron is described by a wavefunction $\psi(z)$, which satisfies Schrödinger's equation,

$$-\frac{\hbar^2}{2m}\frac{d^2}{dz^2}\psi(z) + U(z)\psi(z) = E\psi(z). \tag{1.2}$$

Consider the case of a piecewise-constant potential, as shown in Fig. 1.3. In the classically allowed region, $E > U$, Eq. (1.2) has solutions

$$\psi(z) = \psi(0)e^{\pm ikz}, \tag{1.3}$$

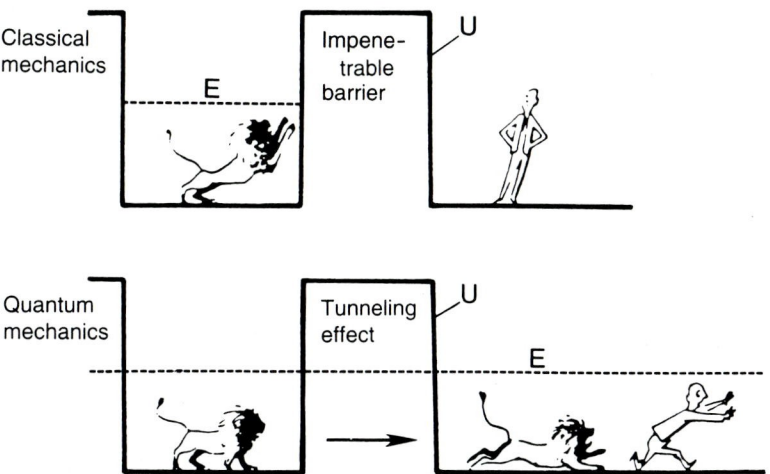

Fig. 1.3. The difference between classical theory and quantum theory. In quantum mechanics, an electron has a nonzero probability of tunneling through a potential barrier. (After Van Vleck; see Walmsley, 1987.)

where

$$k = \frac{\sqrt{2m(E - U)}}{\hbar} \tag{1.4}$$

is the wave vector. The electron is moving (in either a positive or negative direction) with a constant momentum $p_z = \hbar k = [2m(E - U)]^{1/2}$, or a constant velocity $v_z = p_z/m$, the same as the classical case. In the classically forbidden region, Eq. (1.2) has a solution

$$\psi(z) = \psi(0)e^{-\kappa z}, \tag{1.5}$$

where

$$\kappa = \frac{\sqrt{2m(U - E)}}{\hbar} \tag{1.6}$$

is the decay constant. It describes a state of the electron decaying in the $+z$ direction. The probability density of observing an electron near a point z is proportional to $|\psi(0)|^2 e^{-2\kappa z}$, which has a nonzero value in the barrier region, thus a nonzero probability to penetrate a barrier. Another solution, $\psi(z) = \psi(0)e^{\kappa z}$, describes an electron state decaying in the $-z$ direction.

Starting from this elementary model, with a little more effort, we can explain some basic features of metal–vacuum–metal tunneling, as shown in Fig. 1.4. The *work function* ϕ of a metal surface is defined as the minimum energy required to remove an electron from the bulk to the vacuum level. In general, the work function depends not only on the material, but also on the crystallographic orientation of the surface (see Section 4.2). For materials commonly used in STM experiments, the typical values of ϕ are listed in Table 1.1. (The work functions for alkali metals are substantially lower, typically 2–3 eV.) Neglecting the thermal excitation, the *Fermi level* is the upper limit of the occupied states in a metal. Taking the vacuum level as the reference point of energy, $E_F = -\phi$. To simplify discussion, we assume that the work functions of the tip and the sample are equal. The electron in the sample can tunnel into the tip and *vice visa*. However, without a bias voltage, there is no net tunneling current.

Table 1.1. Typical values of work functions. After *Handbook of Chemistry and Physics,* 69th edition, CRC Press (1988).

Element	Al	Au	Cu	Ir	Ni	Pt	Si	W
ϕ (eV)	4.1	5.4	4.6	5.6	5.2	5.7	4.8	4.8

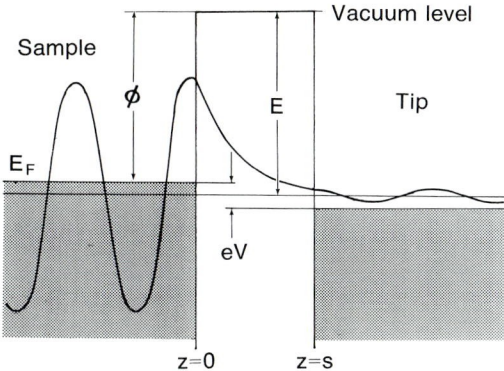

Fig. 1.4. A one-dimensional metal–vacuum–metal tunneling junction. The sample, left, and the tip, right, are modeled as semi-infinite pieces of free-electron metal.

By applying a bias voltage V, a net tunneling current occurs. A sample state ψ_n with energy level E_n lying between $E_F - eV$ and E_F has a chance to tunnel into the tip. We assume that the bias is much smaller than the value of the work function, that is, $eV \ll \phi$. Then the energy levels of all the sample states of interest are very close to the Fermi level, that is, $E_n \approx -\phi$. The probability w for an electron in the nth sample state to present at the tip surface, $z = W$, is

$$w \propto |\psi_n(0)|^2 e^{-2\kappa W}, \tag{1.7}$$

where $\psi_n(0)$ is the value of the nth sample state at the sample surface, and

$$\kappa = \frac{\sqrt{2m\phi}}{\hbar} \tag{1.8}$$

is the decay constant of a sample state near the Fermi level in the barrier region. Using eV as the unit of the work function, and $\text{\AA}^{-1}$ as the unit of the decay constant, the numerical value of Eq. (1.8) is

$$\kappa = 0.51 \sqrt{\phi(\text{eV})} \ \text{\AA}^{-1}. \tag{1.9}$$

In an STM experiment, the tip scans over the sample surface. During a scan, the condition of the tip usually does not vary. The electrons coming to the tip surface, $z = W$, have a constant velocity to flow into the tip. The tunneling current is directly proportional to the number of states on the sample surface within the energy interval eV, which are responsible for the tunneling current. This number depends on the local nature of the sample surface. For metals, it is finite. For semiconductors and insulators, the number is very small or zero. For semimetals, it is in between. By including all the sample states in the energy interval eV, the tunneling current is

$$I \propto \sum_{E_n = E_F - eV}^{E_F} |\psi_n(0)|^2 e^{-2\kappa W}. \tag{1.10}$$

If V is small enough that the density of electronic states does not vary significantly within it, the sum in Eq. (1.10) can be conveniently written in terms of the *local density of states (LDOS)* at the Fermi level. At a location z and energy E, the LDOS $\rho_S(z, E)$ of the sample is defined as

$$\rho_S(z, E) \equiv \frac{1}{\epsilon} \sum_{E_n = E - \epsilon}^{E} |\psi_n(z)|^2, \tag{1.11}$$

for a sufficiently small ϵ. The LDOS is the number of electrons per unit volume per unit energy, at a given point in space and at a given energy. It has a nice feature as follows. The probability density for a specific state, $|\psi_n|^2$, depends on the normalization condition: its integral over the entire space should be 1. As the volume increases, the probability density $|\psi_n|^2$ of a single state decreases; but the number of states per unit energy increases. The LDOS remains a constant. The value of the surface LDOS near the Fermi level is an indicator of whether the surface is metallic or insulating.

The tunneling current can be conveniently written in terms of the LDOS of the sample:

$$\begin{aligned} I &\propto V \rho_S(0, E_F) e^{-2\kappa W} \\ &\approx V \rho_S(0, E_F) e^{-1.025\sqrt{\phi} W}. \end{aligned} \tag{1.12}$$

The typical value of work function is $\phi \approx 4$ eV, which gives a typical value of the decay constant $\kappa \approx 1$ Å^{-1}. According to Eq. (1.12), the current decays about $e^2 \approx 7.4$ times per Å.

The dependence of the logarithm of the tunneling current with respect to distance is a measure of the work function, or the tunneling barrier height (Garcia, 1986; Coombs and Pethica, 1986). In fact, from Eq. (1.12),

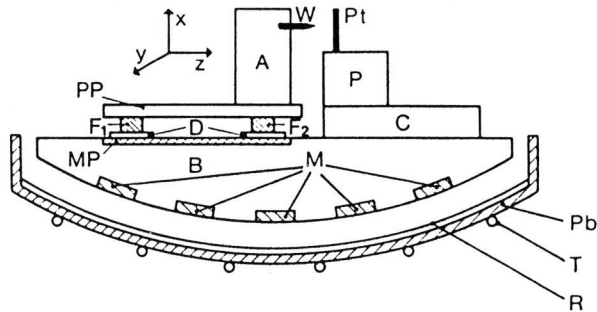

Fig. 1.5. Apparatus for demonstrating vacuum tunneling. The tunneling unit consists of a W tip and a Pt plate. Vibration isolation is achieved by magnetic levitation on a superconducting bowl. (Reproduced from Binnig et al., 1982a, with permission.)

$$\phi = \frac{\hbar^2}{8m}\left(\frac{d\ln I}{dW}\right)^2 \approx 0.95\left(\frac{d\ln I}{dW}\right)^2. \qquad (1.13)$$

Experimentally, this quantity can be conveniently measured by varying the tip–sample distance through the z piezo, often with an ac voltage. The measured values often varies with distance. Eq. (1.13) is the definition of the *apparent barrier height*.

According to Eq. (1.5), at $z = W$, the sample wavefunction is $\psi(W) = \psi(0)\exp(-\kappa W)$. Using the definition of LDOS, Eq. (1.11), the right-hand side of Eq. (1.10) is proportional to the Fermi-level LDOS *of the sample at the tip surface, $z = W$,*

$$\sum_{E_F - eV}^{E_F} |\psi(0)|^2 e^{-2\kappa W} \equiv \rho_S(W, E_F)eV. \qquad (1.14)$$

The tunneling current can be conveniently expressed in terms of this quantity,

$$I \propto \rho_S(W, E_F)V. \qquad (1.15)$$

By scanning the STM tip over the sample surface with the tunneling current kept constant, a topographic STM image is generated. According to this one-dimensional model, a topographic STM image is a constant Fermi-level LDOS contour of the sample surface. As shown in Chapter 5, if the lateral scale of the features of interest on the sample is much larger than a characteristic length related to the decay constant, $\pi/\kappa \approx 3$ Å, at low bias, this elementary model should be approximately valid (Tersoff and Hamann 1983).

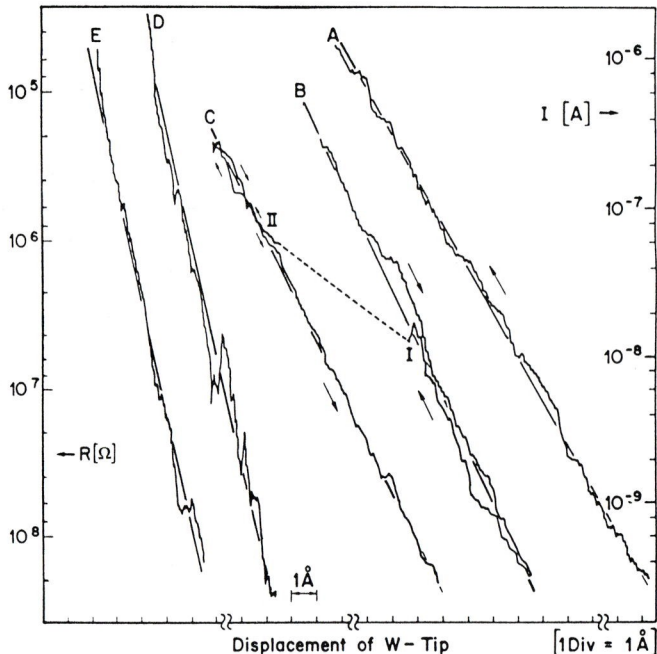

Fig. 1.6. Tunneling through a controllable vacuum gap. The exponential dependence $I \sim V$ is observed over four orders of magnitude. On clean surfaces, an apparent
barrier height of 3.5 eV was observed. (Reproduced from Binnig et al., 1982a, with
permission.)

From Eq. (1.12), it is clear that the tunneling current is extremely sensitive to a minute variance of the tip–sample distance. In order to achieve a
stable metal–vacuum–metal tunneling, the suppression of environmental
vibration is extremely important. In the first successful demonstration of
vacuum tunneling with a controllable gap and scanning capability by Binnig,
Rohrer, Gerber, and Weibel (1982a), the suppression of vibration was achieved
with an elaborate method: magnetic levitation using superconductors working
at liquid helium temperature. Figure 1.5 is a schematic of the apparatus.

Using this apparatus, Binnig et al. demonstrated the exponential distance
dependence of tunneling current over four orders of magnitude. Their first
results are summarized in Fig. 1.6. The value of the work function was found
to depend sharply on the condition of the surfaces. Initially, the measured
values were around $0.6 - 0.7$ eV. After repeated cleaning, the slope became
much steeper. A value of 3.2 eV was obtained, which can last for several
minutes. Being still lower than the value for clean surfaces, $4 - 5$ eV, it was

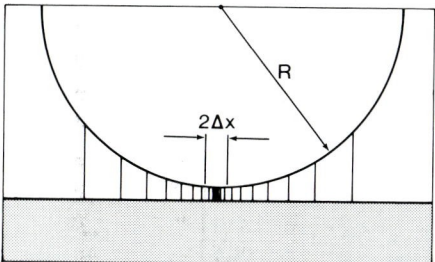

Fig. 1.7. Estimation of the lateral resolution in STM. In 1978, Binnig made an estimation of the possible lateral resolution of STM with a simple spherical-tip model. The tip end, with radius R, is very close to the sample surface. The tunneling current is concentrated in a small region around the origin, $x = 0$. With $r \approx 1000$Å, the radius of the tunneling current column is approximately $\Delta x \approx 45$ Å. (After Quate, 1986.)

attributed to a contamination layer, which was expected to exist under the vacuum conditions of their experiments (10^{-6} Torr).

At the beginning of their experimentation, Binnig and Rohrer anticipated that the realization of controllable vacuum tunneling would result in a lateral resolution much smaller than the radius of the tip end. Their original argument was recorded on Binnig's 1978 laboratory notebook, which was reproduced in a *Physics Today* article by Quate (1986). Schematically, it is shown in Fig. 1.7. If the distance between the tip end and the sample surface is much smaller than the tip radius, near the tip end, the current lines are almost perpendicular to the sample surface. Assuming that at each point the tunneling current density follows the formula for the one-dimensional case, Eq. (1.12), the lateral current distribution is

$$I(\Delta x) = I_0 \exp \left(- 2\kappa \, \frac{\Delta x^2}{2R} \right). \tag{1.16}$$

Typically, $\kappa \approx 1$ Å^{-1}. For $R \approx 1000$ Å, at $\Delta x \approx 45$ Å, the current drops by a factor of e^{-2}, that is, about one order of magnitude. For $R \approx 100$ Å, the current concentrates in a small circle of radius 14 Å. Therefore, with moderate means, a very high lateral resolution is expected.

The actual achievement of STM greatly exceeds this expectation. Details of surface electronic structures with a spatial resolution of 2 Å are now routinely observed. Based on the obtained electronic structure, the atomic structures of surfaces and adsorbates of a large number of systems are revealed. Furthermore, the active role of the STM tip through the tip–sample interactions enables real-space manipulation and control of individual atoms. An era of *experimenting and working on an atomic scale* arises.

1.3 Probing electronic structure at an atomic scale

The invention of the STM realized a dream of physicists and chemists almost two centuries old: to visualize individual atoms and their internal structures in real space. Prior to the invention of the STM, individual atoms could only be imaged under limited circumstances (see Section 1.8).

Since John Dalton published his first volume of *A New System of Chemical Philosophy* in 1808, the concept of atoms and molecules has been the cornerstone of modern physics and chemistry. The upper half of Fig. 1.8 is reproduced from Dalton's book. Even in contemporary chemistry and physics, Dalton's initial dogma remains quite accurate, except that the *atoms* (derived from Greek, ατομος, indivisible), are *divisible*. According to the modern theory of atomic systems, quantum mechanics, an atom consists of a number of electrons in a series of stationary states surrounding an extremely small, positively charged nucleus. Each of these electron states has a well-defined energy level. Apart from electron spin, the electron states are classified according to the quantum numbers l and m. Each individual electronic state has its characteristic probability distribution, or electron density distribution, as shown in the lower half of Fig. 1.8. In molecules or solids, the electronic structure can often be represented as linear combinations of atomic states. However, until very recently, the information about individual atoms and their electronic structures — the individual electronic states — was inferred from indirect measurements. Now, STM has made it technically possible for scientists to probe directly the electronic structures of various materials at an atomic scale (≈ 2 Å).

1.3.1 Semiconductors

The first success of STM was the real-space imaging of the Si(111)-7×7 structure, as shown in Fig. 1.9. The 7×7 reconstruction of the Si(111) surface, one of the most intriguing phenomena in surface science, was discovered by low-energy electron diffraction (LEED) in 1959. The LEED diffraction pattern clearly shows that the unit cell of this reconstructed surface is constituted of 49 silicon atoms on the original Si(111) surface, and exhibits a *p6mm* symmetry. However, the details of the arrangement of the 49 Si atoms in each layer of one unit cell cannot be determined unambiguously from the LEED pattern. In the 1960s and 1970s, a large number of models have been proposed. There was not enough experimental evidence to determine which one was correct. The STM image of that surface taken by Binnig et al. (1983) clearly showed that there are 12 adatoms and one large hole in each unit cell. This experimental fact contradicted all the previous models. Based on the real-space observation, Binnig et al. (1983) proposed the 12-adatom model of the Si(111)-7×7 reconstruction. In addition, they proposed that the surface does

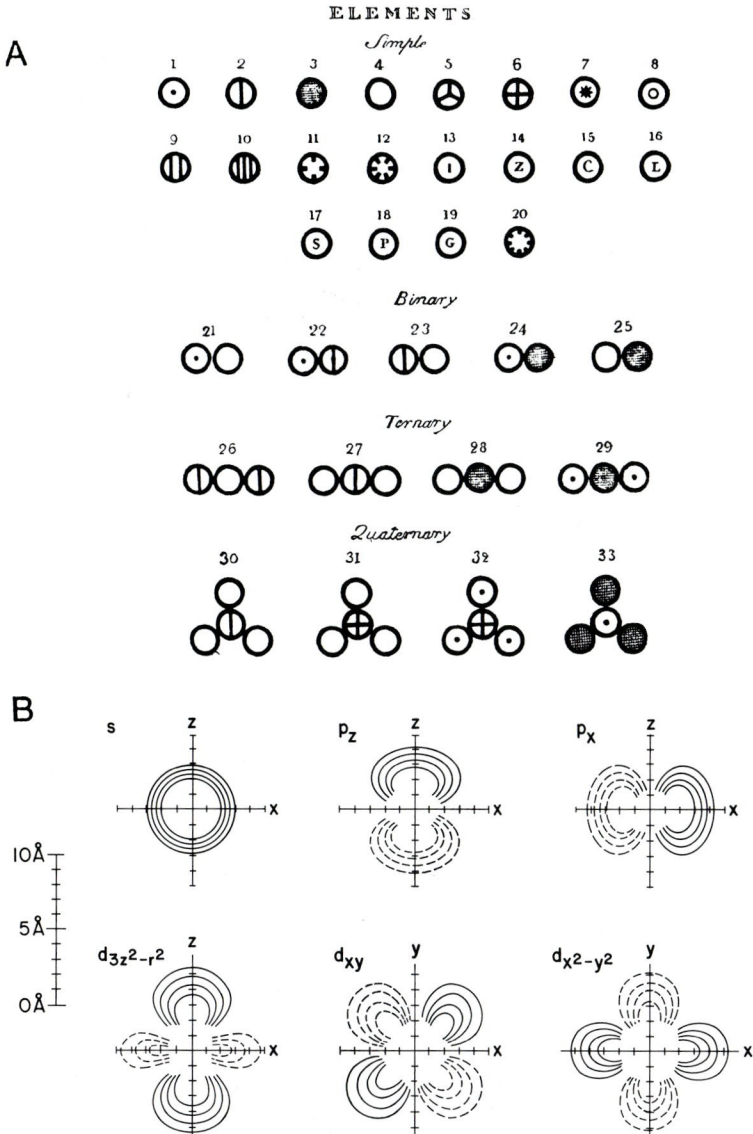

Fig. 1.8. Dalton's atoms and the electronic states in an atom. A, a chart in Dalton's *A New System of Chemical Philosophy,* published in 1808. In modern symbols, these atoms are: 1, H; 2, N; 3, C; 4, O; 5, P; 6, S; 7, Mg; 8, Ca; 9, Na; 10, K; 11, Sr; 12, Ba; 13, Fe; 14, Zn; 15, Cu; 16, Pb; 17, Ag; 18, Pt; 19, Au; 20, Hg. The major modern modification to Dalton's theory is that the atoms are *divisible*. The contour maps in B represent typical electronic states in atoms. The outermost contour on each map represents a density of $10^{-3} Å^{-3}$. The successive contours represent an increase of a factor of 2. The regions with dashed-curve contours have opposite phases in the wavefunction from those with solid-curve contours.

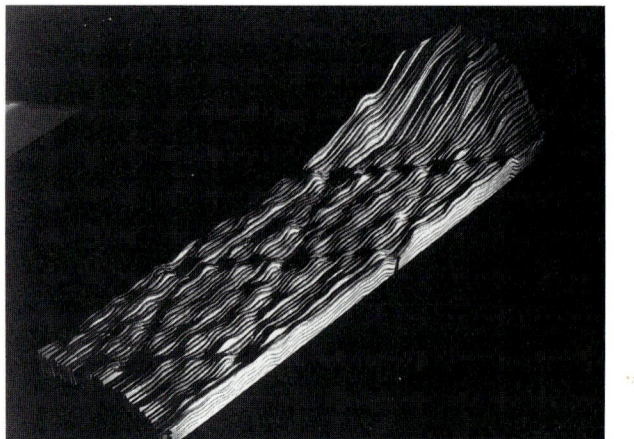

Fig. 1.9. The structure of Si(111)-7×7 resolved in real space. (Reproduced from Binnig, Rohrer, Gerber, and Weibel, 1983, with permission.)

not have a sixfold symmetry. The two triangular halves are different. The true symmetry is *p3m1* instead of *p6mm*. The complete details of the reconstruction were finally worked out by Takayanagi et al. (1985) based on a combination of STM imaging and other surface-probing methods. We will come back to the details of the Takayanagi model later in this section.

To understand the meaning of the STM images of silicon surfaces, we review some basic facts of the crystallography of silicon. We will discuss the simpler Si(111) surfaces first, then the complicated 7×7 reconstruction. In fact, the STM imaging of the simple Si(111) surface is the most elementary case of imaging semiconductors, and perhaps the most instructive one.

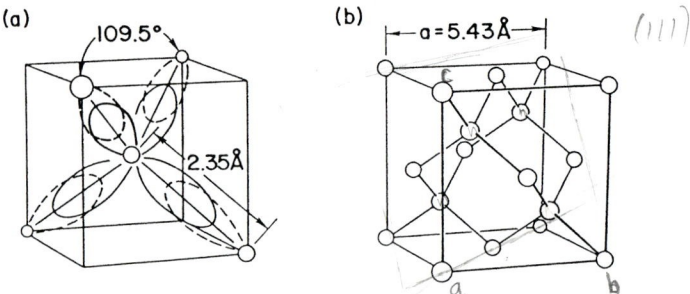

Fig. 1.10. Crystal structure of silicon. (a) The bonding of Si atoms. Each Si atom has four valence electrons, which form four *sp³* orbitals, directed to the four corners of a regular tetrahedron. Each of the *sp³* orbitals is bonded with an *sp³* orbital of another Si atom. (b) The structure of the Si crystal, the so-called diamond structure.

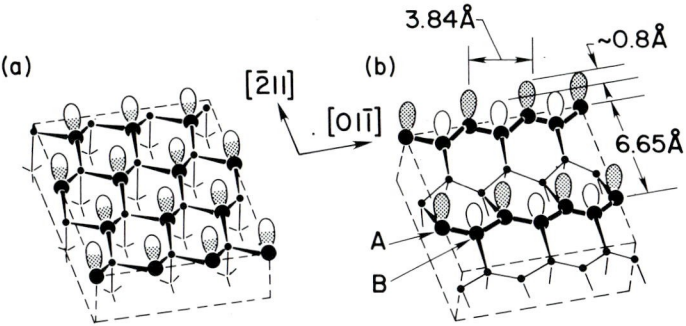

Fig. 1.11. The nascent Si(111) surface and its reconstruction. (a) The nascent Si(111) surface has a threefold symmetry, with nearest-neighbor atomic distance 3.84 Å. (b) The Si(111) surface reconstructs immediately at room temperature to a metastable Si(111)-2×1 surface, which has a lower symmetry. Two rows of dangling bond states are formed: One is filled, another is empty.

The formation of a Si crystal is shown in Fig. 1.10. Aside from the core, each Si atom has four valence electrons: two $3s$ electrons and two $3p$ electrons. To form a Si crystal, one of the $3s$ electrons is excited to the $3p$ orbital. The four valence electrons form four sp^3 hybrid orbitals, each points to a vertex of a tetrahedron, as shown in Fig. 1.10. These four sp^3 orbitals are unpaired, that is, each orbital is occupied by one electron. Since the electron has spin, each orbital can be occupied by two electrons with opposite spins. To satisfy this, each of the directional sp^3 orbitals is bonded with an sp^3 orbital of a neighboring Si atom to form electron pairs, or a *valence bond*. Such a valence bonding of all Si atoms in a crystal form a structure shown in (b) of Fig. 1.10, the so-called *diamond structure*. As seen, it is a cubic crystal. Because all those tetrahedral orbitals are fully occupied, there is no free electron. Thus, similar to diamond, silicon is not a metal.

By inspecting Fig. 1.10, it is obvious that the most natural cleavage plane of a Si crystal is the (111) plane or its equivalent, namely, $(1\bar{1}1)$, $(11\bar{1})$, etc.[1] Right after cleaving, on each of the surface Si atoms, there is a broken bond, or a *dangling bond,* that is perpendicular to the (111) surface. Each of the dangling bond orbitals is half filled, that is, has only one electron. The nascent Si(111) surface is thus *metallic* and exhibits a threefold symmetry, as shown in Fig. 1.11 (a). However, because of the large number of unsaturated bonds, such a surface is unstable. It reconstructs even at room temperature, and loses its threefold symmetry.

---

[1] The symbols for lattice planes, (111) etc., are *Miller indices.* See Ashcroft and Mermin (1976), p. 92.

Fig. 1.13. STM image of the Si(111) surface passivated with H. Taken at a bias of ≈ 2 V, and set current 10 pA. The height range, from white to black, is 0.5 Å. The apparent corrugation is 0.07 Å. The single surface defect in the upper portion of the image is probably an adsorbed molecule. (Reproduced from Higashi et al., 1991, with permission.)

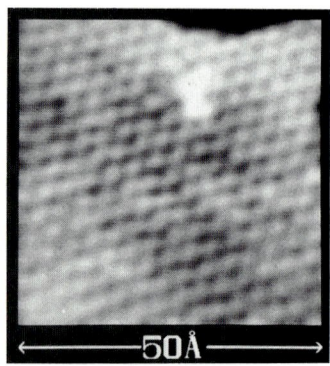

To observe the nascent Si(111) structure, all the dangling bonds have to be capped with something. In silicon technology, this process is called *passivation.* A standard method is to treat it in NH_4F solution, and every unsatisfied dangling bond is capped with a H atom. Higashi et al. (1991) has successfully imaged the hexagonal structure of the passivated Si(111) surface with STM, as shown in Fig. 1.13.

The reconstruction of the Si(111) surface has been the subject of intense research in surface science for many years using various techniques. Based on solid experimental evidence, Pandey (1981, 1982) proposed the π-bond chain model, as shown in Fig. 1.11 (b). The silicon atoms labeled A move up from the second layer to the first layer. The electrons at the silicon atoms labeled B transfer to the A atoms, to make these dangling bonds fully occupied. On the other hand, the dangling bonds at the B atoms become empty states. Again, the surface becomes *semiconducting.* The electronic states in the dangling bonds at the A atoms form a valence band, that is, with energy levels below the Fermi level. The electronic states in the dangling bonds at the B atoms form a conduction band, i. e., with energy levels above the Fermi level. There is an energy gap of about 0.5 eV between them. The local density of states (LDOS) at the Fermi level is virtually zero.

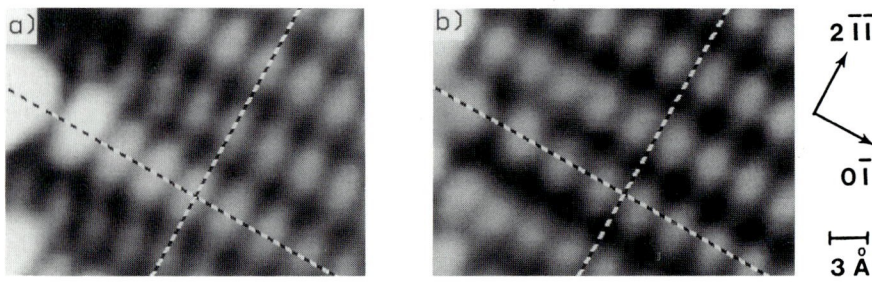

Fig. 1.12. STM image of the Si(111)-2×1 surface. At a different bias, the image is different. (Reproduced from Feenstra and Stroscio, 1987, with permission.)

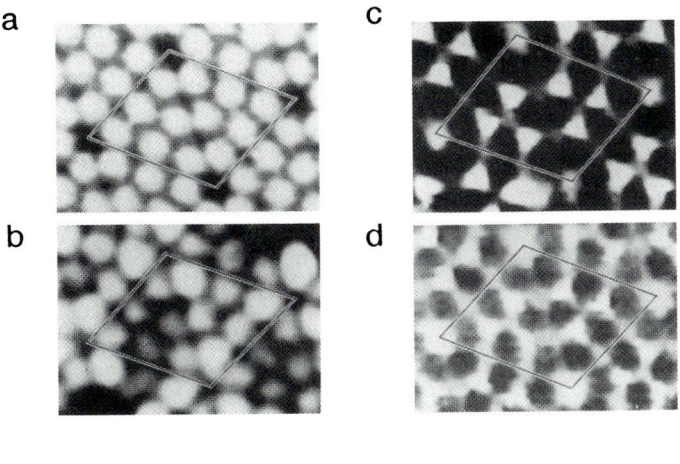

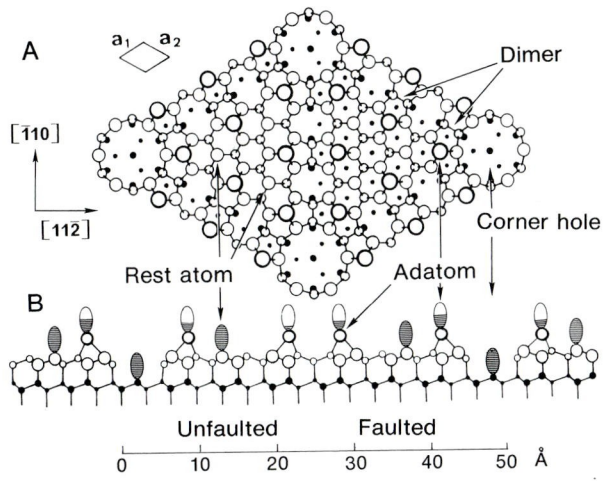

Fig. 1.14. Electronic states and DAS model of Si(111)-7×7. Upper, electronic states on Si(111)-7×7 mapped by STM. (a) Topographic image, taken at a bias of +2 V. (b) Current image at −0.35 V. The two sets of dangling-bond states on the faulted half and the unfaulted half become different. (c) Current image at −0.8 V. The dangling bonds at the rest atoms and the corner holes become apparent. (d) Current image at −1.8 V. The "back-bond" states are observed. (Reproduced from Hamers et al., 1987, with permission.) Below, the DAS model. In each unit cell, there are 9 dimers, 12 adatoms, and a stacking-fault layer. There are 19 dangling bonds in each unit cell. As shown, all 19 dangling bonds and their energy levels have been identified by STM experiments. (After Takayanagi et al., 1985.)

The Si(111)-2×1 surface was first resolved by Feenstra and Stroscio (1987), using the *voltage-dependent imaging* method. By alternatively setting the bias voltage at +1 V and −1 V for consecutive scan lines, two images at opposite bias voltages on the same spatial locations are obtained (Fig. 1.12). The images show that in the $[1\bar{1}0]$ direction, the corrugations of the two STM images are reversed. At the peaks of the image taken at +1 V, there are valleys in the image taken at −1 V. Therefore, *none of the two images represents the atomic structure of the surface.* They are the corrugations of the *surface wavefunctions.* Different types of surface wavefunctions have different energy levels. At different bias voltages, different types of surface wavefunctions are detected. We will provide detailed analysis of these images in Chapters 5 and 6.

By heating the Si(111) surface in ultrahigh vacuum (UHV) to above 400°C, the metastable Si(111)-2×1 is reconstructed to a stable structure. Low energy electron diffraction (LEED) studies showed that the new unit cell has $7 \times 7 = 49$ atoms. The complicated and mysterious structure has been a subject of intensive theoretical and experimental studies for several decades. Combining STM images with other surface analysis techniques, Takayanagi et al. (1985) proposed the DAS model, which has 9 dimers (D), 12 adatoms (A), and a stacking fault layer (S) in each unit cell. Figure 1.14 shows the DAS model. This model is found to be in agreement with all measurements. A review of the history and various models is included in the paper of Takayanagi et al. (1985).

The DAS model has the least number of dangling bonds (19) among all the models ever proposed. There are 12 dangling bonds at the adatoms, 6 at the rest atoms, and 1 at the center atom deep in the corner hole. The 19 dangling bonds are at different energy levels.

By taking images at different biases, or by interrupting the feedback loop temporarily and taking the tunneling current at different biases as the image signal, the different kinds of dangling-bond states as well as other electronic states on the surface can be mapped separately. The STM images in Fig. 1.14 were taken by Hamers et al. (1986). In this figure, (a) is a topographic image of the Si(111)-7×7 surface, taken with a bias +2 V, that is, by tunneling into its unoccupied states. Most of the tunneling current goes into the dangling-bond states at the twelve adatoms. Because these dangling bonds are partially occupied, at a bias of −0.35 V, as shown in (b), the tunneling current from these dangling bonds still dominates the current. However, a distinct asymmetry between the faulted half and the unfaulted half appears: Apparently the energy levels of those two groups of dangling bonds are different. (c) is taken at a bias of −0.8 V. As seen, only the dangling bonds at the rest atoms and the corner holes are present. At a bias of −1.8 V, (d), the "back-bond" states, which are deeply occupied to bind the surface atoms together, become apparent. This set of images shows clearly that STM is capable of imaging

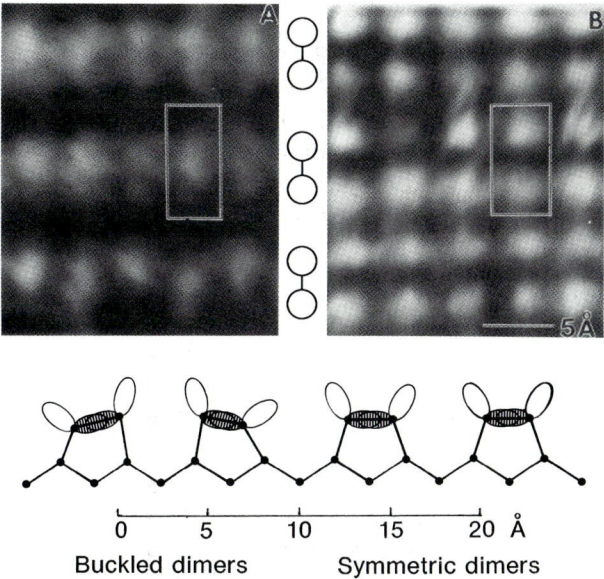

Fig. 1.15. The STM images of Si(100) surface at different bias. (a) At a bias of −1.6 V, the occupied bond between the two atoms in a dimer, is imaged. (b) At a bias +1.6 V, the unoccupied dangling bond on each Si atom is imaged. Notice that the resolution in (b) is much sharper than that in (a). Lower, the dimer model with the dangling bonds is schematically shown. (Reproduced from Hamers, 1989a, with permission.)

individual localized states, both their density distributions in space and their energy levels.

From a technological point of view, Si(001) is one of the most important surfaces. Most integrated circuits are made on it. In the LEED experiments, a (2 × 1) diffraction pattern is observed, which indicates a reconstruction to double the periodicity in one direction. Several models have been proposed. A high-resolution STM study by Tromp et al. (1985) shows unambiguously that the dimer model is correct: In the bulklike terminated surface, each of the top-layer atoms is bonded to two atoms in the second layer, leaving two dangling bonds on each top-layer atom.

Because the bonds between the dimer Si atoms are occupied and the dangling bonds at the top-layer Si atoms are unoccupied, the STM image at positive bias and negative bias should be different. This is indeed observed (Fig. 1.15). As expected, at +1.5 V, that is, by tunneling into the unoccupied states of the Si surface, two well-separated peaks were observed on each dimer. On the other hand, at −1.5 V, that is, by tunneling from the occupied states of the Si surface, only one peak is observed on each dimer. Again, the STM is proved to be sensitive to the localized surface states, or dangling bonds. The very high resolution observed at +1.5 V bias is due to the dangling-bond states, essentially p_z states, having a much sharper charge density distribution than the bond between the two Si atoms in a dimer, when projected on the z plane.

1.3.2 Metals

Atomic resolution has been observed on a large number of clean metal surfaces, including Au(111), Au(001), Al(111), Cu(111), Cu(110), Cu(001), Pt(001), Pt(111), Ru(0001), Ni(110), Ni(001), etc. (see the review article by Behm in *Scanning Tunneling Microscopy and Related Methods,* edited by

Fig. 1.16. Atom-resolved image observed on Cu(111). The atomic distance of Cu(111) is 2.8 Å. A skew dislocation appears on the surface. Near the dislocation, every single Cu atom on the surface is observed. The corrugation amplitude of the flat Cu(111) facet is about 0.3 Å. (Reproduced from Samsavar et al., 1991, with permission.)

Behm et al., 1990). An example is shown in Fig. 1.16, where every single atom on a Cu(111) surface, especially those near a spiral dislocation, is clearly resolved. The observation of atomic resolution on metal surfaces was reported as early as 1987, at the Second International Conference on STM (Chiang et al., 1988; Hallmark et al., 1988). The reported atomic resolution on Au(111) surface, with a corrugation amplitude 0.3 Å, was a pleasant surprise at that time. Later, similar atom-resolved images were observed on virtually every clean metal surface (Behm 1990). The atomic structure of the Au(111)-22× $\sqrt{3}$ surface and the nucleation process on this surface have been one of the most spectacular discoveries by STM, which will be discussed in Chapter 16. An interesting fact to be mentioned is that in many cases, the atomic corrugation is *inverted*. In other words, the atomic sites appear as depressions (minima) on the topographic image, rather than protrusions (maxima). Because such *negative images* sometimes have very high resolution, structural and electronic information can be extracted from those negative images as well. The origin of the negative images will be discussed in Chapter 5.

1.3.3 Organic molecules

A large number of organic molecules have been imaged by STM with atomic resolution. In this sense, the STM opens a new era of organic chemistry. An example is the STM imaging of benzene, adsorbed on the Rh(111) surface with the presence of CO (Fig. 1.17). The benzene molecules form a regular

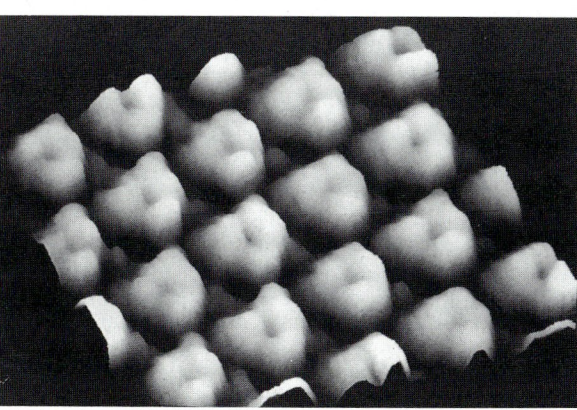

Fig. 1.17. STM image of the benzene molecule. By coabsorbing benzene and CO on Rh(111), a stable hexagonal array of benzene is formed. The CO molecules function as a stabilizing agent. The topographic image is taken with a bias of +10 mV at the sample, and a tunneling current of 2 nA. (Reproduced from Ohtani et al., 1988, with permission.)

hexagonal array on the surface. The center hole of the benzene molecule is clearly visible. The CO molecule in the triangular hole is less visible than the benzene molecule, but make the image of the benzene molecule more triangular than hexagonal.

1.3.4 Layered materials

Many layered materials, such as graphite, $1T$-TaS_2, and $2H$-MoS_2, can easily provide large areas of atomically flat surfaces that are stable in air. On many

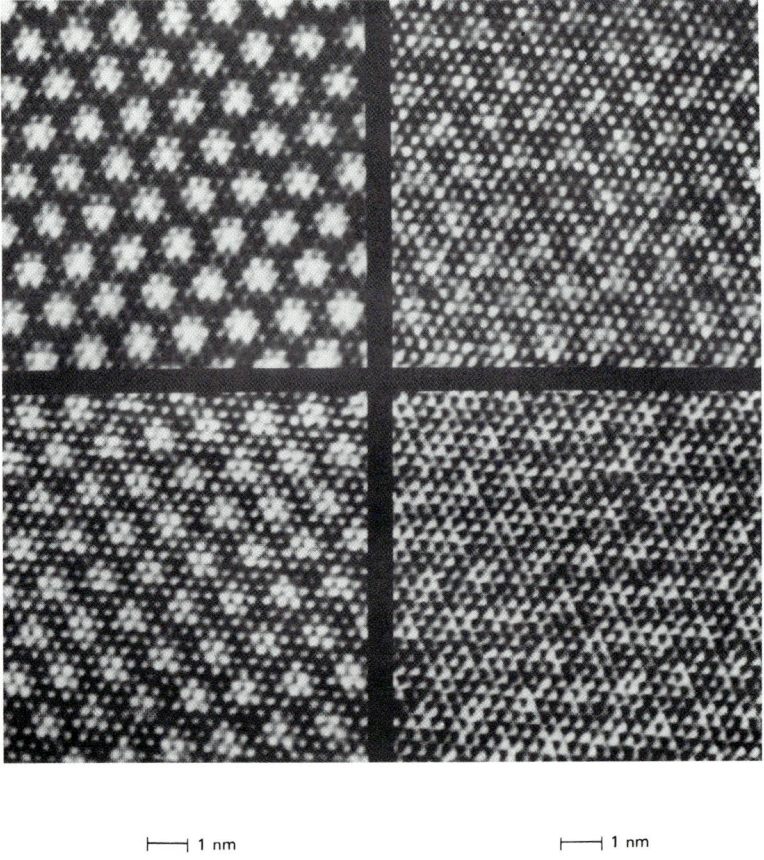

├───┤ 1 nm   ├───┤ 1 nm

Fig. 1.18. Four STM images of 4Hb-TaS$_2$ at 4.2 K. These images were taken during a period of about 2 h on the same area of the surface under identical tunneling conditions (I=2.2 nA, V=25 mV). These images demonstrate the role of tip electronic states on the STM images. (Reproduced from Coleman et al., 1988, with permission.)

layered materials charge-density waves exist. Therefore, layered materials are favorable subjects for STM studies. As with metals, atomic resolution was achieved on virtually every kind of layered-material surfaces, no matter whether its electronic structure was metallic, semimetallic, or semiconducting. An extensive review was published by Coleman et al. (1988).

It is often observed that the actual appearance of the atom-resolved images varies from time to time, and the details are not reproducible. Figure 1.18 shows four STM images of the same area on the same surface, under identical tunneling conditions, within a 2 h period. The difference is due to different electronic states at the tip apex.

1.4 Spatially resolved tunneling spectroscopy

One of the motivations for Binnig et al. (1982a) to pursue tunneling with a controllable gap was to achieve it in a configuration that allows simultaneously spatially resolved tunneling spectroscopy. Historically, tunneling spectroscopy with metal–insulator–metal (MIM) tunneling junctions was first demonstrated by Giaever (1960). His MIM tunneling experiment provided a direct measurement of the energy gap in superconductors, which was a critical evidence for the Bardeen–Cooper–Schrieffer (BCS) theory of superconductivity (1957). Jaklevic and Lambe (1966) further developed a method to study vibrational spectra of molecules sandwiched on the tunneling junction. A wealthy body of concepts has been accumulated in the study of MIM tunneling junctions (see, for example, Hansma, 1982; Duke, 1969; Burstein and Lundquist, 1969; and Wolf, 1985). The concepts developed through the study of MIM tunneling junctions are instrumental in the understanding of the tunneling phenomena in STM, as well as the scanning tunneling spectroscopy (STS). Here, we provide a brief description of the MIM tunneling experiments and the related theoretical concepts.

The realization of MIM tunneling is, to a great extent, owing to a gift from Nature: a superb and easy-to-form insulating film Al_2O_3 (sapphire is Al_2O_3 in crystalline form). By heating a piece of pure Al in air, an extremely stable and strong film of Al_2O_3 is formed. Even as thin as 30 Å, it has exceedingly good electrical properties, and is free from pinholes. By evaporating another metal on the oxide-covered Al piece, an MIM tunneling junction is formed. A typical MIM junction is shown in Fig. 1.19.

The most extensively used theoretical method for the understanding of the MIM tunneling junction is the time-dependent perturbation approach developed by Bardeen (1960). It is sufficiently simple for treating many realistic cases, and has been successfully used for describing a wide variety of effects (Duke, 1969; Kirtley, 1982).

A schematic diagram of the Bardeen approach is shown in Fig. 1.20. Instead of trying to solve the Schrödinger equation of the combined system,

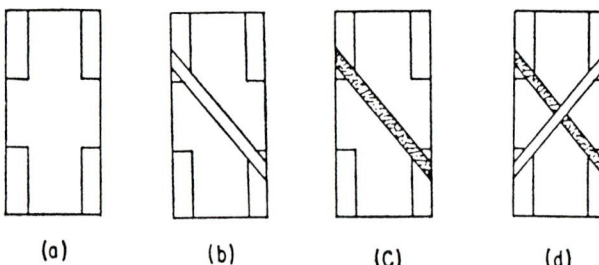

Fig. 1.19. Metal–insulator–metal tunneling junction. The junction is made through the following steps. (a) The substrate, a glass slide with indium contacts. (b) An aluminum strip is vacuum deposited. (c) The aluminum strip is heated in air to form a very thin (~30 Å) aluminum oxide (Al_2O_3). (d) A lead film is deposited across the aluminum strip, forming an Al-Al_2O_3-Pb sandwich. (After Giaever and Megerle, 1961.)

Bardeen considers two separate subsystems first. The electronic states of the separated subsystems are obtained by solving the stationary Schrödinger equations. For many practical systems, those solutions are known. The rate of transferring an electron from one electrode to another is calculated using time-dependent perturbation theory. As a result, Bardeen showed that the amplitude of electron transfer, or the tunneling matrix element M, is determined by the overlap of the surface wavefunctions of the two subsystems at a separation surface (the choice of the separation surface does not affect the results appreciably). In other words, Bardeen showed that the tunneling matrix element M is determined by a surface integral on a separation surface between the two electrodes, $z = z_0$,

$$ M = \frac{\hbar}{2m} \int_{z=z_0} \left(\chi^* \frac{\partial \psi}{\partial z} - \psi \frac{\partial \chi^*}{\partial z} \right) dS, \qquad (1.17) $$

where ψ and χ are the wavefunctions of the two electrodes. The rate of electron transfer is then determined by the Fermi golden rule (Landau and Lifshitz, 1977). Explicitly, the probability w of an electron in the state ψ at energy level E_ψ tunneling to a state χ of energy level E_χ obeys the following equation:

$$ w = \frac{2\pi}{\hbar} |M|^2 \delta(E_\psi - E_\chi), \qquad (1.18) $$

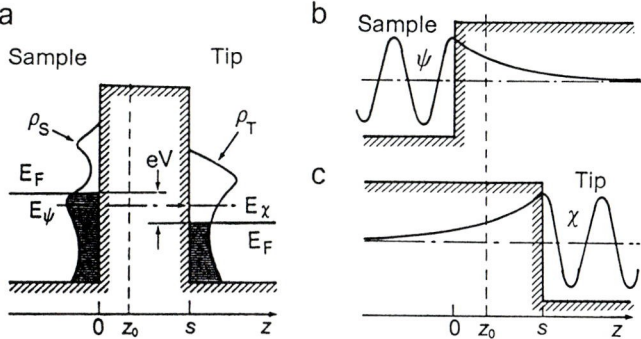

Fig. 1.20. The Bardeen approach to tunneling theory. Instead of solving the Schrödinger equation for the coupled system, a, Bardeen (1960) makes clever use of perturbation theory. Starting with two free subsystems, b and c, the tunneling current is calculated through the overlap of the wavefunctions of free systems using the *Fermi golden rule.*

The δ function in this equation indicates that only the states with the same energy level in both electrodes can tunnel into each other.

   The tunneling current can be evaluated by summing over all the relevant states. At any finite temperature, the electrons in both electrodes follow the Fermi distribution. With a bias voltage V, the total tunneling current is

$$I = \frac{4\pi e}{\hbar} \int_{-\infty}^{\infty} \left[f(E_F - eV + \epsilon) - f(E_F + \epsilon) \right]$$
$$\times \rho_S(E_F - eV + \epsilon)\, \rho_T(E_F + \epsilon) |M|^2\, d\epsilon, \tag{1.19}$$

where $f(E) = \{1 + \exp[(E - E_F)/k_BT]\}^{-1}$ is the Fermi distribution function, and $\rho_S(E)$ and $\rho_T(E)$ are the density of states (DOS) of the two electrodes.

   If k_BT is smaller than the energy resolution required in the measurement, then the Fermi distribution function can be approximated by a step function. In this case, the tunneling current is (see Fig. 1.20):

$$I = \frac{4\pi e}{\hbar} \int_{0}^{eV} \rho_S(E_F - eV + \epsilon)\, \rho_T(E_F + \epsilon) |M|^2\, d\epsilon. \tag{1.20}$$

   In the interpretation of the experiment of Giaever (1960), Bardeen (1960) further assumed that the magnitude of the tunneling matrix element $|M|$ does not change appreciably in the interval of interest. Then, the tunneling current is determined by the convolution of the DOS of two electrodes:

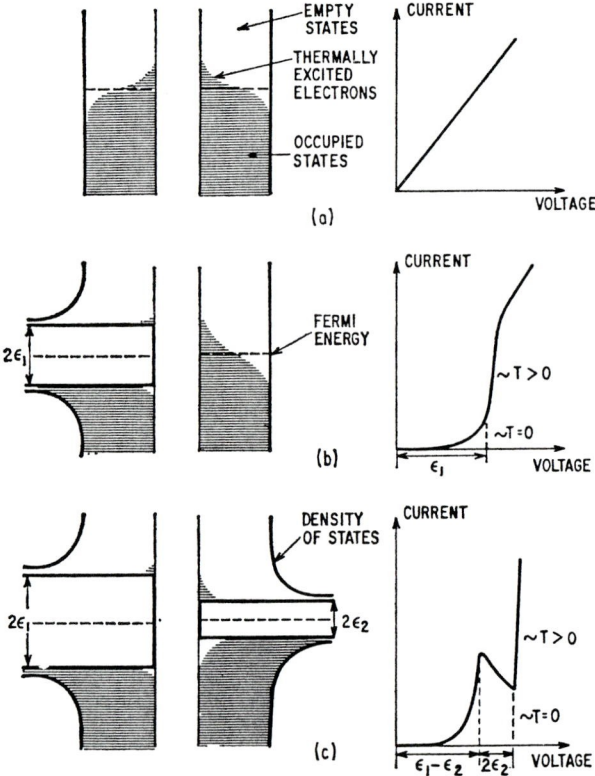

Fig. 1.21. Tunneling spectroscopy in classic tunneling junctions. (a) If both electrodes are metallic, the *I*/*V* curve is linear. (b) If one electrode has an energy gap, an edge occurs in the *I*/*V* curve. (c) If both electrodes have energy gaps, two edges occur. A "negative differential conductance" exists. (After Giaever and Megerle, 1961).

$$I \propto \int_0^{eV} \rho_S(E_F - eV + \epsilon)\, \rho_T(E_F + \epsilon)\, d\epsilon. \tag{1.21}$$

Clearly, according to the Bardeen formula, Eq. (1.21), the electronic structure of the two participating electrodes enters into the formula in a symmetric way. In other words, they are interchangeable. In determining the tunneling current, the DOS of one electrode ρ_S and the DOS of another electrode ρ_T contribute equally in determining the tunneling current. This point was well verified in the classic tunneling junction experiment (Giaever, 1960a; Giaever and Megerle, 1961). As shown in Fig. 1.21, if both electrodes are metals in their normal state, then the $I - V$ curve is a straight line. If one of the metals is superconducting (that is, there is an energy gap and the DOS has

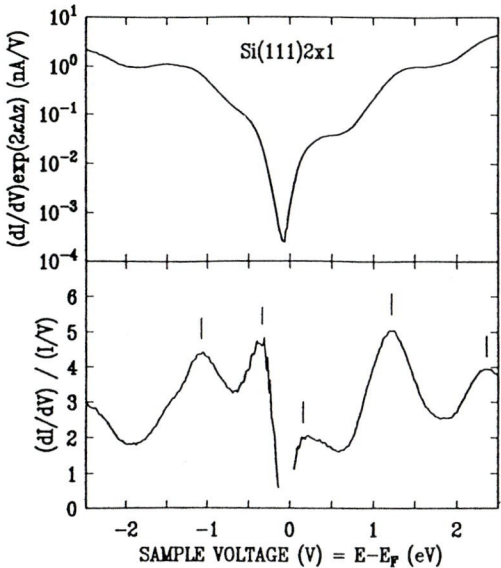

Fig. 1.22. Local tunneling spectrum of Si(111)-2×1. Using the tip treatment procedure developed by Feenstra, Stroscio, and Fein (1987a), reproducible tunneling spectra can be obtained. The tunneling spectrum shown here (Feenstra, 1991) reproduces exactly all the features found in an early measurement (Stroscio, Feenstra, and Fein, 1986). (Reproduced from Feenstra, 1991, with permission.)

a sharp peak), then the $I - V$ curve shows a threshold. If both metals are superconducting (that is, energy gaps exist in the DOS of both), then there are two thresholds with an interval of negative differential conductance.

If in an STS experiment the goal is to obtain the DOS of the sample, one requires a tip with a constant DOS, or a free-electron metal tip. In this case, from Eq. (1.21),

$$\frac{dI}{dV} \propto \rho_S(E_F - eV). \tag{1.22}$$

Thus, with a free-electron metal tip, the dynamic tunneling conductance is proportional to the DOS of the sample.

A systematic method of obtaining local tunneling spectra with STM was developed by Feenstra, Thompson, and Fein (1986). The details of this method will be described in Chapter 14. As expected from the Bardeen formula, the tip DOS plays an equal role as the sample DOS in determining the tunneling spectra. Because the tip is made of transition metals or semiconductors, the tip DOS is usually highly structured. In order to obtain reproducible tunneling spectra of the sample, special tip treatment procedures have to be conducted before the STS experiments are performed. A reproducible recipe of treating tips for spectroscopy was first described by Feenstra, Stroscio, and Fein (1987a). Details will be described in Chapter 14. With such a tip treatment procedure, the tip becomes relatively blunt, and atomic resolution is generally not observed. However, the tunneling spectra obtained by such tips are very reproducible. An example of STS on Si(111)-2×1, by Feenstra (1991), is shown in Fig. 1.22. In the classic tunneling-junction experiments (Giaever, 1960), the observed tunneling spectroscopy is averaged over a large (μm to mm) area. On the other hand, STM allows tunneling spectroscopy to be observed in localized areas on a solid surface.

1.5 Lateral resolution: Early theories

In 1982 and 1983, the unprecedented resolution achieved by STM prompted a number of theoretical studies, aiming at an understanding of the imaging mechanism and an assessment of its intrinsic resolution. Two methods were simultaneously developed. Both theories took the STM image of the profiles of Au(110)-2×1 and Au(110)-3×1 as a testing ground. At that time, atomic resolution was not achieved on gold surfaces. The images were one-dimensional wavy profiles with periodicities 8.15 and 12.23 Å, respectively. One is the scattering approach of Garcia, Ocal, and Flores (1983); and Stoll, Baratoff, Selloni, and Carnevali (1984). Based on it, Stoll (1984) derived an analytic expression of the corrugation function. Another one is the spherical-tip model, or s-wave-tip model of Tersoff and Hamann (1983, 1985).

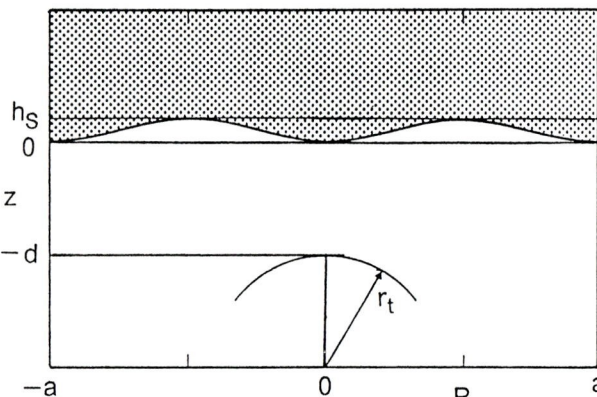

Fig. 1.23. Free-electron-metal model of STM resolution. The sample is modeled as a corrugated free-electron-metal surface. The tip is modeled as a curved free-electron-metal surface with radius r_t at the closest approach to the sample surface. (After Stoll, 1984.)

Both theories predicted similar resolution for the images of the superstructures of the Au(110) surfaces.

1.5.1 The Stoll formula

Based on earlier works of Garcia et al. (1983) and Stoll et al. (1984), Stoll (1984) derived an explicit formula for the corrugation amplitude from a free-electron-metal tip on a corrugated free-electron metal surface. A schematic diagram of the model is shown in Fig. 1.23. The corrugation of the metal surface has a periodicity a and an amplitude h_s. The tip has a radius R near the gap, which has a width d. The tunneling current is calculated directly using a scattering method. The general formalism is complicated, and only numerical results are available. Stoll (1984) showed that when both the periodicity a and the gap width d are large, that is,

$$a \gg \frac{\pi}{\kappa}, \tag{1.23}$$

and

$$d \gg \frac{2}{\kappa}, \tag{1.24}$$

the corrugations of the STM images follow a simple analytic expression,

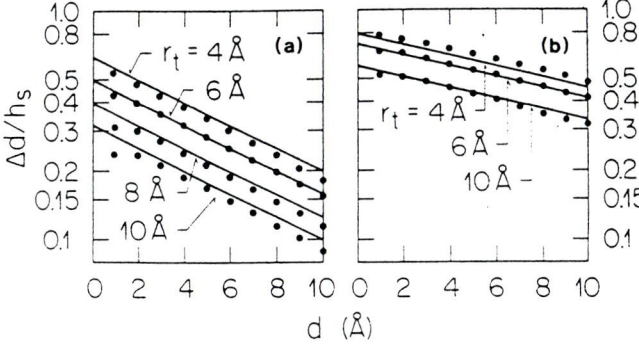

Fig. 1.24. Comparison of Stoll formula with numerical calculations. Solid curves are from the Stoll formula. Dots are results of numerical calculations. Periodicities (a) $a=$ 8.15 Å, (b) $a = 12.23$ Å, corresponding to the case of Au(110)-2×1 and Au(110)-3×1, respectively. (Reproduced from Stoll, 1984, with permission.)

$$\frac{\Delta d}{h_s} = \exp\left(-\frac{\pi^2(R+d)}{\kappa a^2}\right).$$ (1.25)

In these equations, κ is the decay constant, defined by Eq. (1.8), whose typical value is $1\,\text{Å}^{-1}$.

Stoll's formula, Eq. (1.25), was found to be in excellent agreement with the results of numerical calculations for large a and large d, as well as experimental results on Au(110)-2×1 and Au(110)-3×1. For small d, deviations are found, see Fig. 1.24.

Some interesting features of the Stoll formula are worth noting. First, only the sum $(R+d)$ appears in the formula. Therefore, only the distance of the center of curvature of the tip to the sample surface matters. The radius R becomes irrelevant. Second, the STM corrugation decays exponentially with tip–sample separation. By extrapolating the formula to $(R+d)=0$, the corrugation coincides with that of the metal surface. These features are also found to be consistent with experimental results and with the results of the Tersoff–Hamann theory, as described below.

1.5.2 The s-wave-tip model

In the s-wave-tip model (Tersoff and Hamann, 1983, 1985), the tip was also modeled as a protruded piece of Sommerfeld metal, with a radius of curvature R, see Fig. 1.25. The solutions of the Schrödinger equation for a spherical potential well of radius R were taken as tip wavefunctions. Among the numerous solutions of this macroscopic quantum-mechanical problem, Tersoff and Hamann assumed that only the s-wave solution was important. Under such assumptions, the tunneling current has an extremely simple form. At low bias, the tunneling current is proportional to the Fermi-level LDOS *at the center of curvature of the tip* $\mathbf{r}_0$.

$$I \propto \sum_{E_\mu = E_F - eV}^{E_F} |\psi_\mu(\mathbf{r}_0)|^2$$
$$= eV \rho_S(\mathbf{r}_0, E_F),$$ (1.26)

according to the definition of LDOS, Eq. (1.11). Therefore, in the s-wave model, a constant-current STM image is a Fermi-level LDOS contour *of the bare surface,* taken at the center of curvature of the tip $\mathbf{r}_0$. In other words, in the s-wave-tip model, the tip properties can be taken out from the problem, and the STM image reflects the property of the sample *only,* rather than a property of the joint surface–tip system. Furthermore, for free-electron metals, the Fermi-level LDOS contours at a distance from the surface almost coincide

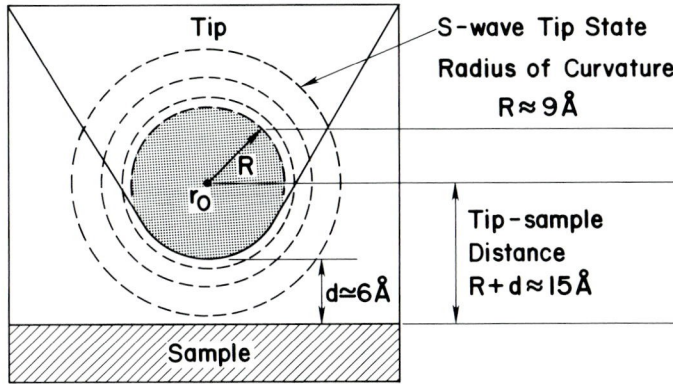

Fig. 1.25. The s-wave-tip model. The tip was modeled as a spherical potential well of radius R. The distance of nearest approach is d. The center of curvature of tip is $\mathbf{r}_0$, at a distance $(R + d)$ from the sample surface. Only the s-wave solution of the spherical-potential-well problem is taken as the tip wavefunction. In the interpretation of the images of the reconstructions on Au(110), the parameters used are: $R = 9\,\text{Å}$, $d = 6\,\text{Å}$. The center of curvature of the tip is 15 Å from the Au surface. (After Tersoff and Hamann, 1983.)

with the contours of the total electron density. Thus, the STM images of free-electron metals are simply the surface charge-density contours (Tersoff and Hamann, 1985).

The Fermi-level LDOS is a well-defined quantity in surface physics. Many of the results of first-principles calculations of the electronic structure of surfaces are presented as LDOS contours at different energy levels, or the contours of integrated LDOS over an energy range. In addition, using an atom-beam scattering technique, the corrugations of the surface LDOS contours at various distances can be measured directly. Therefore, the validity of the s-wave model can be verified directly by comparing the STM images with results of standard numerical calculations of surface electronic structures, and with standard experimental results from helium scattering.

For simple metal surfaces with fundamental periodicity a, the corrugation amplitude of the Fermi-level LDOS as a function of tip–sample distance can be estimated with reasonable accuracy (Tersoff and Hamann, 1985):

$$\Delta z \approx \frac{2}{\kappa} \exp\left[-2\left(\sqrt{\kappa^2 + \frac{\pi^2}{a^2}} - \kappa \right) z \right], \qquad (1.27)$$

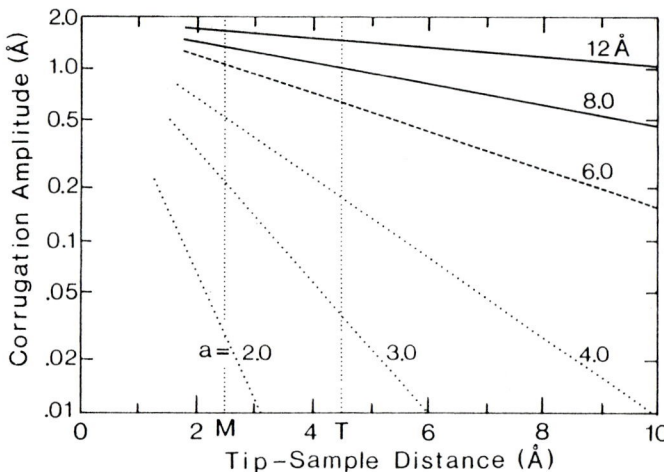

Fig. 1.26. General features of Fermi-level LDOS corrugation amplitude. The corrugation amplitude as a function of the periodicity a and tip–sample distance z, in Å, after Eq. (1.27). The tip–sample distance is defined as the distance from the center of curvature of the tip to the plane of the top-layer nuclei of the sample surface. The point M indicates a sure mechanical contact. The point T indicates the collapse of the tunneling barrier. For profiles of superstructures (solid curves) a is large. The corrugation amplitudes are large and only weakly depend on z. For low-Miller-index surfaces (dotted curves) a is small. The LDOS contours have no detectable corrugation within reasonable tip–sample distance. Dashed curve: intermediate case.

where z is the distance from the top-layer nuclei of the sample surface to the center of curvature of the tip. With a typical value of the work function, $\phi = 4$ eV, a plot of Eq. (1.27) for different structure periodicities is shown in Fig. 1.26. As shown, for different values of structure periodicities a, according to Eq. (1.27), the corrugation amplitude depends dramatically on a. For large periodicities, that is, $a \gg \pi/\kappa$, the corrugation amplitudes are large and approximately independent of the tip–sample distance. Actually, for large superstructures, Eq. (1.27) can be simplified to

$$\Delta z = \text{const.} \times \exp\left(-\frac{\pi^2 z}{a^2 \kappa} \right). \tag{1.28}$$

This expression coincides with the Stoll formula, Eq. (1.25), within a constant factor. Therefore, both theories provide an adequate explanation of the topographic STM images of the superstructures on Au(110) surfaces.

The superstructures of the Au(110) surface, with periodicities 8.15 and 12.23 Å, respectively, are much larger than the resolution of STM observed experimentally. For example, the atomic structure of the Au(110) surface, with atomic spacing 2.9 Å, is clearly resolved by STM (Berndt et al., 1992). Although the s-wave model adequately explains the contours of the superstructures on the Au(110) surface, it cannot explain the atomic resolution observed on low-Miller-index metal surfaces. For most metals (except alkali and alkaline earth metals, which are rarely imaged by STM), the atomic spacing is 2.5–3 Å. According to Eq. (1.27) and Fig. 1.26, the corrugation amplitudes of the Fermi-level LDOS, at any realistic tip–sample distances, are too small to be detected experimentally. Therefore, the s-wave-tip model does not explain the experimentally observed atomic resolution by STM.

The original s-wave-tip model described the tip as a macroscopic spherical potential well, for example, with $r \approx 9$ Å. It describes the protruded end of a free-electron-metal tip. Another incarnation of the s-wave-tip model is the Na-atom-tip model. It assumes that the tip is an alkali metal atom, for example, a Na atom, weakly adsorbed on a metal surface (Lang, 1986; see Section 6.3). Similar to the original s-wave model, the Na-atom-tip model predicts a very low intrinsic lateral resolution.

Because the s-wave tip model predicts a very low intrinsic resolution, several remedies were proposed to explain the observed atomic resolution. The first one is to imagine a tip whose potential and wavefunctions were arbitrarily localized (Hansma and Tersoff, 1987). In other words, the tip is imagined to be a fictitious geometrical point. To explain the observed STM images with atomic resolution, that fictitious geometrical point must be placed at an unphysically short distance from the nuclei on the sample surface, for example, 1.3 Å (Yang et al. 1992). Furthermore, once the radius of the spherical potential well is shrunk to a subatomic dimension, the uncertainty principle becomes effective. The spacings of the energy levels for potential wells with subatomic radii can be as large as a few keV (Chung, Feuchtwang, and Cutler, 1987). Therefore, the point-tip model is not consistent with quantum mechanics, and does not represent a physical reality. Another remedy is to attribute the observed atomic corrugations to a singularity in the band structure of the sample surface. It is called the theory of anomalous corrugation (Tersoff, 1986). For example, on graphite, it was proposed that because the point tip images a single Bloch wave at the Fermi level, the nodes of the Bloch wave provide an infinitely large contrast. However, such an enhancement mechanism depends on a logarithmic singularity, which is extremely unstable. A minute mixing of a non-s-wave tip state will eliminate the enhancement completely, as shown by Lawanmi and Payne (1990). A careful analysis by Tersoff and Lang (1990) showed that the s-wave-tip model breaks down in the case of graphite, and the STM images depend dramatically on tip electronic states. Hence, the theory of anomalous corrugation is not valid. The analysis of Tersoff and Lang (1990) will be presented in Section 5.6.

1.6 Origin of atomic resolution in STM

As we have discussed in Section 1.3, experimentally, atomic resolution has been observed on literally every clean surfaces of metals and semiconductors. Today, atomic resolution on rigid surfaces has become a "must" in STM operation (Rohrer, 1992). In order to resolve single atoms, a lateral resolution of 2 Å is required. The importance of the STM — the feature that sets it apart from other instruments — is that it can resolve details in the vicinity of a single atom, otherwise it would not have created the excitement that now surrounds it (Quate, 1986). Here, we briefly discuss the origin of its atomic resolution.

A review of the early observations of the atomic resolution on metals by STM is provided by Behm (1990). It was first observed on Au(111) (Chiang et al., 1988; Hallmark et al., 1988; Wöll et al., 1989). Subsequently, atomic resolution is also observed on Al(111), Ni(100), Ru(0001), Cu(100), Cu(111), Au(110), Ag(111), Ni(110), and Pt(111), etc. No exception is reported.

An early systematic experimental study on the imaging mechanism was conducted on Al(111) (Wintterlin et al., 1989). The observed corrugation amplitude was more than one order of magnitude larger than the Fermi-level LDOS corrugation. Aluminum is a textbook example of simple metals. The electronic states on the Al(111) surface have been studied thoroughly.

The surface charge density of Al(111) has been well characterized by first-principles calculations as well as helium scattering experiments. The asymptote of the corrugation amplitude Δz of equal-LDOS surface contours follows an exponential law, as obtained from a first-principles calculation of the electronic structure of the Al(111) surface (Mednick and Kleinman, 1980):

$$\Delta z \approx 3.0 \, e^{-1.56z}, \tag{1.29}$$

where z is the distance from the plane of the top layer nuclei of the Al surface. The units are in Å. At a distance of 3 Å, the corrugation amplitude is about 0.03 Å. This is confirmed experimentally by atom-beam diffraction. A helium atom can approach to within 3 Å of the top-layer nuclei. An STM tip cannot approach closer than a helium atom. If the STM image is the contour of equal LDOS surface, then the corrugation amplitude in any STM image should be much smaller than 0.03 Å.

STM experiments have routinely demonstrated corrugation amplitudes more than one order of magnitude greater than the Fermi-level LDOS corrugation. As reported by Wintterlin et al. (1989), With an appropriate tip-sharpening procedure, a corrugation amplitude of 0.3 Å can be routinely obtained. The largest corrugation observed was 0.8 Å. A complete record of the dependence of corrugation amplitude with distance is shown in Fig. 1.27. The corrugation amplitudes are larger than 0.1 Å over a distance range of 2 Å.

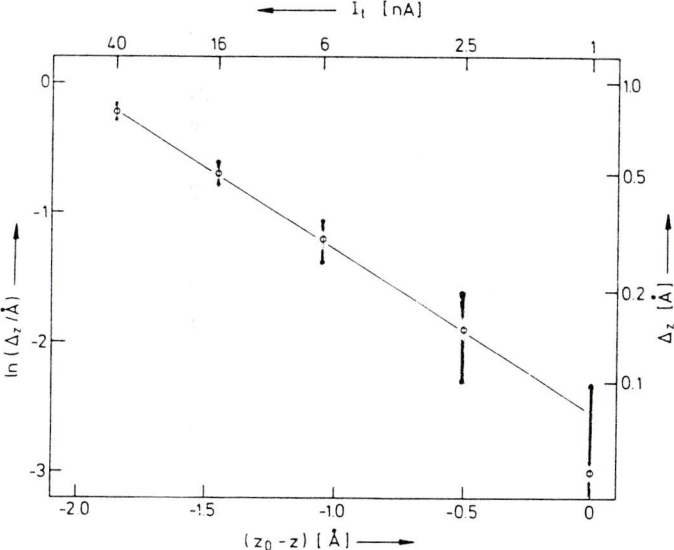

Fig. 1.27. Quantitative results from STM images of Al(111). An exponential relation between the corrugation amplitude and the tip–sample distance is observed. The best corrugation observed is more than 20 times greater than the maximum corrugation amplitude expected from the Fermi-level LDOS contour. (Reproduced from Wintterlin et al., 1989, with permission.)

An exponential dependence of corrugation amplitude with distance is clearly observed.

The atomic resolution in STM can be understood in terms of tip electronic states and tip–sample interactions. We will discuss the effect of tip electronic states in this section, and the tip–sample interactions in the next section.

Soon after the first STM experiments on Si(111)-7×7 (Binnig et al., 1983), Baratoff (1984) proposed that the atomic resolution in STM is probably due to a single dangling bond protruding from the tip. Many transition-metal surfaces, such as W(100) and Mo(100), have a strong tendency to form highly localized surface states (see Weng et al., 1978, and references therein). Especially, the d_z-like surface state on W(001) surface is located at the Fermi level. Those localized surface states were discovered experimentally by Swanson and Crouser (1966, 1967), and studied extensively by Posternak et al. (1980). We will discuss them in detail in Chapter 4.

To investigate the STM imaging mechanism further, Ohnishi and Tsukada (1989) made a first-principles calculation of the electronic states for a number of W clusters. From the calculations, they found that on the apex atom of many W clusters, there is a d_z-like state protruding from the apex atom, energetically very close to the Fermi level. Using Green's function methods,

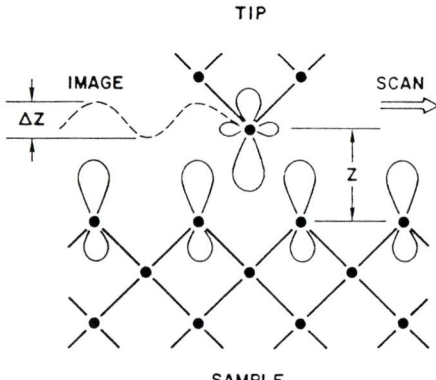

Fig. 1.28. Microscopic view of STM imaging mechanism. An atomic state at the tip end, exemplified by a d_{z^2} state protruding from the apex of a W tip, interacts with a two-dimensional array of atomic states, exemplified by sp^3 states on the Si surface. This results in a highly corrugated tunneling current distribution. (Reproduced from Chen, 1991, with permission.)

they also found that the tunneling current is predominately contributed by this d state. Ohnishi and Tsukada (1989) proposed that such an orbital would be advantageous for a sharp STM image. Fig. 1.29 shows the electronic states near the Fermi level on W_4 and W_5 clusters.

Demuth et al. (1988) analyzed the effect of electronic states on the tip based on a number of experimental facts. They emphasized that the tip is one half of the STM experiment, since the tunneling current is determined by the convolution of the tip state and the sample state. Unlike the surface, the tip tends to be more difficult to control than the sample. Experimentally, even the best prepared clean tungsten tips usually do not immediately produce the highest resolution on the Si surface. Most often the highest resolution is achieved after long periods of scanning or controlled tip crashing. When there is no atomic resolution, an effective procedure to achieve atomic resolution is to collide the tip mildly with the Si surface. After such a controlled crashing, a crater is found on the Si surface, which shows that a Si cluster has been picked up by the tip. Atomic resolution is then often achieved. Demuth et al. (1988) proposed that at the end of the Si cluster, there is a p_z like dangling bond protruding from the tip end, which provides atomic resolution.

The effect of p_z or d_{z^2} dangling bonds on STM resolution can be understood in the light of the *reciprocity principle* (Chen, 1988), which is the fundamental microscopic symmetry between the tip and the sample: By interchanging the "acting" electronic state of the tip and the sample state under observation, the image should be the same.

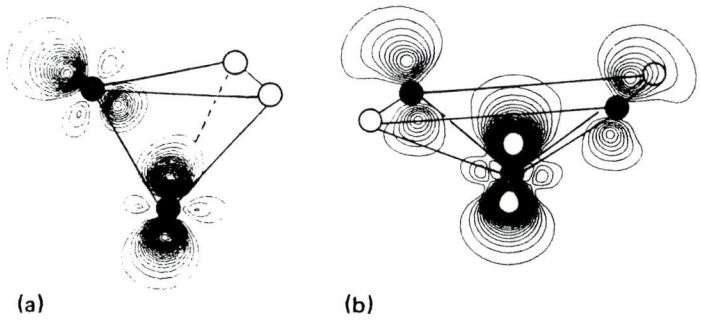

(a)   **(b)**

Fig. 1.29. Electronic states on W clusters. The electronic states near the Fermi level on tungsten clusters, W_4 and W_5, were calculated by Ohnishi and Tsukuda (1989). At low bias, these d like tip states contribute more than 90% of the tunneling current. (Reproduced from Tsukada, Kobayashi, and Ohnishi, 1990, with permission.)

The effect of a dangling bond at the tip apex can be illustrated by experimental facts and the reciprocity principle as follows. Consider the STM images of the Si(100) surface on different biases, Fig. 1.15. The two images of the same area are taken with the same atomlike state on the tip. It can be imagined as the image of one sample state with an array of tip states. Clearly, if the tip state is the p_z like dangling bond, the images should be much sharper than the images with the bond between the dimer as the tip state.

The discrepancy between the sharp STM image and the low corrugation of the charge density on Al(111) can be explained in light of the reciprocity principle as well. Figure 1.30 shows a qualitative explanation of the effect of a d_z tip state. For an s-wave tip state, the STM image of a metal surface is the charge-density contour, which can be evaluated using *atomic-charge superposition* (that is, as a sum of the charge densities of individual atoms, each made of s-states). According to the reciprocity principle (see Fig. 1.30), with a d_z tip state, the tip no longer traces the contour of the Fermi-level LDOS. Instead, it traces the charge-density contour of a *fictitious surface* with a d_z state on each atom. Obviously, this contour exhibits much stronger atomic corrugation than that of the Fermi-level LDOS.

The importance and effect of tip electronic states were discussed by many authors (Tersoff, 1990; Tersoff and Lang, 1990; Doyen et al., 1990; Behm, 1990; Lawunmi and Payne, 1990; Sacks and Noguera, 1991). Doyen et al. (1990) made a first-principles calculation of a realistic W tip. They found that the electronic structure of the W tip exhibits a $5d_z$ resonance near the Fermi level, which is the most possible origin of atomic resolution. Sacks and Noguera (1991) noted that s and d states dominate the DOS of the W surface near the Fermi level, and p states could arise from an adsorbed foreign atom. They have also derived the necessary formalism to account for the effect of p

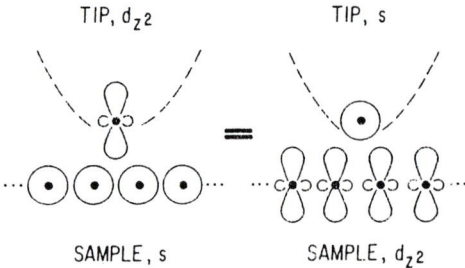

Fig. 1.30. Origin of atomic resolution on metal surfaces. According to the reciprocity principle, the image taken with a d_z tip state (which exists on a W tip) on a free-electron-metal surface is equivalent to an image taken with a point tip on a fictitious sample surface with a d_z state on each top-layer atom, which obviously has a strong corrugation. (Reproduced from Chen, 1990, with permission.)

and d tip states. Tersoff and Lang (1990) made a careful calculation of the electronic states of five different tip atoms, Na, Ca, Si, Mo, and C. Although the results for Na and Ca resemble the conclusions of the Tersoff–Hamann theory, they noted that real tips are neither Na nor Ca, but rather transition metals, possibly contaminated with atoms from the surface (for example, Si and C are common sample materials). For Si, with an s^2p^2 atomic configuration, the $m = 1$ contribution is almost as large as the $m = 0$ (which includes both s and p_z). And for Mo, while the p contribution is reduced, this is more than compensated by the large contribution from states of d like symmetry.

Some recent experiments further illustrated the importance of tip electronic states. Barth et al. (1990) found that the corrugation of STM image of the Au(111) surface is often inverted, that is, the atom sites appears as depressions instead of protrusions. It is also reported that with a sudden change of the tip, the STM image switches suddenly from inverted to noninverted, which indicates a tip-state effect. Yang et al. (1992) performed an atomic charge superposition calculation of STM images of glycine and alanine, and found that their images resembled the LDOS contour of a tip–sample distance of 1.3 Å, which is unphysical. By checking at experimental conditions, they concluded that the only interpretation was the existence of a d-type localized state on the tip.

1.7 Tip–sample interaction effects

In order to achieve atomic resolution, even with p and d tip states, the tip–sample distance should be very short. At such short distances, the tip–sample interactions are strong. There are two kinds of interaction effects to be considered. The first is the modification of the sample wavefunction due to the proximity of the tip. The second is the force between the tip and the sample. Both are important to consider.

The necessity of considering the modification of sample wavefunctions was already discussed in the early years of STM by Binnig et al. (1984), Baratoff (1984), and Garcia (1986). Because of the effect of the imaging force, the top of the potential barrier is much lower than the vacuum level. At the optimum distance for observing atomic resolution, the barrier collapses. From the semiclassical point of view, it is no longer tunneling. If Bardeen's tunneling theory (1960) is going to be applied, the modification of the sample wavefunction due to the proximity of the tip must be taken into account, and vice versa, as shown in Fig. 1.31.

A series of first-principles calculations of the *combined system,* that is, the tip *and* the sample, has been carried out by many authors, for example, Ciraci, Baratoff, and Batra (1990, 1990a). The three-dimensional shape of the potential barrier as well as the force between the tip and the sample are calculated. Three systems have been studied: graphite–carbon, graphite–aluminum, and aluminum–aluminum. All those studies reached the same conclusion: The top of the potential barrier between the tip and the sample is either very close to or lower than the Fermi level within the normal tip–sample distances of STM.

The problem of wavefunction modification can be treated by a stationary-state perturbation theory. With such modifications considered, Bardeen's time-dependent perturbation approach becomes adequately accurate. This two-step perturbation approach is similar to the perturbation treatment of the hydrogen molecular ion by Holstein (1955), which is presented in the quantum-mechanics textbook of Landau and Lifshitz (1977). In the case of two planar electrodes and the case of the hydrogen molecular ion, the modification of the bare wavefunctions is a numerical constant. Actually, since the image force is a long-range interaction, even at large tip–sample distances, a modification of the bare wavefunctions is necessary. The accuracy of the perturbation approach is verified by comparing with an analytically soluble case — the hydrogen molecular ion problem. It reproduces the asymptotes of the exact solution with very high accuracy.

The effect of atomic forces in STM was observed repeatedly by direct experiments. Before the invention of the STM, Teague (1978) already reported the observation of the attractive and repulsive forces in a gold–vacuum–gold tunneling experiment. We will discuss this in Section 1.8. Shortly after the invention of the STM, the importance of repulsive force in STM experiments was suggested by Coombs and Pethica (1986). Soler et al. (1986) proposed that the force and deformation form the primary mechanism behind the observed giant corrugation on graphite. Dürig et al. (1988, 1990) made a direct measurement of the force in STM. Their method is to make *simultaneous* measurements of the force *gradient* and the tunneling

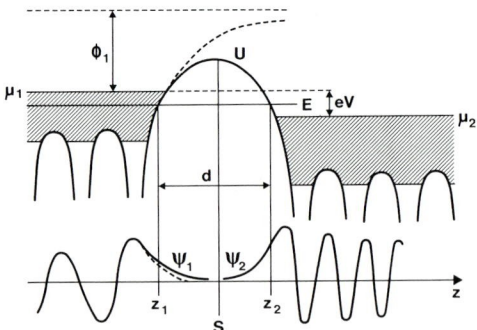

Fig. 1.31. The effect of potential barrier lowering on wavefunctions. At the normal tip–sample distance of STM operation, due to the interaction between the tip and the sample, the top of the potential barrier becomes much lower than the vacuum level. The wavefunctions of both parties in the middle of the gap are different from the "free" wavefunctions. (Reproduced from Baratoff, 1984, with permission.)

conductance in an STM experiment. By integrating the force gradient, the equilibrium point is determined. The tunneling conductance as a function of the tip–sample distance (with respect to the equilibrium point) is directly measured. It was determined that the normal tip–sample distance of STM operation is 1–4 Å from the equilibrium point.

To summarize, the existence and role of force in STM is now a well-established scientific fact. At a relatively large absolute distance, for example, 5 Å, the force between these two parties is attractive. (By absolute distance we mean the distance between the nucleus of the apex atom of the tip and the top-layer nuclei of the sample surface.) At very short absolute distances, for example, 1.5 Å, the force between these two parts is repulsive. Between these two extremes, there is a well-defined position where the net force between the tip and the sample is zero. It is the *equilibrium distance*. On the absolute distance scale, the equilibrium distance is about 2–2.5 Å. Therefore, the tip–sample distance of normal STM operation is 3–7 Å on the absolute distance scale. In this range, the attractive atomic force dominates, and the distortion of wavefunctions cannot be disregarded. Therefore, any serious attempt to understand the imaging mechanism of STM should consider the effect of atomic forces and the wavefunction distortions.

1.8 Historical remarks

All major scientific discoveries have prior arts, and STM is no exception. Imaging of individual atoms had been achieved many years before the invention of STM by field-ion microscopy (FIM) and field-emission microscopy (FEM); see Tsong (1990) and Brodie (1978); also, for more limited situations, by high-resolution transmission electron microscopy (TEM); see Crewe et al. (1970) and Spence (1988). Vacuum tunneling has been demonstrated using the topografiner by Young, ward, and Scire (1971); and by Teague (1978). We are presenting those methods not only for historical interest. FIM and FEM are still among the most useful methods for characterizing the STM tip. The discovery of the role of atomic forces in vacuum tunneling by Teague (1978) is important in the understanding of the STM imaging mechanism. It also inspired the invention of atomic force microscopy (AFM); see Binnig, Quate, and Gerber (1986).

1.8.1 Imaging individual atoms

Two important methods of imaging individual atoms before STM were both invented by one scientist — Erwin Müller: the FEM in 1936, and the FIM in 1951. There are books describing these methods (Müller and Tsong, 1969; Tsong, 1990). A detailed history of the inventions together with a biography of Erwin Müller are given by Drechsler (1978).

Field-ion microscopy

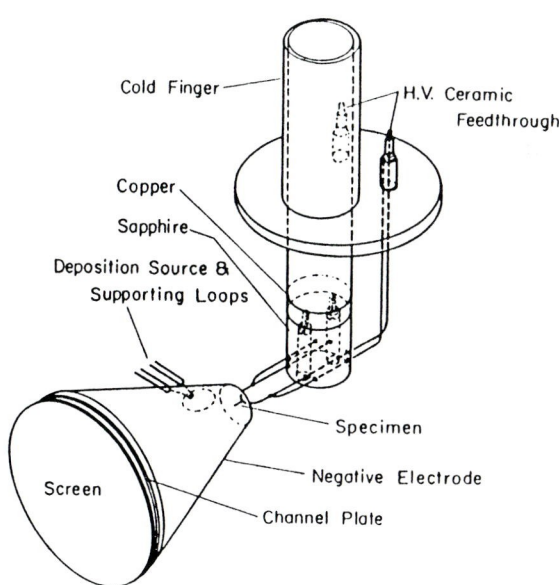

Fig. 1.32. Schematics of a field-ion microscope (FIM). The sample, a tip of radius ≈ 100 Å is at a high positive voltage relative to a fluorescent screen, placed in a chamber filled with a few millitorrs of He. The He ions generated at the tip surface projects an atomic image of the tip. (Reproduced from Tsong, 1990, with permission.)

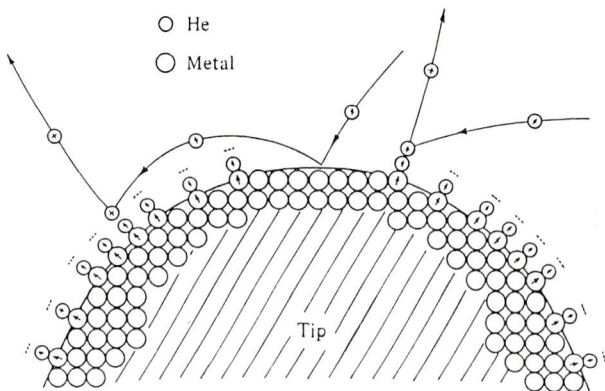

Fig. 1.33. Image mechanism of field-ion microscopy. A sharp tip, usually made of refractory metal, is placed in a chamber filled with He at about 10^{-4} torr. With an electric field of a few V/Å, each of those atoms in the more protruding positions will have a He atom adsorbed on it. The He atom may be ionized to form a He ion, then be accelerated by the electric field to form an image of this surface atom. (Reproduced from Tsong, 1990, with permission.)

Field-ion microscopy (FIM) was the first method to enable the visualization of single atoms at certain metal surfaces. It is shown schematically in Fig. 1.32. A vacuum chamber with basic vacuum better than 10^{-6} torr is filled with a gas, for example, 5×10^{-3} torr of helium. A tip of small radius R is kept at a high positive voltage of the order of 10 kV, opposite to a grounded fluorescent screen at a distance L. Due to the highly nonuniform electrical field near the tip end, the helium atoms are polarized and attracted to the tip. The He atoms are ionized by the very high electrical field at the surface of the tip, and then accelerated towards the screen. The trajectories of the helium ions are strongly affected by the local variation of electrical field at the tip end. An image of the field distribution at the tip end is formed on the fluorescence screen, then observed visually or photographed. (See Fig. 1.33.)

A rough estimation of the magnification M can be obtained from a heuristic geometrical argument to yield

$$M = \frac{L}{R}. \tag{1.30}$$

Typically, $L = 10$ cm, $R = 100$ Å, and $M = 10^7$. An atom at the tip end, with a typical diameter 3 Å, creates a 3-mm-size spot on the screen. A deviation up to a factor of 3 from the prediction of Eq. (1.30) was observed experimentally (Müller and Tsong, 1969). Nevertheless, it gives the correct order of magnitude.

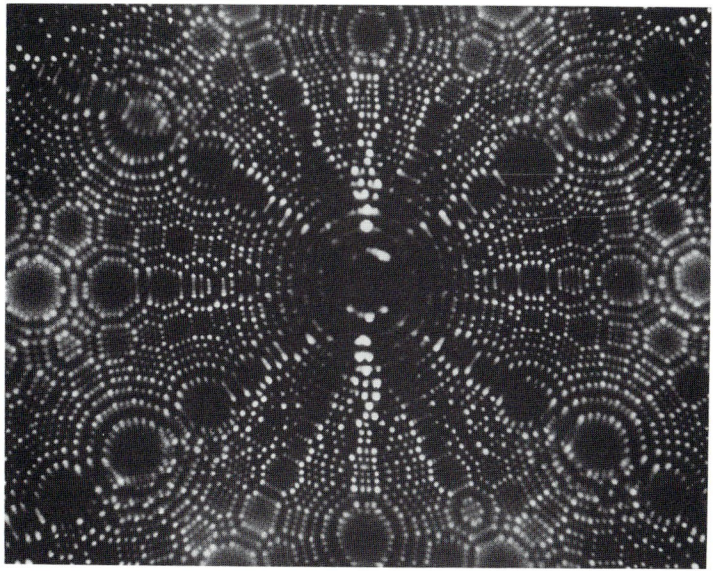

Fig. 1.34. FIM image of a W tip. (Courtesy of T. T. Tsong.)

To obtain sharper images, the tip is usually cooled below liquid nitrogen temperature. However, atomic features can be observed at room temperature without cooling, which is more convenient for in situ characterization of the STM tip.

The most frequently studied samples with FIM are refractory metal tips, such as W, Mo, Pt, Ir, etc. The field evaporation threshold for refractory metals is appreciably higher than the field to ionize helium atoms, which is 4.5 V/Å. Field evaporation is also used for forming and cleaning the FIM sample, which is the tip end, to make it a sharp end and to remove adsorbed exotic atoms. A typical FIM image is shown in Fig. 1.34.

Tsong (1987) made an extensive comparison between FIM and STM. The major points are as follows:

1. The samples in FIM are limited to the end of a tip, with an image area $<(80 \text{ Å})^2$. The samples in STM have no shape and size limitation in principle.

2. FIM can only be operated in vacuum at cryogenic temperatures, typically liquid nitrogen and liquid hydrogen temperatures. STM can be operated in vacuum, gas, and liquid, from millikelvins to a few hundred degrees centigrade.

3. The surfaces prepared for FIM reflect the field evaporation process itself, whereas the surfaces studied by STM are the thermal equilibrium surfaces.

4. The FIM images are nonlinear and difficult to calibrate. The STM images are basically linear and easy to calibrate.

5. The FIM images do not contain information about the chemical identity of the atoms. By combining with tunneling spectroscopy, limited chemical information can be extracted from STM images.

6. By combining with a time-of-flight atom-probe method, FIM is capable of providing information about the chemical identity and even the isotope number of the surface atoms, when it is evaporated from the surface (Tsong, 1990). In this application, FIM is unique.

The fact FIM only images a small area at the apex of the tip makes it insensitive to vibration and to thermal drift. Because of this, FIM has been used to study the motion of a single atom on a tip over a long period of time. By tracing the trajectory of a single atom, the surface diffusion coefficient and the rate of directional walk of single atoms was directly measured.

The directional walk of a single atom on the tip surface is not only an interesting observation, but is also relevant to STM experiments. Many tip treatment methods for STM rely on the application of a high electric field near the tip end, which will induce directional walks of atoms at the tip surface, among other effects. Figure 1.35 is an example of the directional walk of a W atom on the surface of a W tip (Tsong, 1990).

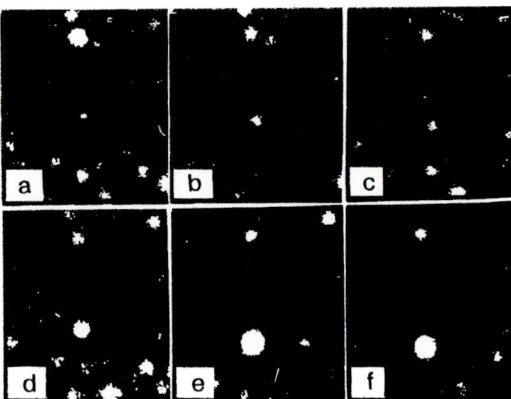

Fig. 1.35. Observation of directional walk of a W atom by FIM. In the absence of an electric field, a W atom on a W surface walks randomly. Under the influence of an electric field, the W atom is polarized and walks toward the direction of high field intensity. (Reproduced from Tsong and Kellogg, 1975, with permission.)

Field-emission microscopy

Although field-emission microscopy (FEM) was invented as early as 1936 (see Drechsler, 1978), atomic resolution was not achieved until the late 1950s. (A brief description of the field-emission phenomenon is included in Section 1.8.2.) In FEM, instead of atoms, electrons are used as the probing particles. Because the de Broglie wavelength of low-energy electrons is much longer than that for atoms, the intrinsic resolution of FEM should be lower than FIM. In the 1950s, it was believed that the resolution of FEM was about 10 Å. Atomic resolution, even 4 Å, was impossible. However, Brodie (1978) showed that with a high screen voltage ($\approx$ 10 kV) and an extremely small tip radius ($\approx$ 100 Å), atomic resolution was possible. Especially, since some atoms can alter the local work function dramatically, for example, alkali metal and alkaline earth atoms. Those atoms are possible candidates of atomic resolution with FEM. For example, by evaporating barium onto a nickel tip, numerous bright spots on the fluorescence screen are clearly observed, because the Ba atoms lower the local barrier height (see Brodie, 1978). This makes FEM an important tool for the characterization of STM tips, especially the tips for tunneling spectroscopy studies (Binh et al., 1992).

High-resolution transmission electron microscopy

The scattering cross section of electrons by heavy atoms is substantially higher than by light atoms. Using an extremely thin carbon film, single heavy atoms, for example, uranium atoms or single molecules containing uranium, can be imaged by transmission electron microscopy. This was demonstrated by Crewe et al. (1970). However, this kind of single-atom observation is much more limited than FIM and especially STM.

1.8.2 Metal–vacuum–metal tunneling

The Topografiner

The first attempt to demonstrate metal–vacuum–metal tunneling with a controllable gap was conducted by Young et al. (1972a) using the *topografiner*.[2] It is probably the instrument closest to the STM in basic principle and operation. A schematic of the topografiner is shown in Fig. 1.36. A tungsten tip is spot-welded to a filament, then attached to a mounting block. The block is fixed to the x and y piezos, which drive a raster scan. The sample is attached to the z piezo, driven by the feedback circuit. The tip is positioned at about 1000 Å

---

2 The word topografiner was coined by Young et al. (1971, 1972) from the Greek word τοπογραφεῖν, which means "to describe a place".

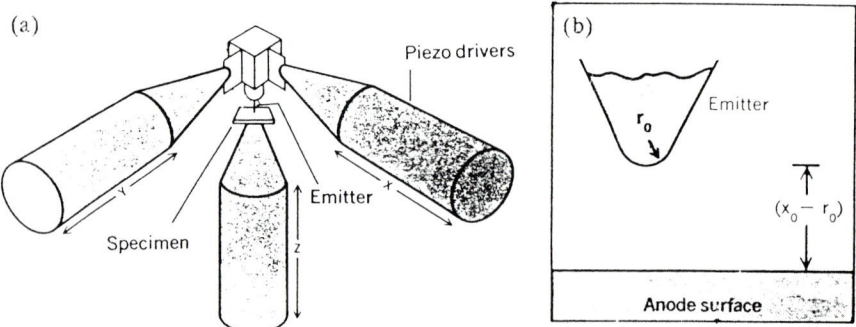

Fig. 1.36. The topografiner. An instrument developed by Young, Ward, and Scire in the late 1960s, which is the closest ancestor of the STM. (a) The tip is driven by the x and y piezos, and the sample is mounted on the z piezo. By applying a high voltage between the tip and the sample, a field-emission current is induced. Using the field-emission current as the feedback signal, topography of the sample surface is obtained. (b) Close-up of the tip and the sample. The end of the tip has a small radius, typically a few hundred Å. The typical tip–sample distance is a few thousand Å. (After Young, 1971.)

from the sample surface. By applying a high positive voltage on the sample, typically a few kV, electrons are stripped off from the tip and flow to the sample surface. This phenomenon, *field emission,* was discovered at the turn of this century, and explained by Fowler and Nordheim (1928) in terms of quantum mechanics. By using the field-emission current intensity as a feedback signal, topographic images of the sample surface were obtained (Young, 1971; Young, Ward, and Scire, 1972). An example of the images is shown in Fig. 1.37.

Field emission is a *tunneling phenomenon* in solids and is quantitatively explained by quantum mechanics. Also, field emission is often used as an auxiliary technique in STM experiments (see Part II). Furthermore, field-emission spectroscopy, as a *vacuum-tunneling spectroscopy* method (Plummer et al., 1975a), provides information about the electronic states of the tunneling tip. Details will be discussed in Chapter 4. For an understanding of the field-emission phenomenon, the article of Good and Müller (1956) in *Handbuch der Physik* is still useful. The following is a simplified analysis of the field-emission phenomenon based on a semiclassical method, or the Wentzel–Kramers–Brillouin (WKB) approximation (see Landau and Lifshitz, 1977).

As shown in Fig. 1.36, in field-emission experiments, the tip is always made very sharp. In other words, the end of the tip has a small radius r_0, typically on the order of 1000 Å. By applying a voltage V of a few kV, the field

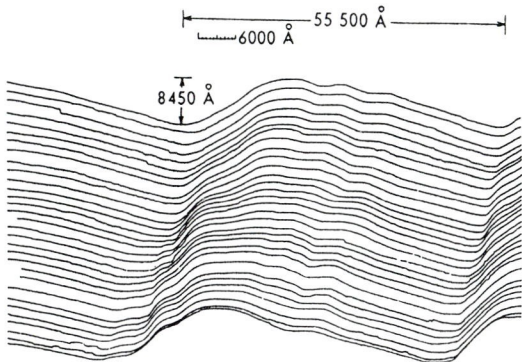

Fig. 1.37. Topographic image of a grating obtained by the topografiner. The vertical resolution is about 30 Å, and the lateral resolution is about 4000 Å. (Reproduced from Young, Ward, and Scire, 1972, with permission.)

intensity near the tip end is high. The field F can be estimated from the applied voltage and the tip radius (see Fig. 1.36),

$$F \approx V \left(\frac{1}{r_0} - \frac{1}{x_0} \right). \tag{1.31}$$

The typical field intensity near the tip is 0.1 to 1 V/Å.

The potential curve for the electrons near the tip surface is shown in Fig. 1.38. The relevant dimensions are much smaller than the radius of the tip end. Therefore, a one-dimensional model is adequate. In the metal, the energy level of the electrons is lower than the vacuum level by the value of the work function ϕ. From the point of view of classical mechanics, the electrons cannot escape from the metal even with a very high external field, that is, the potential barrier is impenetrable. From the point of view of quantum mechanics, there is always a finite probability that the electrons can penetrate the potential barrier. In the semiclassical (WKB) approximation, the transmission coefficient for a general potential barrier is (Landau and Lifshitz, 1977):

$$T \propto \exp \left(-\frac{2}{\hbar} \int_0^{z_0} \{2m[U(z) - E]\}^{1/2} dz \right), \tag{1.32}$$

where m is the electron mass and $U(z)$ is the potential. The potential barrier in field emission is (see Fig. 1.38)

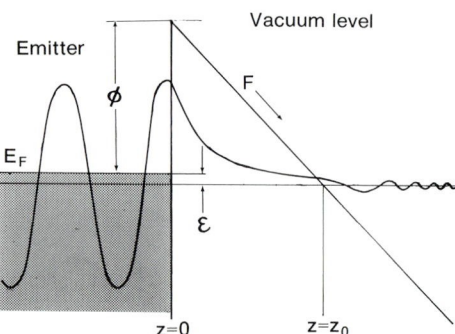

Fig. 1.38. The Fowler–Nordheim equation for field emission. The relevant dimensions are much smaller than the tip radius. Therefore, a one-dimensional model is adequate. Neglecting the image-force effect, the potential $U(z)$ outside the metal surface is linear with respect to distance z. The relevant parameters are: work function of the material, ϕ, and the field intensity near the surface, F.

$$U(z) - E = \phi - eFz, \tag{1.33}$$

where F is the field intensity near the surface. The point z_0 is where the classical momentum vanishes, that is, $z_0 = \phi/eF$. An elementary integration gives

$$T \propto \exp\left(-\frac{4\sqrt{2m}}{3e\hbar}\frac{\phi^{3/2}}{F}\right). \tag{1.34}$$

Eq. (1.34) is the simplified version of the Fowler–Nordheim equation. In the original Fowler–Nordheim equation, there is an algebraic prefactor to the exponential factor. Experimentally, the exponential factor always dominates the functional dependence, leaving the existence and the specific form of the algebraic factor hardly distinguishable (see Fowler and Nordheim, 1928; Good and Müller, 1956). In practical units (work function ϕ in eV, and the field intensity F in V/Å), Eq. (1.34) becomes

$$T \propto \exp\left(-0.68\frac{\phi^{3/2}}{F}\right). \tag{1.35}$$

The standard method for treating field-emission data is to make a plot of the logarithm of field-emission current $\ln I$ versus $1/V$ or $1/F$, then fit the data points with a straight line. This plot is called a Fowler–Nordheim plot (Good and Müller, 1956).

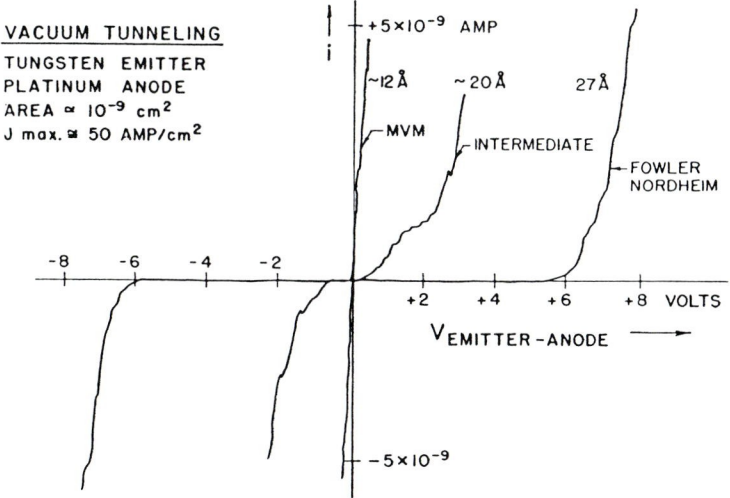

Fig. 1.39. Transition from field emission to vacuum tunneling. (Reproduced from Young, Ward, and Scire, 1971a, with permission.)

As seen from Fig. 1.36 and Eq. (1.31), the field-emission current is not very sensitive to the tip–sample distance x_0. Using the topografiner, the best vertical resolution is found to be 30 Å, and the best horizontal resolution is 4000 Å (Young 1971, Young, Ward, and Scire, 1972).

By bringing the tip very close to the sample surface, and applying a bias voltage smaller than the work function ϕ, metal–vacuum–metal tunneling should occur. Using the topografiner, Young et al. (1971) observed the transition from field emission to metal–vacuum–metal tunneling. Figure 1.39 shows their experimental results. When the tip–sample separation is substantially reduced, a linear dependence $I \propto V$ is observed. They estimated that the tip–sample distance was about 12 Å. Their results indicate that the implementation of metal–vacuum–metal tunneling with scanning is feasible. It is regrettable that their research project was discontinued at that time.

The experiments of Teague

A few years before the invention of STM, Teague (1978) conducted an extensive study of the tunneling between two gold spheres through a vacuum gap ranging from about 20 Å to a point where the two electrodes touched. A schematic of the experiment is shown in Fig. 1.41. The original report, a thesis unpublished otherwise until 1986, contains both a theoretical analysis and a series of measurements. Although Teague's setup does not scan and does not have a feedback circuit, much understanding of the effects of image potential and the effect of forces in vacuum tunneling processes was achieved.

In Teague's theoretical analysis of the tunneling phenomenon at very small tunneling gap widths, the many-body effects are taken into account through the image potential. Details of the many-body effect on surface potentials are discussed in Section 4.2. The residual many-body effects beyond the image potential are estimated to be less than 1%, which is negligible. The potential between two planar free-electron metals, including multiple image potentials, is shown in Fig. 1.40 As shown, for $W \geq 4$ Å, the potential barrier is essentially a square barrier with the top following the $1/W$

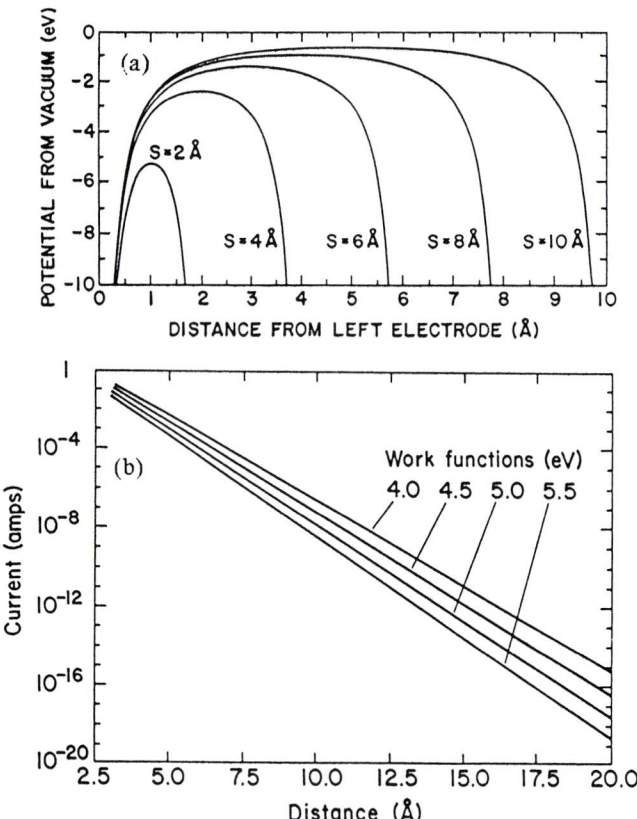

Fig. 1.40. Potential curves and tunneling characteristics. (a) The potential curves in the tunneling gap. The many-body effects are taken into account by introducing multiple image potential; see Sections 2.2 and 4.2. (b) Dependence of tunneling current with gap width, resulting from a numerical calculation. The tunneling current shows an accurate exponential dependence on tunneling gap width, up to a mechanical contact. The apparent barrier height is almost equal to the nominal barrier height at large gap widths over the entire range. (After Teague, 1978.)

law of the classical image potential. For $W = 2$ Å, the barrier top becomes lower than the Fermi level (-5 eV), and the square-barrier model is no longer accurate.

The tunneling current as a function of gap width and bias voltage were studied using numerical methods. The tunneling current resulting from numerical calculations shows an almost strict exponential dependence to the separation s from 20 Å down to 3 Å. The barrier height obtained from the slope of the ln $I{\approx}s$ curve almost accurately equal to the nominal work function, as shown in Fig. 1.40.

A schematic of the heart of Teague's experimental apparatus is shown in Fig. 1.41. To maintain a clean gold surface, care was taken. The electrodes were made from 1.25 mm gold wire by repeated chemical cleaning and electron-bombardment melting of its end in high vacuum. The tunneling experiment was conducted in an oil-free high-vacuum chamber with a two-stage thermostat enclosing the tunneling apparatus (Fig. 1.41). In a sturdy rectangular aluminum block (1), the two gold electrodes (2) are placed along the axis of a cylindrical assembly. On one end, the gold wire is attached to a pair of membranes (3). A stack of piezoelectric plates (4) actuates the fine displacement. Coarse displacement is actuated by a differential screw (5). The thermostat was so effective that no drift was observed through the I/V curve during the experiments, although there is no feedback mechanism.

Typical $I{\approx}s$ curves are shown in Fig. 1.42. As shown, the exponential dependence was observed over five orders of magnitude.

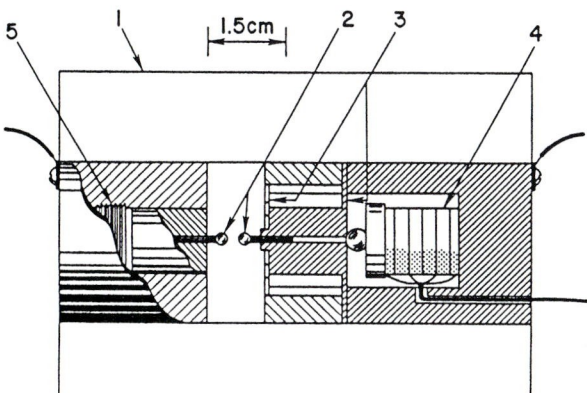

Fig. 1.41. Micropositioning assembly in the MIM tunneling experiment. In an aluminum block (1), two gold electrodes (2) are contained. The right electrode is attached to a pair of membranes (3), and actuated by a stack of piezoelectric plates (4). The left electrode is actuated by a differential screw (5) for coarse positioning. The entire assembly is enclosed in a thermostat and then a vacuum chamber. (After Teague, 1978.)

Fig. 1.42. Measured relation between the logarithm of current and electrode spacing. The dashed line corresponds to a work function of approximately 3eV, indicating that the gold electrodes have some organic contaminants. The electrode separation is measured from the voltage applying on the piezoelectric stack. For small apparent separations, the actual electrode separation change is smaller than the apparent change due to the repulsive force and the deformation of the electrodes. The 100mV curve was measured immediately after the 10mV curve. Similar characteristics were obtained for negative bias voltages. After Teague (1978).

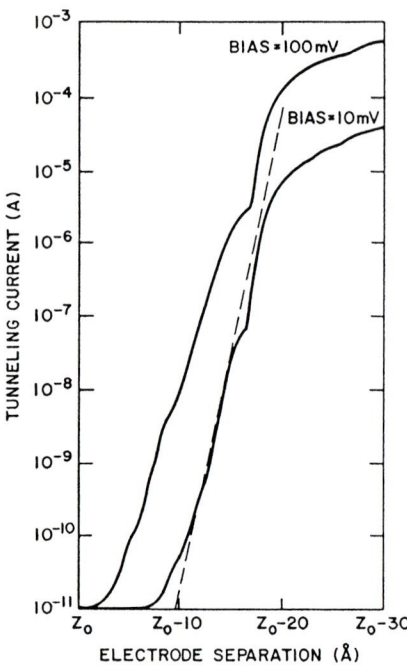

From the observed dependence of the tunneling current on electrode separations, Teague (1978) found that the experimental data cannot be properly explained unless the *attractive force* between the electrodes is taken into account. The attractive force causes a minute deformation of the ends of the gold electrodes. Because the tunneling current is extremely sensitive to the gap width, a minute deformation is enough to change the observed tunneling current significantly. At very short spacings, the apparent barrier height *increases* instead of decreases, then drops off steeply to almost zero. In other words, as a result of the attractive and repulsive forces, when the gap width is small, the gap displacement measured by the piezo voltage does not equal the actual gap displacement. Therefore, the apparent barrier height obtained from the piezo displacement does not equal the apparent barrier height calculated based on rigid electrodes. The nature and the effect of the attractive atomic force in STM will be discussed in Chapters 7 and 8.

PART I

IMAGING MECHANISM

The first part of the growth of a physical science consists in the discovery of a system of quantities on which its phenomena may be conceived to depend. The next stage is the discovery of the mathematical form of the relations between these quantities. After this, the science may be treated as a mathematical science, and the verification of the laws is effected by a theoretical investigation of the conditions under which certain quantities can be most accurately measured, followed by an experimental realization of these conditions, and actual measurement of the quantities.

On the Mathematical Classification of Physical Quantities.

James Clerk Maxwell
Distinguished Professor of
Experimental Physics
Cambridge University

Bardeen's perturbation approach (1960) to tunneling problems has been the backbone of the theoretical understanding of Giaever's classical tunneling junction experiments (1960), and has played an important role in the early years of STM. Because of the close proximity of the tip and the sample, Bardeen's approach (BA), in its original form, is not appropriate. Yet the attractiveness of Bardeen's concept, the perturbation approach, is irresistible: It may provide analytic expressions for the tunneling current, which may facilitate conceptual understanding as well as a thorough comparison with experimental measurements.

The deficiencies of BA can be corrected by including the modifications of the wavefunctions of one party (for example, sample) due to the existence of another party (for example, tip). Actually, the concept and method for such modifications were proposed by Holstein (1955) and Herring (1962) in the treatment of *atomic forces*. By including such corrections, a *modified Bardeen approach* (MBA) is established.[3] It reproduces the analytical results for the square-barrier problem with high accuracy, even when the top of the barrier lies below the energy level. By applying the MBA to the only tunneling problem in Nature that has an exact analytical solution, the hydrogen molecular ion, a quantitative description of the exchange interaction as well as high-accuracy expressions for the asymptotic potential curves is found. At least for those soluble cases, the accuracy of the MBA is well verified. For many practical cases in STM, the integrals appearing in the MBA result in simple analytical forms. The correction to the wavefunctions, in many practical cases in STM and AFM, can be managed analytically. The two origins of the MBA, Bardeen's theory of tunneling and Holstein's theory of atomic force, result in a natural connection between tunneling conductance and attractive atomic force. The consequences of this equivalence can be verified experimentally.

In spite of its simplicity, the MBA yields results that account for much of the physics in STM, and give reasonably accurate quantitative agreement with measurements, including topographic and current images, tunneling spectra, tunneling barrier measurements, and the effect of force between the tip and the sample. Of course, it does not mean that the first-principles calculations are not important. For any practical system of interest, both as the tip and as the sample, an all-electron first-principles calculation, instead of simplified models, is essential for achieving a thorough understanding.

---

3 Acronyms often cause confusion in the scientific literature. Here we have a simple mnemonics: A person with a Bachelor of Arts (BA) degree, after further study, might earn a Master of Business Administration (MBA).

CHAPTER 2

ATOM-SCALE TUNNELING

2.1 Introduction

STM experiments with atomic resolution are always performed with very short tip–sample distances. Direct measurements have shown that the normal tip–sample distances in STM are 1–4 Å before a mechanical contact, as shown in Fig. 2.1 (Dürig, Züger, and Pohl, 1988). The experiments were *simultaneous* measurements of the force gradient and the tunneling conductance in an STM. By integrating the force gradient, the point of a mechanical contact is determined. The tunneling conductance as a function of the tip–sample distance (with respect to the point of a mechanical contact) is directly measured. At a tunneling conductance of 1 MΩ, the distance is found to be about 1 Å from a mechanical contact. The tunneling current varies about one order of magnitude per ångström. Therefore, the normal tip–sample distance in an STM experiment, typically with a tunneling resistance of 1 MΩ to 1 GΩ, is 1–4 Å from a mechanical contact.

The transport process in the distance range of normal STM operation is very different from the tunneling process in the classical tunneling junction (Giaever, 1960), where the thickness of the insulator between the conducting electrodes is typically 20–30 Å. First, because of the overlap of the surface potentials of the tip and the sample, the top of the potential barrier in the gap region is substantially lower than the vacuum level. In many cases, the top of the potential barrier is even lower than the Fermi level. Second, the local atomic structure and the local electronic structure near the gap region play a dominant role. Third, the forces between the tip and the sample, both attractive and repulsive, are important. The forces deform the tip and the sample in the vicinity of the gap region. A tangible effect on the tunneling current is often observed.

We start our discussion with a definition of the tip–sample distance. Because the STM deals with atoms, the basic definition should be microscopic. If at least part of the sample surface is atomically flat, a natural definition of the tip–sample distance is the distance from the plane of the topmost nuclei of the sample to the nucleus of the apex atom of the tip. Throughout the book, this microscopic distance is denoted by a lower-case letter, d, z_0, or

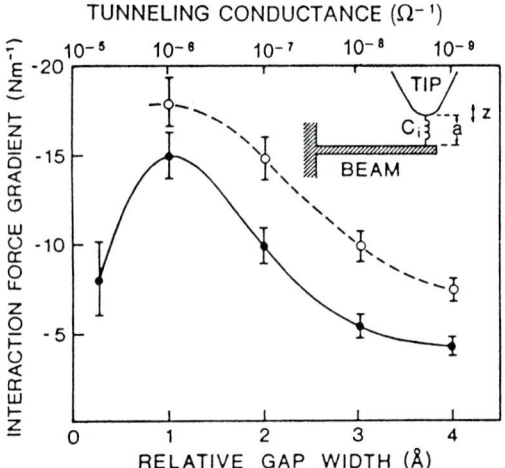

Fig. 2.1. Correlation between tunneling conductance and force. The tunneling conductance and the force gradient are measured simultaneously. The origin of the abscissa is the equilibrium point, where the net force is zero. It corresponds to a tunneling conductivity of $10^{-5}\ \Omega^{-1}$. The experimental setup is schematically shown in the inset. (Reproduced from Dürig et al., 1988, with permission.)

s. As a convention, the norm out of the sample surface is denoted as the positive z direction. From an experimental point of view, there is another zero point that is directly measurable: the equilibrium distance or equilibrium point. We may consider an STM as a giant molecule consisting of two parts, the tip and the sample. At a relatively large (absolute) distance, for example, 5 Å, the force between these two parts is attractive. At very short distances, for example, 2 Å, the force between these two parts is repulsive. Between these two extremes, there is a well-defined position where the net force between the tip and the sample is zero. We define it as the *equilibrium distance*. On the microscopic scale, the equilibrium distance is approximately equal to the inter-atomic distance in the material under investigation, which is 2–3 Å. Thus, the normal tip–sample distance in STM experiments is 4–7 Å on the microscopic distance scale.

In some cases, macroscopic models are used for simplified discussions of certain phenomena without atomic resolution. A macroscopic tip–sample distance should be defined. To avoid confusion, we use the term *barrier thickness* instead. Throughout the book, the barrier thickness is always denoted by a upper-case letter, such as W or L. In the Sommerfeld model of the free-electron metals, the barrier thickness is the distance between the surface of the metal pieces. In the jellium model (see Chapter 4), the barrier thickness is defined as the distance between the image-force planes.

Because of the force and deformation, the displacement measured by the voltage applied on the z piezo may different from the true tip–sample distance. We will denote it as the *apparent displacement* whenever it appears.

2.2 The perturbation approach

In this Chapter, we discuss a perturbation theory for STM, the modified Bardeen approach (MBA). The illustrate the concept of a perturbation approach, let us consider the following four regimes of interactions (z_0 denotes the microscopic tip–sample distance):

1. When the tip–sample distance is large, for example, $z_0 > 100$ Å, the mutual interaction is negligible. By applying a large electrical field between them, field emission may occur. Those phenomena can be described as the interaction of one electrode with the electrical field, without considering any interactions from the other electrode.

2. At intermediate distances, for example, $10 < z_0 < 100$ Å, a long-range interaction between the tip and the sample takes place. The wavefunctions of both the tip and the sample are distorted, and a van der Waals force arises. The van der Waals interaction follows a power law, with an order of magnitude of a few meV per atom.

3. At short distances, for example, $3 < z_0 < 10$ Å, the electrons may transfer from one side to another. The exchange of electrons gives rise to an attractive interaction, that is, the *resonance energy* of Heisenberg (1926) and Pauling (1977), which is the origin of the chemical bond. If a bias is applied between them, a *tunneling current* occurs. The resonance energy follows an exponential law, and can be much larger than the van der Waals interaction, for example, of a good fraction of an eV per atom.

4. At extremely short distances, for example, $z_0 < 3$ Å, the repulsive force becomes dominant. It has a very steep distance dependence. The tip–sample distance is virtually determined by the short-ranged repulsive force. By pushing the tip farther toward the sample surface, the tip and sample deform accordingly.

The polarization, or the van der Waals interaction, can be accounted for by a stationary-state perturbation theory, effectively and accurately. The exchange interaction or tunneling can be treated by time-dependent perturbation theory, following the method of Oppenheimer (1928) and Bardeen (1960). In this regime, the polarization interaction is still in effect. Therefore, to make an accurate description of the tunneling effect, both perturbations must be considered simultaneously. This is the essence of the MBA.

To elucidate the basic concepts and to provide a test case for the accuracy of MBA, we discuss a one-dimensional case first. In Chapter 7, we will discuss another elementary case, the hydrogen molecular ion, which provides another test case for the accuracy of MBA. The close relation between STM and AFM is established then.

2.3 The image force

In this section, we discuss the concept of classical image force. The validity of this concept has been verified using quantum mechanics in a many-body formalism (Bardeen, 1936; Lang and Kohn, 1970; Appelbaum and Hamann, 1972; Herring, 1992). We will present it in Chapter 4.

The concept of image force was introduced by Schottky (1923) in his early attempt to interpret field emission. Nordheim (1928) then studied its effect in the quantum-mechanical treatment of the same problem (Fowler and Nordheim, 1928). The origin of image force can be understood from elementary electrostatics, as shown in Fig. 2.2. An electron located outside a metal surface induces positive charge on the metal surface. The positive surface charge attracts the electron. For an electron at a distance z from the metal surface, the effect of the surface charge is equivalent to the effect of a positive "image charge" inside the metal at a distance z from the surface. Coulomb's law gives the attractive force on the electron:

$$f = -\frac{e^2}{(2z)^2} .$$

(2.1)

To integrate the force over z, we find the potential

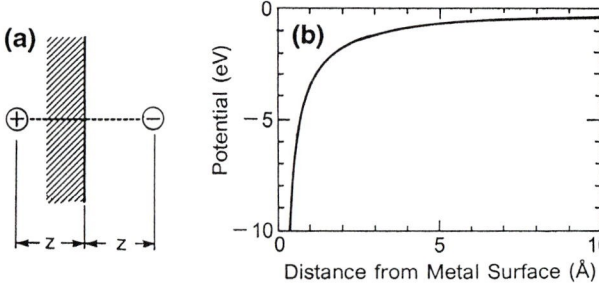

Fig. 2.2. The image potential of an electron near a metal surface. The electron induces positive charge at the metal surface. (a) The effect of the positive surface charge is equivalent to a fictitious image charge behind the metal surface. (b) The distance dependence of the image potential.

$$U = - \int f \, dz$$

$$= - \frac{e^2}{4z} = - \frac{3.60}{z} \ (\text{eV Å}). \tag{2.2}$$

Here, the vacuum potential is chosen to be zero — when the electron is far from the metal.

When an electron is located inside a tunneling barrier, positive charges are induced on both metal surfaces. The total force can be treated by a series of image charges induced by both metal surfaces. The force acting on the electron inside the barrier is the sum of forces from all the image charges, as shown in Fig. 2.3,

$$f = - \frac{e^2}{4z^2} + \sum_{n=0}^{\infty} \left(\frac{e^2}{4(nW - z)^2} - \frac{e^2}{4(nW + z)^2} \right). \tag{2.3}$$

The potential can be obtained by integration. A convenient zero point is chosen by the condition that when $W \gg z$, only the left-hand surface is effective, and the effect of the right-hand metal surface should disappear. We obtain (Simmons, 1969)

$$U = - \frac{e^2}{4z} - \frac{e^2}{2} \sum_{n=0}^{\infty} \left(\frac{nW}{n^2 W^2 - z^2} - \frac{1}{nW} \right). \tag{2.4}$$

The potential is symmetric with respect to both metal surfaces. In fact, by direct substitution, it is easy to show that the potential at $W - z$ equals that at z. At the middle of the barrier, $z = W/2$, the potential is

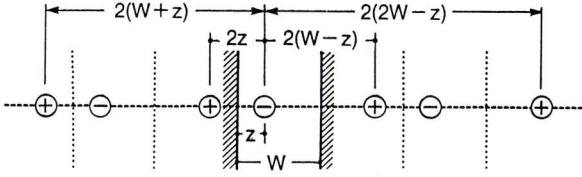

Fig. 2.3. The image potential of an electron inside a tunneling barrier. The electron induces positive charge at both metal surfaces. The effect of positive surface charges is equivalent to a series of image charges behind the two metal surfaces.

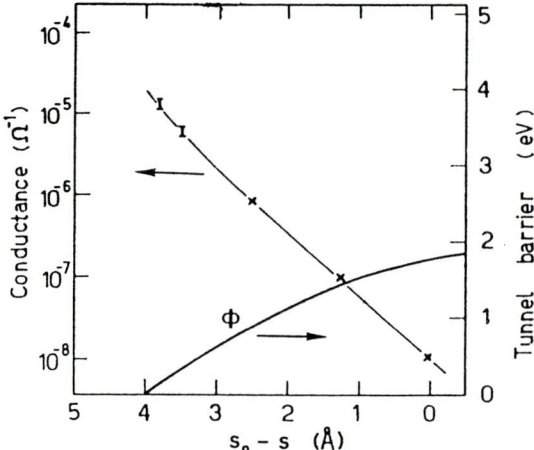

Fig. 2.4. Measured tunneling conductance and the potential barrier. The experimental tunneling conductance is measured with a tunneling current $I=10$ nA and a
varying bias. The slope of the logarithm of the measured tunneling conductance yields
a constant apparent barrier height 3.2 eV. Due to the image force, the barrier height ϕ
becomes negative whereas the measured apparent barrier height remains constant. Near
the point of the potential barrier collapsing, the apparent barrier height *increases*
instead of decreases. This is due to the *attractive force* between the tip and the sample.
(Reproduced from Binnig et al., 1984, with permission.)

$$U = -\frac{e^2}{2W}\left(1 + \sum_{n=1}^{\infty}\frac{1}{n(4n^2-1)}\right)$$
$$= -\frac{e^2\ln 2}{W}$$
$$\approx -\frac{9.97}{W}\ (\text{eV Å}).$$

$$(2.5)$$

   Detailed plots of the potential between two metal electrodes were published by Teague (1978), as shown in Fig. 1.40. It seems that the image
potential would reduce the apparent barrier height observed in STM at reduced
tip–sample distances. However, experimentally, this has never been observed.
A ubiquitous experimental observation is that the apparent barrier height is
almost constant up to a mechanical contact, with a value close to the nominal
work function, as exemplified in Fig. 2.4 (Binnig et al., 1984). Using numerical methods, Teague (1978) showed that the image potential does not affect
the apparent barrier height, as shown in Fig. 1.40. A simple explanation of the
observed constant apparent barrier height is provided in the following section.

Furthermore, Binnig et al. (1984) reported that near the mechanical contact, the observed apparent barrier height *increases* rather than decreases, as shown in Fig. 2.4. It is due to the *attractive force* between the tip and the sample, which causes a deformation of the tip and the sample near the barrier region. The same observations were reported earlier by Teague (1978).

2.4 The square-barrier problem

In this section, we will analyze an elementary problem in quantum mechanics, the square barrier. The purpose is twofold. First, such an analysis can provide physical insight into the process, to gain a conceptual understanding. Second, analytically soluble models are indispensable for assessing the accuracy of approximate methods, such as the MBA.

From a semiclassical point of view (Landau and Lifshitz, 1977), there are two qualitatively different cases, as shown in Fig. 2.5. When the top of the barrier is higher than the energy level, the barrier is *classically forbidden,* and the process is called *tunneling.* When the top of the barrier is lower than the energy barrier, the barrier is *classically allowed,* and the process is called *channeling* or *ballistic transport.* We will show that for square potential barriers of atomic scale, the distinction between the classically forbidden case and the classically allowed case disappears. There is only one unified phenomenon, *quantum transmission.*

The derivation of the transmission coefficients for a square barrier can be found in almost every textbook on elementary quantum mechanics (for example, Landau and Lifshitz 1977). However, the conventions and notations are not consistent. Figure 2.5 specifies the notations used in this book. To make it consistent with the perturbation approach later in this chapter, we take the reference point of energy at the vacuum level.

In the problem of quantum transmission, three traveling waves are involved: the incoming wave $\exp(iqx)$, the reflected wave $A \exp(-iqx)$, and the transmitted wave $B \exp(iqx)$, where

$$q = \frac{\sqrt{2m(E - U_0)}}{\hbar} \approx 0.51\sqrt{E - U_0} \tag{2.6}$$

is the wave vector of the traveling waves. Practical units are used here: eV for energy, Å for distance, and $Å^{-1}$ for wave vector. There are two different conventions in defining the transmission coefficient according to the normalization of the initial state. One is to normalize it with respect to the incoming traveling wave, $D = |B|^2$. Another is to normalize it with respect to the quasistanding wave on the left-hand side:

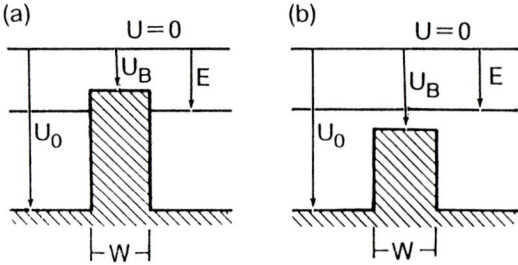

Fig. 2.5. Classically forbidden and classically allowed potential barriers. (a) A classically forbidden potential barrier, where the top of the potential barrier is higher than the energy level. From a semiclassical point of view, the transport process across such a barrier is called *tunneling*. (b) A classically allowed potential barrier, where the top of the potential barrier is lower than the energy level. From a semiclassical point of view, the transport process across such a barrier is called *channeling* or *ballistic transport*.

$$T = \frac{|B|^2}{1 + |A|^2}. \tag{2.7}$$

This convention more conveniently connects with perturbation theory, which is developed later in this chapter. If the barrier is classically forbidden, i. e., $E < U_B$ or $|E| > |U_B|$, the transmission coefficient is

$$T = \left[1 + \frac{1}{2}\left(\frac{\kappa}{q} + \frac{q}{\kappa}\right)^2 \sinh^2(\kappa W)\right]^{-1}, \tag{2.8}$$

where

$$\kappa = \frac{\sqrt{2m(U_B - E)}}{\hbar} \approx 0.51\sqrt{U_B - E}. \tag{2.9}$$

If the barrier is classically allowed, that is, $E > U_B$ or $|E| < |E_B|$, the transmission coefficient is

$$T = \left[1 + \frac{1}{2}\left(\frac{k}{q} - \frac{q}{k}\right)^2 \sin^2(kW)\right]^{-1}, \tag{2.10}$$

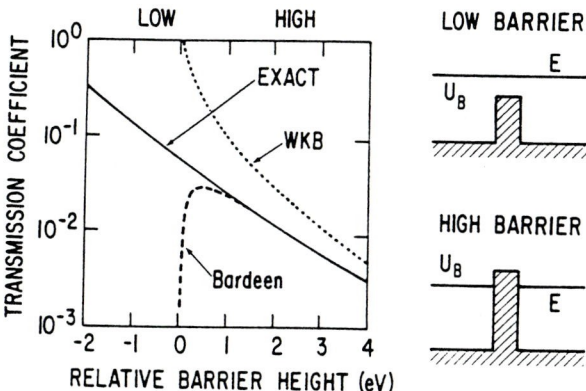

Fig. 2.6. Quantum transmission through a thin potential barrier. From the semi-classical point of view, the transmission through a high barrier, *tunneling*, is qualitatively different from that of a low barrier, *ballistic transport*. Nevertheless, for a thin barrier, here $W = 3$ Å, the logarithm of the exact quantum mechanical transmission coefficient (solid curve) is nearly linear to the barrier height from 4 eV above the energy level to 2 eV below the energy level. As long as the barrier is thin, there is no qualitative difference between tunneling and ballistic transport. Also shown (dashed and dotted curves) is how both the semiclassical method (WKB) and Bardeen's tunneling theory become inaccurate for low barriers.

where

$$k = \frac{\sqrt{2m(E - U_B)}}{\hbar} \approx 0.51\sqrt{E - U_B} \ . \tag{2.11}$$

Taking the semiclassical limit, these two cases become very different. For thick, classically forbidden barriers, Eq. (2.8) becomes

$$T \approx 2\left(\frac{\kappa}{q} + \frac{q}{\kappa}\right)^{-2} e^{-2\kappa W}. \tag{2.12}$$

We recovered here the usual WKB formula for tunneling probability, which exhibits an exponentially decaying behavior. On the other hand, from Eq. (2.10), we observed immediately that *resonances* occur when the thickness of the barrier equals integer multiples of one half of the de Broglie wavelength in the barrier region,

$$kW = n\pi, \quad n = 1, 2, 3, \ldots \tag{2.13}$$

At those values of the thickness, $T = 1$, as if the barrier does not exist. Such a quantum resonance phenomenon is currently the subject of intense theoretical and experimental research (see for example, Escapa and Garcia, 1990).

Asymptotically, the behaviors of the two regimes are radically different. One is decaying; another one is oscillating. However, the oscillatory behavior only happens when the barrier thickness is comparable to or larger than one half of the de Broglie wavelength. For a classically allowed potential barrier, if its thickness is shorter then one half of the de Broglie wavelength, the transmission coefficient is a *direct continuation* of the classically forbidden case. Figure 2.6 shows the dependence of the transmission coefficient with respect to the relative barrier height, $(U_B - E)$, for a potential barrier with thickness $W=3$ Å. As shown, the logarithm of the transmission coefficient varies almost linearly from 4 down to -2 eV.

This trend can be further illustrated by expanding the logarithm of the transmission coefficient, Eq. (2.8) or (2.10), into a Taylor series. The result is

$$\ln T = \ln(1 + \frac{\eta}{2}) + \frac{1 + \eta/6}{1 + \eta/2} \epsilon +$$
$$+ \left(\frac{1}{2\eta} + \frac{1}{12} + \frac{\eta}{45} - \frac{\eta^2}{360} \right) \epsilon^2 + O(\epsilon^3), \tag{2.14}$$

where the dimensionless quantities are: $\epsilon = 8m(U_B - E)W^2/\hbar^2$, and $\eta = 8m(E - U_0)W^2/\hbar^2$. The energy dependence of the transmission coefficient is not only continuous, but also analytic.

2.4.1 Apparent barrier height

The definition of apparent barrier height, Eq. (1.13), is based on the exponential decaying wavefunction in the barrier region. Experimentally, as the tip is moving toward the sample, the barrier height and the tip–sample distance are changing at the same time. Therefore, the *measured* apparent barrier height no longer equals the nominal barrier height in the energy diagram. Binnig et al. (1984) show that by expanding the image term into binomials, the first term in the variation of the apparent barrier height with distance vanishes. Experimentally, almost every author reported the observation of a constant apparent barrier height down to almost a mechanical contact, in spite of the effect of the image force (for example, Teague, 1978; Binnig et al., 1984; Payne and Inkson, 1985; and the recent extensive study of Schuster et al., 1992.) Here, we provide a simple explanation based on the exact transmission coefficient for the square-potential-barrier problem. Assuming that the barrier height varies according to the image force law,

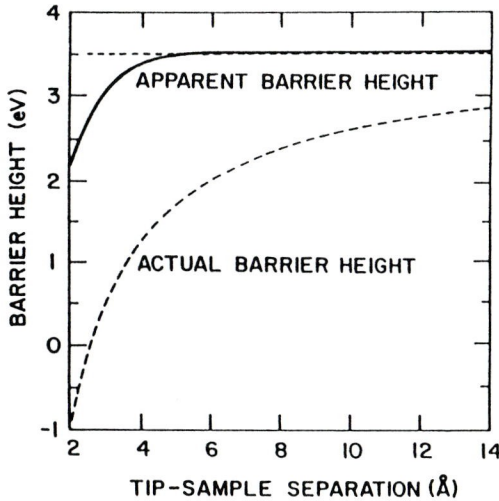

Fig. 2.7. Apparent barrier height calculated from the exact solution. Variation of the apparent barrier height $0.95(d \ln I/dz)^2$ with barrier thickness, as calculated from the exact solution of the square-potential-barrier problem. The actual barrier height (dashed curve) drops dramatically because of the image force potential. The apparent barrier height (solid curve) almost always equals the nominal value of barrier height. (Parameters used: $U_0 = 3.5$ eV, $E_k = 7.5$ eV.)

$$U = U_0 - \frac{9.97}{W} \text{ (eV Å)}. \qquad (2.15)$$

By definition, the apparent barrier height is

$$\phi_{ap} = 0.95 \left\{ \frac{\partial}{\partial W} \ln \left[1 + \frac{1}{2} \left(\frac{\kappa}{q} + \frac{q}{\kappa} \right)^2 \sinh^2(2\kappa W) \right] \right\}^2 . \qquad (2.16)$$

In terms of Å and eV, $\kappa \approx 0.51\sqrt{U}$, $q = 0.51\sqrt{E_k}$. The result of a direct numerical calculation is illustrated in Fig. 2.7. As shown, the apparent barrier height is almost a constant up to the point the barrier collapses. Actually, when the barrier is very thin, a square barrier following Eq. (2.15) becomes a poor approximation for the effect of the image force. The actual barrier is more slim than the square barrier, and the current is growing even faster. The apparent barrier height should not drop, as in the left edge of Fig. 2.7. Rather, it remains almost constant, as also shown by numerical calculations of Teague (1987) (see Fig. 1.40).

2.4.2 Uncertainty-principle considerations

The continuity of tunneling and ballistic transport for atom-scaled potential barriers is a straightforward consequence of the uncertainty principle. It can be shown from the following two different points of view:

The first view is based on the uncertainty relation between time and energy. When the electron is in the region of the barrier, its velocity, as determined by the kinetic energy, is:

$$v = \sqrt{\frac{2(E - U_B)}{m}} \, .\tag{2.17}$$

The time for an electron to pass the barrier is

$$\Delta t = W \sqrt{\frac{m}{2(E - U_B)}} \, .\tag{2.18}$$

The energy uncertainty for an electron in the barrier region is then

$$\Delta E > \frac{\hbar}{\Delta t} = \frac{\hbar}{W} \sqrt{\frac{2(E - U_B)}{m}} \, .\tag{2.19}$$

At $E - U_B = 2 \, eV$ and $W = 3 \, \%$ Å, the energy uncertainty is $\Delta E \approx 2.5$ eV. In other words, *the energy uncertainty is larger than the absolute value of the kinetic energy.* Therefore, for atom-scale barriers, the distinction between tunneling and ballistic transport disappears.

The second view is based on the uncertainty relation between coordinate and momentum. In the region of a classically allowed barrier, the de Broglie wavelength of an electron is

$$\lambda = \frac{2\pi\hbar}{\sqrt{2m(E - U_B)}} \approx \frac{12.3}{\sqrt{E - U_B}} \, ,\tag{2.20}$$

in Å and eV. For example, an electron in the barrier region with kinetic energy 2 eV has a de Broglie wavelength 8.7 Å. In STM, this occurs when the thickness of the barrier is less than 3 Å. Therefore, the position uncertainty of this electron is much greater than the barrier thickness. In other words, for an electron near the center of the barrier, at any time, it is uncertain whether the electron is on one or another side of the barrier, or right on top of the barrier.

2.5 The modified Bardeen approach

In this section, we present a step-by-step derivation of the MBA. First, we show that by evaluating Bardeen's tunneling matrix element from properly *modified* wavefunctions, the perturbation theory is accurate even if the barrier *collapses*. Next, we show that for a relatively flat surface, the modification is essentially a constant multiplier. It provides a simple explanation of the observed constant apparent barrier height. In Chapter 7, we will show that the modified Bardeen integral equals the resonance energy of Heisenberg (1926) and Pauling (1977) in the problem of interatomic forces. The wavefunction modification for the simplest molecule, the hydrogen molecular ion, is again a constant multiplier (Holstein, 1955). By comparing with the exact solution of the H_2^+ problem, we show that the modified Bardeen integral provides an accurate expression for the asymptotic potential curve, again verifying the accuracy of the MBA.

2.5.1 The transition probability

Figure 2.8 is the energy schematic of the combined system. As the tip and the sample approach each other with a finite bias V, the potential U in the barrier region becomes different from the potentials of the free tip and the free sample. To make perturbation calculations, we draw a separation surface between the tip and the sample, then define a pair of subsystems with potential surfaces U_S and U_T, respectively. As we show later on, the exact position of the separation surface is not important. As shown in Fig. 2.8, we *define* the potentials of the individual systems to satisfy two conditions. First, the sum of the two potentials of the individual systems equals the potential of the combined system, that is,

$$U_S + U_T = U. \qquad (2.21)$$

Second, the product of the two potentials is zero throughout the entire space,

$$U_S U_T = 0. \qquad (2.22)$$

The reference point of energy, the vacuum level, is well defined in the STM problem. The entire system is neutral. Therefore, at infinity, there is a well-defined vacuum potential. In the vicinity of the apex of the tip, the potential barrier in the gap is substantially lowered. However, the barrier lowering is confined in a small region near the tip end. Outside the interaction region, the potential in the space equals the vacuum level. This condition, Eq. (2.22), minimizes the error estimation term of Oppenheimer (1928).

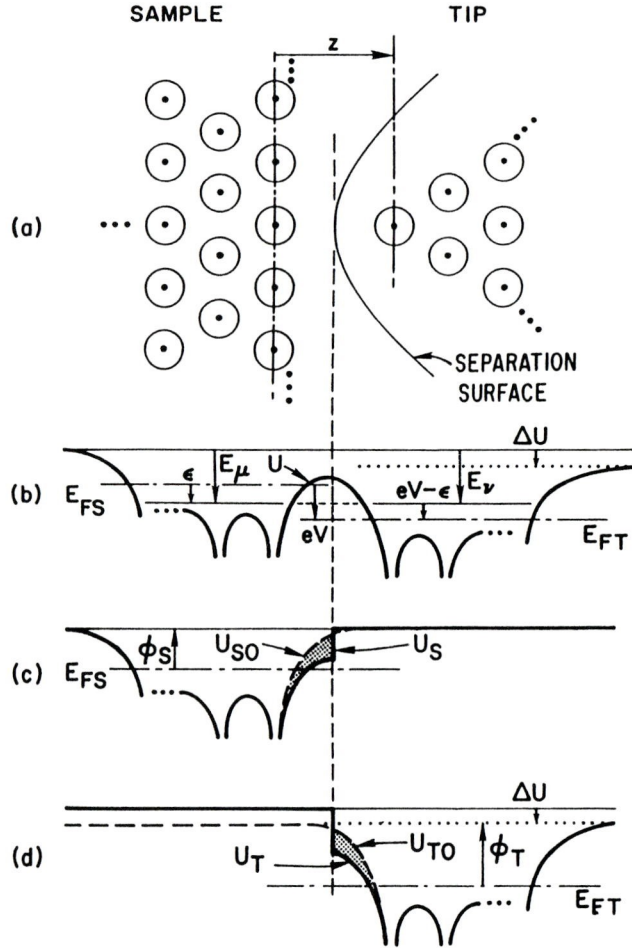

Fig. 2.8. Perturbation approach for quantum transmission. (a) A separation surface is drawn between the two subsystems. The precise location is not critical. (b) The potential surface of the system. (c) and (d) The potential surfaces of the subsystems. (Reproduced from Chen, 1991b, with permission.)

As seen from Fig. 2.8, the potentials U_S and U_T are different from the potentials of the free tip, U_{T0}, and the free sample, U_{S0}, respectively. The effect of the differences ($V_S = U_{S0} - U_S$ and $V_T = U_{S0} - U_S$), can be taken into account by time-independent perturbation methods.

At $t < 0$, U_T is turned off. The Schrödinger equation gives the stationary states of the sample:

$$(T + U_S)\psi_\mu = E_\mu \psi_\mu , \tag{2.23}$$

where $T = -(\hbar^2/2m)\nabla^2$ is the kinetic energy. At $t > 0$, the potential in the tip body is turned on. The sample state starts to evolve according to the time-dependent Schrödinger equation

$$i\hbar \frac{\partial \Psi}{\partial t} = (T + U_S + U_T)\Psi . \tag{2.24}$$

On the other hand, the modified wavefunctions of the tip are defined by the following Schrödinger equation,

$$(T + U_T)\chi_\nu = E_\nu\chi_\nu . \tag{2.25}$$

We expand the time-dependent wavefunction Ψ in terms of χ_ν,

$$\Psi = \sum_\nu a_\nu(t) \, \chi_\nu \, e^{-iE_\nu t/\hbar} . \tag{2.26}$$

For a state that is ψ_μ at $t = 0$, let

$$a_\nu(t) = (\chi_\nu, \psi_\mu) \, e^{-i(E_\mu - E_\nu) t/\hbar} + c_\nu(t) , \tag{2.27}$$

with $c_\nu(0) = 0$, then

$$\Psi = \psi_\mu \, e^{-iE_\mu t/\hbar} + \sum_\nu c_\nu(t) \, \chi_\nu \, e^{-iE_\nu t/\hbar} . \tag{2.28}$$

Note that the expansion is done in terms of the tip wavefunctions, which form a complete and orthogonal set of states. Substituting Eq. (2.28) for Eq. (2.24), we obtain the exact equation for $c_\nu(t)$:

$$i\dot{c}_\nu(t) = (\chi_\nu, U_T \psi_\mu) e^{-iE_\mu t/\hbar} + \sum_\lambda c_\lambda(t) \, (\chi_\nu, U_S \chi_\lambda) e^{-iE_\lambda t/\hbar} . \tag{2.29}$$

The transition probability of an electron from ψ_μ to χ_ν in first-order perturbation theory is then given by the Fermi golden rule,

$$w_{\mu\nu}^{(1)} = \frac{2\pi}{\hbar} \, |M_{\mu\nu}|^2 \, \delta(E_\nu - E_\mu), \tag{2.30}$$

where the matrix element is

$$M_{\mu\nu} = \int_{\Omega_T} \chi^*_\nu \, U_T \, \psi_\mu \, d\tau. \tag{2.31}$$

Since U_T is nonvanishing only in the tip body, the matrix element $M_{\mu\nu}$ is evaluated only in the volume of the tip, Ω_T.

2.5.2 The modified Bardeen integral

In this subsection, we show that using Schrödinger's equations, the tunneling matrix element can be converted to a surface integral similar to Bardeen's. Using Schrödinger's equation for the tip states, Eq. (2.25), the matrix element is converted into

$$\begin{aligned} M_{\mu\nu} &= \int_{\Omega_T} \psi_\mu \, U_T \chi^*_\nu \, d\tau \\ &= \int_{\Omega_T} \psi_\mu (E_\nu - T) \, \chi^*_\nu \, d\tau. \end{aligned} \tag{2.32}$$

The delta function factor in Eq. (2.30), that is, the condition of elastic transition, requires that $E_\mu = E_\nu$. Using Schrödinger's equation for the sample states, Eq. (2.23), and noticing that in the tip body, Ω_T, the potential of the sample U_S is zero, the transition matrix element is converted into

$$\begin{aligned} M_{\mu\nu} &= \int_{\Omega_T} \left(\chi^*_\nu \, E_\mu \, \psi_\mu - \psi_\mu \, T\chi^*_\nu \right) d\tau \\ &= \int_{\Omega_T} \left(\chi^*_\nu \, T \psi_\mu - \psi_\mu \, T\chi^*_\nu \right) d\tau. \end{aligned} \tag{2.33}$$

By writing down the kinetic energy term $T = -(\hbar^2/2m)\nabla^2$ explicitly, and using Green's theorem, the transition matrix element is finally converted into a surface integral similar to Bardeen's, in terms of *modified wavefunctions*:

$$M_{\mu\nu} = -\frac{\hbar^2}{2m} \int_\Sigma \left(\chi^*_\nu \, \nabla \psi_\mu - \psi_\mu \, \nabla \chi^*_\nu \right) \cdot d\mathbf{S}. \tag{2.34}$$

The matrix element has the dimension of *energy*. In Chapter 7, we will show that the physical meaning of Bardeen's matrix element is the energy lowering due to the interplay of the two states. In other words, it is the *resonance energy* of Heisenberg (1926) and Pauling (1977).

By integrating over all the states in the tip and the sample, taking into account the occupation probabilities, the tunneling current is

$$I = \frac{4\pi e}{\hbar} \int_{-\infty}^{\infty} \left[f(E_F - eV + \epsilon) - f(E_F + \epsilon) \right]$$
$$\times \rho_S(E_F - eV + \epsilon)\, \rho_T(E_F + \epsilon)\, |M|^2 \, d\epsilon, \tag{2.35}$$

where $f(E) = \{1 + \exp[(E - E_F)/k_B T]\}^{-1}$ is the Fermi distribution function, $\rho_S(E_F)$ is the density of states (DOS) of the sample, and $\rho_T(E_F)$ is the DOS of the tip. If $k_B T$ is much smaller than the feature sizes in the energy spectrum of interest, then

$$I = \frac{4\pi e}{\hbar} \int_0^{eV} \rho_S(E_F - eV + \epsilon)\, \rho_T(E_F + \epsilon)\, |M|^2 \, d\epsilon. \tag{2.36}$$

2.5.3 Error estimation

It is important to know how accurate the transition probability from first-order time-dependent perturbation theory is. Following the method of Oppenheimer (1928), we show that our choice of unperturbed potentials minimizes the error estimation term. Therefore, it is the optimum choice.

To make an error estimation, we evaluate the second-order transition probability by taking into account the second term in Eq. (2.29),

$$w_{\mu\nu}^{(2)} = \frac{2\pi}{\hbar} \left| \sum_{\lambda} \frac{(\chi_\nu, U_S\, \chi_\lambda)\, (\chi_\lambda, U_T\, \psi_\mu)}{E_\lambda - E_\nu} \right|^2 \delta(E_\mu - E_\nu). \tag{2.37}$$

Oppenheimer (1928) estimated the error as follows. Since E_μ, E_ν are always several eV below vacuum level, the tip wavefunctions with energy levels close to that of the sample decay quickly outside the tip body. Thus, the matrix elements $(\chi_\nu, U_S\, \chi_\lambda)$ for $E_\lambda \leq E_\mu$ are much smaller than those with $E_\lambda \approx 0$. On the other hand, while $E_\lambda > 0$, the tip wavefunction oscillates quickly, and the matrix elements become small again. Therefore, the numerator must have a sharp maximum somewhere at $E_\lambda \approx \bar{E}$. By replacing E_λ by $\bar{E}$ in the denominator of Eq. (2.37), we obtain the Oppenheimer error estimation term:

$$w_{\mu\nu}^{(2)} \approx \frac{2\pi}{\hbar} \left| \sum_{\lambda} \frac{(\chi_{\nu}, U_S \chi_{\lambda})(\chi_{\lambda}, U_T \psi_{\mu})}{\bar{E} - E_{\nu}} \right|^2 \delta(E_{\mu} - E_{\nu})$$

$$= \frac{2\pi}{\hbar} \left| \frac{(\chi_{\nu}, U_S U_T \psi_{\mu})}{\bar{E} - E_{\nu}} \right|^2 \delta(E_{\mu} - E_{\nu}) = 0$$

(2.38)

because of Eq. (2.22). Thus, the accuracy of the first-order perturbation is justified.

2.5.4 Wavefunction correction

The wavefunctions in Eq. (2.34) are different from the wavefunctions of the free tip and free sample. The effect of the distortion potential ($V = U_S - U_{S0}$ and $V = U_S - U_{S0}$), can be evaluated through time-independent perturbation. In the following, we present an approximate method based on the Green's function of the vacuum (see Appendix B). To first order, the distorted wavefunction ψ is related to the undistorted one, ψ_0, by

$$\psi(\mathbf{r}) = \psi_0(\mathbf{r}) + \int G(\mathbf{r}, \mathbf{r}')V(\mathbf{r}')\psi_0(\mathbf{r}') \, d^3\mathbf{r}'.$$

(2.39)

The Green's function is defined by

$$\left(-\frac{\hbar^2}{2m}\nabla^2 + U_{S0} - E \right)G(\mathbf{r}, \mathbf{r}') = -\delta(\mathbf{r} - \mathbf{r}').$$

(2.40)

Using the properties of the Green's function (see Appendix B), the evaluation of the effect of distortion to transmission matrix elements can be greatly simplified. First, because of the continuity of the wavefunction and its derivative across the separation surface, only the *multiplier of the wavefunctions at the separation surface* is relevant. Second, in the first-order approximation, the effect of the distortion potential is additive [see Eq. (2.39)]. Thus, to evaluate the multiplier, a simpler undistorted Hamiltonian might be used instead of the accurate one. For example, the Green's function and the wavefunction of the *vacuum* can be used to evaluate the distortion multiplier. Consider the one-dimensional case. From Eq. (2.40),

$$\psi(z_0) = e^{-\kappa z_0} - \frac{m}{\kappa\hbar^2} \int e^{-\kappa|z_0 - z|} e^{-\kappa z} V(z) \, dz.$$

(2.41)

In the region $z > z_0$, the integrand vanishes rapidly. The integration in region $z < z_0$ gives

$$\psi(z_0) = e^{-\kappa z_0}\left(1 - \frac{m}{\kappa\hbar^2}\int_{-\infty}^{z_0} V(z)\,dz\right). \tag{2.42}$$

For $z \geq z_0$, the distorted wavefunction has the same exponential dependence on z as the free wavefunction. Thus, $\partial\psi/\partial z$ gains the same factor. The tip wavefunction χ gains a similar factor. Therefore, the transmission probability becomes

$$T = T_0\left(1 - \frac{m}{\kappa\hbar^2}\int_{-\infty}^{z_0} V(z)\,dz\right)^2\left(1 - \frac{m}{\kappa\hbar^2}\int_{z_0}^{\infty} V(z)\,dz\right)^2. \tag{2.43}$$

We will test this method with the exact solutions of the square-barrier problem in the following subsection.

2.5.5 Comparison with exact solutions

As the semiclassical tunneling theory and Bardeen's original approach become inaccurate for potential barriers close to or lower than the energy level, the validity range of the MBA is much wider. In this subsection, the accuracy of the MBA is tested against an exactly soluble case, that is, the one-dimensional transmission through a square barrier of thickness $W=2$ Å (see Fig. 2.9).

The exact transmission coefficient is presented in Equations Eq. (2.8) and (2.10). The logarithm of the exact transmission coefficient varies almost linearly with barrier height, all the way from 4 eV above the energy level to 2 eV below the energy level.

The transmission coefficient of the MBA can be evaluated analytically. For the case of $L = W/2$, the result is

$$T = \frac{8\kappa_0^2 q^2}{\left[\kappa^2\left(\sinh\kappa W + \dfrac{\kappa_0}{\kappa}\cosh\kappa W\right)^2 + q^2\left(\cosh\kappa W + \dfrac{\kappa_0}{\kappa}\sinh\kappa W\right)^2\right]^2}, \tag{2.44}$$

where $\kappa_0 = (2mU_0)^{1/2}/\hbar$.

As shown in Fig. 2.10, the transmission coefficient of the MBA remains accurate even with a barrier 2 eV below the energy level. By taking $L = W/3$, a similar result is found, as shown in Fig. 2.9. Therefore, the accuracy of MBA is approximately independent of the choice of the separation surface. The transmission coefficient of the original BA can be obtained by letting

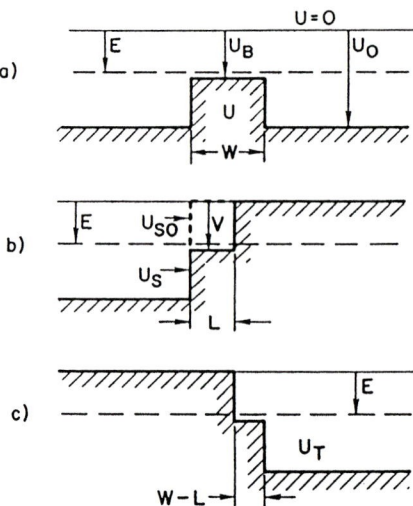

Fig. 2.9. Perturbation treatment for the square-barrier problem. (a), Original problem. (b) and (c) The potentials of the subsystems for a perturbation treatment.

$\kappa_0 = \kappa$ in Eq. (2.44). As shown in Fig. 2.9, the BA becomes inaccurate when the barrier top is close to or lower than the energy level. The accuracy of semiclassical theory is even worse than the BA, as shown in Fig. 2.10. As the top of the barrier comes close to the energy level, the WKB transmission coefficient quickly becomes unity. The MBA with corrections with Green's function is also shown. For the case of Fig. 2.9, the values of the integrals are $(-U_B L)$ and $[-U_B(W-L)]$, respectively. As shown, the result is fairly accurate.

The MBA provides another simple explanation of why the image potential is not observable. According to Eq. (2.42), as long as the integral of the distortion potential is a constant (that is, the shaded area in Fig. 2.8 remains constant while varying the tip–sample distance), the effect of distortion is a constant independent of the barrier thickness. Therefore, the effect of barrier lowering due to image force is not observable.

2.6 Effect of image force on tunneling

In this section, we will derive the correction factor for the tunneling current due to the image force. For a free metal surface, the image potential pertinent to this surface, Eq. (2.2), is always present. The the simple image potential is always an essential part of the free sample, and is always implied in any first-

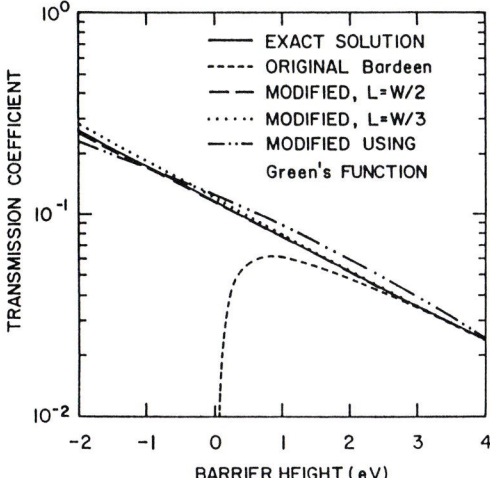

Fig. 2.10. Different approximate methods for the square barrier problem. (Parameters used: $W= 2$ Å; $E= 4$ eV; $U_0= 16$ eV.) The original Bardeen theory breaks down when the barrier top comes close to the energy level. The modified Bardeen tunneling theory is accurate with separation surfaces either centered ($L = W/2$) or off-centered ($L = W/3$). By approximating the distortion of wavefunctions using Green's functions, the error in the entire region is only a few percent.

principles calculations. Therefore, a correction due to image force in tunneling means the correction due to the second term in Eq. (2.4); that is,

$$V(z) = -\frac{e^2}{2} \sum_{n=0}^{\infty} \left(\frac{nW}{n^2 W^2 - z^2} - \frac{1}{nW} \right).$$ (2.45)

We choose the bisection plane in the barrier as the separation surface. The correction for tunneling current can be obtained from the correction for the wavefunction on the bisection plane. Using the Green's function method, following Eq. (2.42), the correction factor for the wavefunction at $z = W/2$ is

$$\psi = \psi_0 \left(1 - \frac{m}{\kappa \hbar^2} \int_0^{W/2} V(z)\, dz \right).$$ (2.46)

Substituting Eq. (2.45) for Eq. (2.46), performing the elementary integrals, noticing that

$$a_0 = \frac{me^2}{\hbar^2}$$ (2.47)

is the Bohr radius, 0.529 Å, and the value of the fast-converging numerical sum is

$$\sum_{n=0}^{\infty} \left(\frac{1}{2} \ln \frac{2n+1}{2n-1} - \frac{1}{2n} \right) \approx 0.02898, \tag{2.48}$$

we find

$$\psi = \psi_0 \left(1 + \frac{0.02898}{\kappa a_0} \right). \tag{2.49}$$

At $\phi = 3.5$ eV, $\kappa \approx 0.95$ Å^{-1}. The numerical factor in Eq. (2.49) is

$$\psi = 1.057\psi_0. \tag{2.50}$$

The wavefunction of another electrode gets the same factor. The tunneling current is proportional to the square of the matrix element. Thus,

$$I = 1.25 I_0. \tag{2.51}$$

Again, the barrier thickness W does not show up in the correction factor. In other words, the correction factor is independent of the barrier thickness. This result is not intuitively obvious, but reasonable, because the image potential is a long-range interaction. In Chapter 7, we will see that for the exchange energy of interatomic forces, the polarization correction is also a constant multiplier independent of the interatomic distance, and the value of the multiplier for hydrogen atoms is larger than the multiplier for the planar tunneling junction.

CHAPTER 3
TUNNELING MATRIX ELEMENTS

3.1 Introduction

In Chapter 2, we showed that the tunneling current can be determined with a perturbation approach. The central problem is to calculate the matrix elements. Those are determined by the modified Bardeen surface integral, evaluated from the wavefunctions of the tip and the sample (with proper corrections) on a separation surface between them, as shown in Fig. 3.1:

$$M = -\frac{\hbar^2}{2m} \int_{\Sigma} (\chi^* \nabla \psi - \psi \nabla \chi^*) \bullet d\mathbf{S}, \qquad (3.1)$$

where ψ_μ is the wavefunction of the sample, and χ_ν is the wavefunction of the tip, respectively. These matrix elements are usually called *tunneling matrix elements*. As we have shown in Chapter 2, after proper correction, they are also suitable for describing processes where the top of the potential barrier becomes lower than the energy level. As shown in Eq. (3.1), only the values of these wavefunctions at the separation surface, which is located roughly in the middle of the gap, are relevant. Certain corrections to the free wavefunctions are required. As we showed in Chapter 2, for the case of a flat surface, the correction due to the image force is essentially a constant multiplier. For a single atom, the correction is also essentially a constant multiplier, although with a different value, as we will discuss in Chapter 7. The matrix element has the dimension of *energy*. In Chapter 7, we will show that the physical meaning of Bardeen's matrix element is the energy lowering due to the overlap of the two states. In other words, it is the *resonance energy* of Heisenberg (1926) and Pauling (1977), which is the dominant term in the force between the tip and the sample under normal STM operating conditions. Therefore, an explicit form for the tunneling matrix elements described by Eq. (3.1) is important in the quantitative understanding of STM and AFM.

The problem of evaluating the tunneling matrix elements has been investigated by many authors (Tersoff and Hamann, 1983, 1985; Baratoff, 1984; Chung, Feuchtwang, and Cutler, 1987; Chen 1988, 1990, 1990a; Lawunmi and

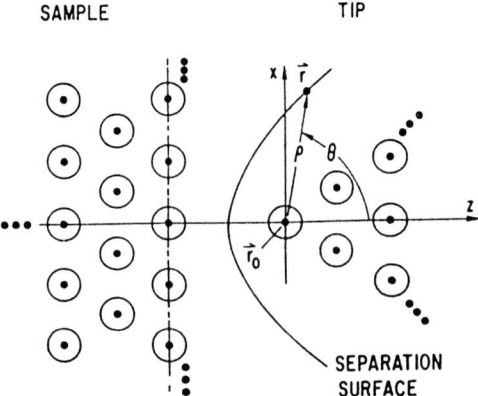

Fig. 3.1. Derivation of the tunneling matrix elements. A separation surface is placed between the tip and the sample. The exact position and the shape of the separation surface is not important. The coordinates for the Cartesian coordinate system and spherical coordinate system are shown, except y and ϕ. (Reproduced from Chen, 1990a, with permission.)

Payne, 1990; Sacks and Noguera, 1991). All those authors used a spherical-harmonic expansion to represent tip wavefunctions in the gap region, which is a natural choice. The spherical-harmonic expansion is used extensively in solid-state physics as well as in quantum chemistry for describing and classifying electronic states. In problems without a magnetic field, the real spherical harmonics are preferred, as described in Appendix A.

In this Chapter, we present step-by-step derivations of the explicit expressions for matrix elements based on the spherical-harmonic expansion of the tip wavefunction in the gap region. The result — *derivative rule* is extremely simple and intuitively understandable. Two independent proofs are presented. The mathematical tool for the derivation is the spherical modified Bessel functions, which are probably the simplest of all Bessel functions. A concise summary about them is included in Appendix C.

3.2 Tip wavefunctions

From Eq. (3.1), the tunneling matrix elements are determined by the wavefunctions of the tip and the sample at the separation surface, which is located roughly in the middle of the vacuum gap, as shown in Fig. 3.1. For both tip states and sample states near the Fermi level, the wavefunctions on and beyond the separation surface satisfy Schrödinger's equation in the vacuum,

$$(\nabla^2 - \kappa^2)\chi(\mathbf{r}) = 0, \tag{3.2}$$

where $\kappa = (2m\phi)^{1/2}\hbar^{-1}$ is the decay constant, determined by the work function ϕ. Near the center of the apex atom, Eq. (3.2) is not valid. Nevertheless, for calculating the tunneling matrix element, we only need the tip wavefunction near the separation surface, which always satisfies Eq. (3.2).

The tip wavefunctions can be expanded into the spherical-harmonic components, $Y_{lm}(\theta, \phi)$, with the nucleus of the apex atom as the origin. Each component is characterized by quantum numbers l and m. In other words, we are looking for solutions of Eq. (3.2) in the form

$$\chi(\mathbf{r}) = \sum_{l, m} C_{lm} f_{lm}(\kappa\rho) \, Y_{lm}(\theta, \phi), \tag{3.3}$$

where $\rho = |\mathbf{r} = \mathbf{r}_0|$, and $\mathbf{r}_0$ is the center of the apex atom. Substituting Eq. (3.3) into Eq. (3.2), we obtain the differential equation for the radial functions $f_{lm}(u)$,

$$\frac{d}{du}\left(u^2 \frac{d f(u)}{du}\right) - \left[u^2 + l(l+1)\right] f(u) = 0. \tag{3.4}$$

As seen, the radial functions depend only on l.

The standard linear-independent solutions for Eq. (3.4) are the spherical modified Bessel functions, $i_l(u)$ and $k_l(u)$ (Arfken, 1968). A brief introduction to them is provided in Appendix C. These so-called special functions are actually *elementary functions*:

$$i_l(u) = u^l\left(\frac{d}{u \, du}\right)^l \frac{\sinh u}{u}, \tag{3.5}$$

and

$$k_l(u) = (-1)^l \, u^l\left(\frac{d}{u \, du}\right)^l \frac{\exp(-u)}{u}. \tag{3.6}$$

Obviously, the functions $i_l(u)$ diverge at large u, which are not appropriate to represent tip wavefunctions. The functions $k_l(u)$ are regular at large u, which satisfies the desired boundary condition. Therefore, a component of tip wavefunction with quantum numbers l and m has the general form

$$\chi_{lm}(\mathbf{r}) = C \, k_l(\kappa\rho) \, Y_{lm}(\theta, \phi). \tag{3.7}$$

Table 3.1. Vacuum tails of tip wavefunctions

State	Wavefunction
s	$C \dfrac{1}{\kappa\rho} e^{-\kappa\rho}$
p_z	$C \left[\dfrac{1}{\kappa\rho} + \dfrac{1}{(\kappa\rho)^2} \right] e^{-\kappa\rho} \cos\theta$
p_x	$C \left[\dfrac{1}{\kappa\rho} + \dfrac{1}{(\kappa\rho)^2} \right] e^{-\kappa\rho} \sin\theta \cos\phi$
p_y	$C \left[\dfrac{1}{\kappa\rho} + \dfrac{1}{(\kappa\rho)^2} \right] e^{-\kappa\rho} \sin\theta \sin\phi$
d_{z^2}	$C \left[\dfrac{1}{\kappa\rho} + \dfrac{3}{(\kappa\rho)^2} + \dfrac{3}{(\kappa\rho)^3} \right] e^{-\kappa\rho} \left(\cos^2\theta - \dfrac{1}{3} \right)$
d_{xz}	$C \left[\dfrac{1}{\kappa\rho} + \dfrac{3}{(\kappa\rho)^2} + \dfrac{3}{(\kappa\rho)^3} \right] e^{-\kappa\rho} \sin 2\theta \cos\phi$
d_{yz}	$C \left[\dfrac{1}{\kappa\rho} + \dfrac{3}{(\kappa\rho)^2} + \dfrac{3}{(\kappa\rho)^3} \right] e^{-\kappa\rho} \sin 2\theta \sin\phi$
d_{xy}	$C \left[\dfrac{1}{\kappa\rho} + \dfrac{3}{(\kappa\rho)^2} + \dfrac{3}{(\kappa\rho)^3} \right] e^{-\kappa\rho} \sin^2\theta \sin 2\phi$
$d_{x^2-y^2}$	$C \left[\dfrac{1}{\kappa\rho} + \dfrac{3}{(\kappa\rho)^2} + \dfrac{3}{(\kappa\rho)^3} \right] e^{-\kappa\rho} \sin^2\theta \cos 2\phi$

In the absence of magnetic field, it is convenient to write those tip wavefunctions in real form, as listed in Table 3.1. The coefficients C are determined by comparing with first-principles calculations of actual tip states.

3.3 Green's function and tip wavefunctions

The Green's function for the Schrödinger equation, Eq. (3.2), is defined by the differential equation (Arfken, 1968)

$$\left[\nabla^2 - \kappa^2 \right] G(\mathbf{r} - \mathbf{r}_0) = -\delta(\mathbf{r} - \mathbf{r}_0). \tag{3.8}$$

With the boundary condition that it is regular at $|\mathbf{r} - \mathbf{r}_0| \to \infty$, the explicit form of the Green's function is (see Appendix B):

$$G(\mathbf{r} - \mathbf{r}_0) = \frac{\exp(-\kappa|\mathbf{r} - \mathbf{r}_0|)}{4\pi|\mathbf{r} - \mathbf{r}_0|}, \tag{3.9}$$

which can be verified by direct substitution.

Denoting $\rho = |\mathbf{r} - \mathbf{r}_0|$, the Green's function can be written in terms of a spherical modified Bessel function of the second kind,

$$G(\mathbf{r} - \mathbf{r}_0) = \frac{\kappa}{4\pi} k_0(\kappa\rho). \tag{3.10}$$

Therefore, the s-wave tip wavefunction is equal to the Green's function up to a constant, *with the center of the apex atom taken as* $\mathbf{r}_0$,

$$\chi_s(\mathbf{r}) = \frac{4\pi C}{\kappa} G(\mathbf{r} - \mathbf{r}_0). \tag{3.11}$$

By taking the derivative of both sides of Eq. (3.10) with respect to z_0, and using the relation (see Appendix C):

$$\frac{d}{du} k_0(u) = -k_1(u) \tag{3.12}$$

we obtain

$$\frac{\partial}{\kappa \partial z_0} G(\mathbf{r} - \mathbf{r}_0) = \frac{\kappa}{4\pi} \frac{z - z_0}{\rho} k_1(\kappa\rho). \tag{3.13}$$

The second fraction in Eq. (3.13) is $\cos\theta$. Therefore, the tip wavefunction for the p_z state is

$$\chi_{pz}(\mathbf{r}) = \frac{4\pi C}{\kappa} \frac{\partial}{\kappa \partial z_0} G(\mathbf{r} - \mathbf{r}_0). \tag{3.14}$$

Similarly, we have

$$\chi_{px}(\mathbf{r}) = \frac{4\pi C}{\kappa} \frac{\partial}{\kappa \partial x_0} G(\mathbf{r} - \mathbf{r}_0), \tag{3.15}$$

$$\chi_{py}(\mathbf{r}) = \frac{4\pi C}{\kappa} \frac{\partial}{\kappa \partial y_0} G(\mathbf{r} - \mathbf{r}_0). \tag{3.16}$$

By taking derivative with respect to x_0 on both sides of Eq. (3.13), and noticing that

$$\frac{d}{du}\left(\frac{k_1(u)}{u}\right) = -\frac{k_2(u)}{u},$$ (3.17)

we obtain

$$\frac{\partial^2}{\kappa^2\,\partial x_0\,\partial z_0}\,G(\mathbf{r} - \mathbf{r}_0) = \frac{\kappa}{4\pi}\,\frac{x - x_0}{\rho}\,\frac{z - z_0}{\rho}\,k_2(\kappa\rho).$$ (3.18)

Using the geometrical relation

$$\frac{x - x_0}{\rho}\,\frac{z - z_0}{\rho} = \cos\theta\,\sin\theta\,\sin\phi,$$ (3.19)

and comparing with Table 3.1, we obtain

$$\chi_{dxz}(\mathbf{r}) = \frac{4\pi C}{\kappa}\,\frac{\partial^2}{\kappa^2\,\partial x_0\,\partial z_0}\,G(\mathbf{r} - \mathbf{r}_0).$$ (3.20)

Similar results can be obtained for the wavefunctions of d_{yz} and d_{xy} states.

By taking the derivative with respect to z_0 on both sides of Eq. (3.13), an extra term containing $k_1(\kappa\rho)/\rho$ is generated:

$$\frac{\partial^2}{\kappa^2\,\partial z_0^2}\,G(\mathbf{r} - \mathbf{r}_0) = \frac{\kappa}{4\pi}\left(\frac{(z - z_0)^2}{\rho^2}\,k_2(\kappa\rho) - \frac{1}{\rho}\,k_1(\kappa\rho)\right).$$ (3.21)

Remembering that (see Appendix C)

$$\frac{3k_1(u)}{u} = k_0(u) - k_2(u),$$ (3.22)

and combining with Eq. (3.11), we obtain

$$\chi_{dz^2}(\mathbf{r}) = \frac{4\pi C}{\kappa}\left(\frac{\partial^2}{\kappa^2\,\partial z_0^2}\,G(\mathbf{r} - \mathbf{r}_0) - \frac{1}{3}\,G(\mathbf{r} - \mathbf{r}_0)\right).$$ (3.23)

For the wavefunction of the $d_{x^2-y^2}$ tip state, the extra term generated by taking the second derivative cancels. Therefore,

$$\chi_{d(x^2 - y^2)}(\mathbf{r}) = \frac{4\pi C}{\kappa} \left(\frac{\partial^2}{\kappa^2 \partial x_0^2} - \frac{\partial^2}{\kappa^2 \partial y_0^2} \right) G(\mathbf{r} - \mathbf{r}_0). \tag{3.24}$$

We will use these relations to evaluate tunneling matrix elements in the following section.

3.4 The derivative rule: individual cases

As shown in Eq. (3.1), the transmission matrix elements for different tip states are determined by the Bardeen integral on a surface separating the sample and the tip with one of the tip states. For an s-wave tip state, using Eq. (3.11),

$$\begin{aligned} M_s &\equiv \frac{\hbar^2}{2m_e} \int_\Sigma \left(\chi_s \nabla \psi - \psi \nabla \chi_s \right) \cdot d\mathbf{S} \\ &= \frac{2\pi C \hbar^2}{\kappa m} \int_\Sigma \left[G(\mathbf{r} - \mathbf{r}_0)\nabla \psi - \psi \nabla G(\mathbf{r} - \mathbf{r}_0) \right] \cdot d\mathbf{S}. \end{aligned} \tag{3.25}$$

Using Green's theorem, it can be converted into a volume integral over Ω_T, the tip side from the separation surface. Noticing that the sample wavefunction ψ satisfies Schrödinger's equation, Eq. (3.2), in Ω_T, and that the Green's function satisfies Eq. (3.8), we obtain immediately

$$\begin{aligned} M_s &= \frac{2\pi C \hbar^2}{\kappa m} \int_{\Omega_T} \left[G(\mathbf{r} - \mathbf{r}_0)\nabla^2 \psi - \psi \nabla^2 G(\mathbf{r} - \mathbf{r}_0) \right] d\tau \\ &= \frac{2\pi C \hbar^2}{\kappa m} \psi(\mathbf{r}_0), \end{aligned} \tag{3.26}$$

which was derived first by Tersoff and Hamann (1983, 1985) using the two-dimensional Fourier transform method. As seen, the simplicity of this result is because the s-wave tip wavefunction *is* the Green's function of the Schrödinger equation in vacuum.

Taking the derivative with respect to z_0 on both sides of Eq. (3.25), and noticing that z_0 is a parameter in the integral (which does not involve the process of evaluating the integration), and the expression of the p_z tip wavefunction, Eq. (3.14), we find

$$\frac{2\pi C\hbar^2}{\kappa m} \frac{\partial \psi(\mathbf{r}_0)}{\partial z_0}$$

$$= \frac{2\pi C\hbar^2}{\kappa m} \frac{\partial}{\partial z_0} \int_\Sigma [G(\mathbf{r} - \mathbf{r}_0)\nabla\psi - \psi\nabla G(\mathbf{r} - \mathbf{r}_0)]\cdot d\mathbf{S}$$

$$= \frac{2\pi C\hbar^2}{\kappa m} \int_\Sigma \left(\frac{\partial G}{\partial z_0} \nabla\psi - \psi\nabla \frac{\partial G}{\partial z_0} \right)\cdot d\mathbf{S} \qquad (3.27)$$

$$= \frac{\hbar^2}{2m} \int_\Sigma [\chi_{pz}(\mathbf{r})\nabla\psi - \psi\nabla\chi_{pz}(\mathbf{r})]\cdot d\mathbf{S}$$

$$= M_{pz}.$$

Therefore, the tunneling matrix element for a p_z tip state is proportional to the z derivative of the sample wavefunction at the center of the apex atom.

The tunneling matrix elements from the rest of the nine tip wavefunctions can be derived using the relation between the tip and Green's functions established in the previous section. For example, for the d_{xz} tip state,

$$\frac{2\pi C\hbar^2}{\kappa m} \frac{\partial^2 \psi(\mathbf{r}_0)}{\partial x_0 \partial z_0}$$

$$= \frac{2\pi C\hbar^2}{\kappa m} \frac{\partial^2}{\partial x_0 \partial z_0} \int_\Sigma [G(\mathbf{r} - \mathbf{r}_0)\nabla\psi - \psi\nabla G(\mathbf{r} - \mathbf{r}_0)]\cdot d\mathbf{S}$$

$$= \frac{2\pi C\hbar^2}{\kappa m} \int_\Sigma \left(\frac{\partial^2 G}{\partial x_0 \partial z_0} \nabla\psi - \psi\nabla \frac{\partial^2 G}{\partial x_0 \partial z_0} \right)\cdot d\mathbf{S} \qquad (3.28)$$

$$= \frac{\hbar^2}{2m} \int_\Sigma [\chi_{dxz}(\mathbf{r})\nabla\psi - \psi\nabla\chi_{dxz}(\mathbf{r})]\cdot d\mathbf{S}$$

$$= M_{dxz}.$$

Similarly, we can obtain the tunneling matrix elements for all the tip states listed in Table 3.1. Table 3.2 is a summary of the results.

The results can be summarized as the *derivative rule:* Write the angle dependence of the tip wavefunction in terms of x, y, and z. Replace them with the simple rule,

$$x \longrightarrow \frac{\partial}{\kappa\partial x} ; \quad y \longrightarrow \frac{\partial}{\kappa\partial y} ; \quad z \longrightarrow \frac{\partial}{\kappa\partial z} ; \qquad (3.29)$$

and act on the sample wavefunction. The term r^2 corresponds to ∇^2. The Schrödinger equation (Eq. (3.2)) implies $\nabla^2\psi = \kappa^2\psi$. Thus, all tunneling matrix elements listed in Table 3.2 follow the derivative rule.

Table 3.2. Tunneling matrix elements

Tip state	Matrix element
s	$\dfrac{2\pi C\hbar^2}{\kappa m}\,\psi(\mathbf{r}_0)$
p_z	$\dfrac{2\pi C\hbar^2}{\kappa m}\,\dfrac{\partial\psi}{\partial z}(\mathbf{r}_0)$
p_x	$\dfrac{2\pi C\hbar^2}{\kappa m}\,\dfrac{\partial\psi}{\partial x}(\mathbf{r}_0)$
p_y	$\dfrac{2\pi C\hbar^2}{\kappa m}\,\dfrac{\partial\psi}{\partial y}(\mathbf{r}_0)$
d_{zx}	$\dfrac{2\pi C\hbar^2}{\kappa m}\,\dfrac{\partial^2\psi}{\partial z\,\partial x}(\mathbf{r}_0)$
d_{zy}	$\dfrac{2\pi C\hbar^2}{\kappa m}\,\dfrac{\partial^2\psi}{\partial z\,\partial y}(\mathbf{r}_0)$
d_{xy}	$\dfrac{2\pi C\hbar^2}{\kappa m}\,\dfrac{\partial^2\psi}{\partial x\,\partial y}(\mathbf{r}_0)$
$d_{z^2}-\frac{1}{3}r^2$	$\dfrac{2\pi C\hbar^2}{\kappa m}\left(\dfrac{\partial^2\psi}{\partial z^2}-\dfrac{1}{3}\kappa^2\psi\right)(\mathbf{r}_0)$
$d_{x^2-y^2}$	$\dfrac{2\pi C\hbar^2}{\kappa m}\left(\dfrac{\partial^2\psi}{\partial x^2}-\dfrac{\partial^2\psi}{\partial y^2}\right)(\mathbf{r}_0)$

3.5 The derivative rule: general case

In this section, we present an alternative proof of the derivative rule, which provides an expression for the transmission matrix element from an arbitrary tip state expanded in terms of spherical harmonics. In the previous sections, we have expanded the tip wavefunction *on the separation surface* in terms of spherical harmonics. In general, the expansion is

$$\chi = \sum_{l,\,m}\beta_{lm}\,k_l(\kappa\rho)Y_{lm}(\theta,\phi)\,. \tag{3.30}$$

The coefficients β_{lm} are determined by fitting the tip wavefunction on and beyond the separation surface. Inside the tip body, the actual tip wavefunction does not satisfy Eq. (3.2), and the expansion (Eq. (3.30)) does

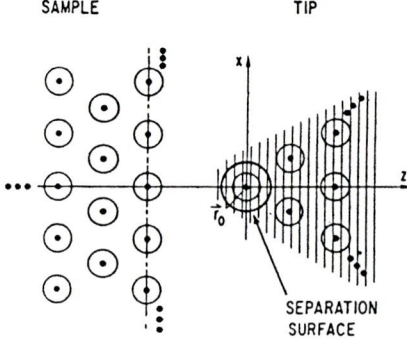

Fig. 3.2. Derivation of the derivative rule: general case. In the shaded region, the tip wavefunction does not satisfy the Schrödinger equation in the vacuum. However, the expansion in Eq. (3.30) satisfies the Schrödinger equation in the vacuum except at the nucleus of the apex atom. Thus the surface on which the Bardeen integral is evaluated can be deformed to be any surface that encloses the nucleus of the apex atom.

not represent the actual tip wavefunction. Instead, it represents the *vacuum continuation* of the vacuum tail of the tip wavefunction into the tip body. Actually, except for the center of the apex atom, that is, the origin of the spherical harmonics, the expansion Eq. (3.30) satisfies the Schrödinger equation in the vacuum, Eq. (3.2).

On the other hand, the sample wavefunction in the entire volume of the tip body should satisfy the Schrödinger equation, Eq. (3.2). Especially, it should be regular at the origin. Therefore, in the tip body, the sample wavefunction must have the form

$$\psi = \sum_{l, m} \alpha_{lm}\, i_l(\kappa\rho) Y_{lm}(\theta, \phi)\,, \tag{3.31}$$

where the coefficients α_{lm} are determined by the sample wavefunction in the vacuum.

To proceed with the proof, we first state a property of the Bardeen integral: If both functions involved, ψ and χ, satisfy the same Schrödinger equation in a region Ω, then the Bardeen integral J on the surface enclosing a closed volume ω within Ω vanishes. Actually, using Green's theorem, Eq. (3.2) becomes

$$J = \frac{\hbar^2}{2m} \int_{\Omega} (\chi^* T\psi - \psi T\chi^*)\, d\Omega. \tag{3.32}$$

Because ψ and χ satisfy the same Schrödinger equation with the same energy eigenvalue, Eq. (3.32) becomes

$$J = -\frac{\hbar^2}{2m} \int_\omega (\chi^* U\psi - \psi U\chi^*) \, d\Omega \tag{3.33}$$
$$= 0.$$

As we have shown, except at the nucleus of the apex atom, or the origin of the spherical harmonics, the expansion form of the tip wavefunction χ, Eq. (3.30), and the sample wavefunction ψ satisfy the same Schrödinger equation, Eq. (3.8). Therefore, we can take any surface enclosing the nucleus of the apex atom to evaluate the transmission matrix element, Eq. (3.1), especially, a sphere of arbitrary radius r_0 centered at the nucleus of the apex atom. Substituting Equations (3.30) and (3.31) into Eq. (3.1),

$$M = \frac{\hbar^2}{2m} \sum_{l,m} \sum_{l',m'} \alpha_{lm} \beta^*_{l'm'}$$
$$\times \left(\frac{di_l(\kappa r)}{dr} k_{l'}(\kappa r) - \frac{dk_{l'}(\kappa r)}{dr} i_l(\kappa r) \right) \tag{3.34}$$
$$\times r^2 \int \sin\theta \, d\theta \, d\phi \, Y_{lm}(\theta, \phi) Y^*_{l'm'}(\theta, \phi).$$

The integral in Eq. (3.34) is $\delta_{ll'}\delta_{mm'}$ due to the orthonormal property of the spherical harmonics. The second line is the Wronskian of the spherical modified Bessel functions,

$$i_l(u) k_l'(u) - i_l'(u) k_l(u) = -\frac{1}{u^2} . \tag{3.35}$$

The r^{-2} factor in the Wornskian cancels the r^2 factor of the surface integral. As expected, the value of the surface integral is independent of the actual location of the surface. Therefore, we obtain a general expression for the tunneling matrix elements, the *sum rule,*

$$M = \frac{\hbar^2}{2m\kappa} \sum_{l,m} \alpha_{lm} \beta^*_{lm} . \tag{3.36}$$

The steps from Eq. (3.34) to Eq. (3.36) simply mean that for each component of the tip wavefunction with angular dependence characterized by l and m, the tunneling matrix element is proportional to the corresponding component of the sample wavefunction with the same angular dependence.

In the following, we show that the coefficients α_{lm} in Eq. (3.31) are related to the derivatives of the sample wavefunction ψ with respect to x, y, and z at the nucleus of the apex atom in an extremely simple way. (To simplify the notation, we take the nucleus of the apex atom as the origin of the coordinate system, i.e., $x_0 = 0$, $y_0 = 0$, and $z_0 = 0$.) This is similar to the well-known case that the expansion coefficients for a power series are simply related to the derivatives of the function at the point of expansion, the so-called Taylor series or MacLaurin series. We will then obtain the derivative rule again, from a completely different point of view.

The key of the proof is the properties of the spherical modified Bessel function of the first kind, $i_l(u)$. For small values of u, the function $i_l(u)$ has the following form:

$$i_l(u) = \frac{(u)^l}{(2l + 1)!!} \left[1 + O(u^2) \right]. \tag{3.37}$$

By multiplying $i_l(\kappa\rho)$ with a spherical harmonics of order l, this term becomes a homogeneous polynomial of x, y, and z of order l. By taking a partial derivative of $i_l(\kappa r)$ with respect to x, y, and z with (summed) order n, and taking the value at r=0, all the terms with powers of x, y, and z of $l > n$ drop off. In particular, for the cases of l=0 and 1, there is only one term left in the derivative at ρ=0, that is, a term containing only one coefficient in Eq. (3.36). For l=2, the derivative may contain the second term in $i_0(\kappa\rho)$, which should be subtracted off to obtain the coefficient for an l=2 component.

We start our derivation by writing down the explicit form of the vacuum asymptote of a tip wavefunction (as well as its vacuum continuation in the tip body). As we have explained in Section 5.3, for the simplicity of relevant mathematics, the rather complicated normalization constants of the spherical harmonics are absorbed in the expression of the sample wavefunction. Up to l=2, we define the coefficients of the expansion by the following expression:

$$\chi = \beta_{00}\, k_0(\kappa\rho) + \left(\beta_{10}\, \frac{z}{\rho} + \beta_{11}\, \frac{x}{\rho} + \beta_{12}\, \frac{y}{\rho} \right) k_1(\kappa\rho)$$

$$+ \left[\beta_{20} \left(\frac{z^2}{\rho^2} - \frac{1}{3} \right) + \beta_{21}\, \frac{xz}{\rho^2} + \beta_{22}\, \frac{yz}{\rho^2} \right. \tag{3.38}$$

$$\left. + \beta_{23}\, \frac{xy}{\rho^2} + \beta_{24} \left(\frac{x^2}{\rho^2} - \frac{y^2}{\rho^2} \right) \right] k_2(\kappa\rho).$$

Similarly, for the sample wavefunction, up to the lowest significant term in the power expansion of the spherical modified Bessel function of the first kind, $i_l(\kappa\rho)$,

$$
\begin{aligned}
4\pi\psi = \alpha_{00} &\left[1 + \frac{1}{6}(\kappa\rho)^2 \right] + \alpha_{10}\,\kappa z + \alpha_{11}\,\kappa x + \alpha_{12}\,\kappa y \\
&+ \alpha_{20}\,\frac{3\kappa^2}{4}\left(z^2 - \frac{1}{3}\rho^2 \right) + \alpha_{21}\,(\kappa^2 xz) + \alpha_{22}\,(\kappa^2 zy) \\
&+ \alpha_{23}\,(\kappa^2 xy) + \alpha_{24}\left[\kappa^2(x^2 - y^2) \right] + O(\rho^3).
\end{aligned} \tag{3.39}
$$

The factor 4π is introduced for convenience. Now, it is straightforward to obtain a relation between the coefficients α_{lm} and the derivatives of sample wavefunctions. For example, because of $\rho^2 = x^2 + y^2 + z^2$, we have

$$
\frac{4\pi}{\kappa^2}\frac{\partial^2}{\partial z^2}\psi(\mathbf{r}_0) = \alpha_{20} + \frac{1}{3}\alpha_{00}. \tag{3.40}
$$

Noticing that

$$
4\pi\psi(\mathbf{r}_0) = \alpha_{00}, \tag{3.41}
$$

we obtain

$$
\alpha_{20} = 4\pi\left(\frac{1}{\kappa^2}\frac{\partial^2}{\partial z^2} - \frac{1}{3} \right)\psi(\mathbf{r}_0). \tag{3.42}
$$

The derivations for other components are very straightforward. Therefore, the tunneling matrix element for an arbitrary tip state, up to $l = 2$, is

$$
\begin{aligned}
M = \frac{2\pi\hbar^2}{m\kappa}\Bigg[& \beta_{00}\,\psi + + \beta_{10}\,\frac{\partial\psi}{\kappa\,\partial z} + \beta_{11}\,\frac{\partial\psi}{\kappa\,\partial x} + \beta_{12}\,\frac{\partial\psi}{\kappa\,\partial y} \\
&+ \beta_{20}\left(\frac{\partial^2\psi}{\kappa^2\,\partial z^2} - \frac{1}{3}\psi \right) + \beta_{21}\,\frac{\partial^2\psi}{\kappa^2\,\partial x\,\partial z} + \beta_{22}\,\frac{\partial^2\psi}{\kappa^2\,\partial y\,\partial z} \\
&+ \beta_{23}\,\frac{\partial^2\psi}{\kappa^2\,\partial x\,\partial y} + \beta_{24}\left(\frac{\partial^2\psi}{\kappa^2\,\partial x^2} - \frac{\partial^2\psi}{\kappa^2\,\partial y^2} \right) \Bigg].
\end{aligned} \tag{3.43}
$$

The values of the derivatives are taken at $\mathbf{r}_0$.

Again, we obtain the "derivative rule" from a completely different point of view. The second derivation gives a general formula to calculate the tunneling matrix element for an arbitrary tip wavefunction, with its vacuum tail expanded in terms of spherical harmonics.

The present approach is convenient for treating a single atomic state at the tip. If the tip structure is more complicated, the expansion, Eq. (3.30) may converge very slowly or not converge at all. An example is the case of two tip atoms at almost the same vertical distance to the sample surface, but with a large horizontal distance in x,y. This is a commonly occurring pathological condition, the double tip. In this case, the total tunneling current can be considered as the sum of the two components of tunneling current, each from one of these two tip atoms, evaluated separately using the derivative rule.

3.6 An intuitive interpretation

In this section, we present an intuitive interpretation for the derivative rule, which we have derived with mathematical rigor in the previous sections. Consider first the effect of a p_z tip state centered at $x = x_0$. The p_x state has two lobes with same weight but opposite phase. If at a point of the positive lobe, $x_0 + \Delta x$, the tunneling matrix element picks up a value proportional to $\psi(x_0 + \Delta x)$, then at the corresponding point $x_0 - \Delta x$ the tunneling matrix element picks up a value proportional to $-\psi(x_0 - \Delta x)$. The total contribution is then

$$M \propto \left[\psi(x_0 + \Delta x) - \psi(x_0 - \Delta x) \right]$$
$$\propto \frac{\partial \psi(x_0)}{\partial x}. \tag{3.44}$$

A similar interpretation can be applied to other cases. For example, the state d_{xy} has four lobes, as shown in Fig. 1.8. For a d_{xy} tip state centered at x_0, y_0, any tunneling matrix element must have the following form:

$$M \propto \left[\psi(x_0 + \Delta x, y_0 + \Delta y) - \psi(x_0 + \Delta x, y_0 - \Delta y) \right.$$
$$\left. + \psi(x_0 - \Delta x, y_0 - \Delta y) - \psi(x_0 - \Delta x, y_0 + \Delta y) \right]$$
$$\propto \frac{\partial^2 \psi(x_0, y_0)}{\partial x \, \partial y}. \tag{3.45}$$

As we have discussed in Chapter 2, a direct consequence of the Bardeen tunneling theory (or the extension of it) is the *reciprocity principle:* If the electronic state of the tip and the sample state under observation are interchanged, the image should be the same. An alternative wording of the same principle is, an image of microscopic scale may be interpreted either as by probing the sample state with a tip state or by probing the tip state with a sample state. The validity of this criterion, as applied to the derivative rule, can be tested by direct calculation. Actually, if one of the two states is an s state, the reciprocity principle is a direct consequence of the Green's-function representation of the other tip states.

CHAPTER 4
WAVEFUNCTIONS AT SURFACES

4.1 Types of surface wavefunctions

As we have shown in Chapters 2 and 3, under the normal operating conditions of STM, the tunneling current can be calculated from the wavefunctions a few Å from the outermost nuclei of the tip and the sample. The wavefunctions at the surfaces of solids, rather than the wavefunctions in the bulk, contribute to the tunneling current. In this chapter, we will discuss the general properties of the wavefunctions at surfaces. This is to fill the gap between standard solid-state physics textbooks such as Kittel (1986) and Ashcroft and Mermin (1985), which have too little information, and monographs as well as journal articles, which are too much to read. For more details, the book of Zangwill (1988) is helpful.

At crystalline surfaces, there are three types of wavefunctions as shown in Fig. 4.1. (1) The Bloch states are terminated by the surface, which become evanescent into the vacuum but remain periodic inside the bulk. (2) New states created at the surfaces in the energy gaps of bulk states, which decay both into the vacuum and into the bulk, the so-called *surface states*. (3) Bloch states in the bulk can combine with surface states to form *surface resonances*, which have a large amplitude near the surface and a small amplitude in the bulk as a Bloch wave.

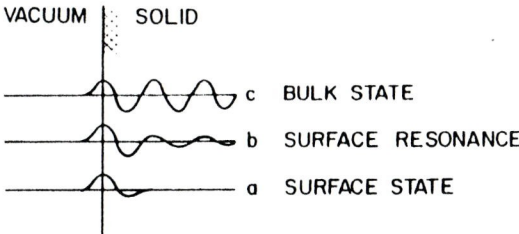

Fig. 4.1. Three types of wavefunctions at surfaces. (Reproduced from Himpsel, 1983, with permission.)

In the bulk of a perfect crystal, the electronic states follow the Bloch theorem, which have the form

$$\psi_{\mathbf{k}}(\mathbf{r}) = e^{i\mathbf{k}\bullet\mathbf{r}}\, u_{\mathbf{k}}(\mathbf{r}), \tag{4.1}$$

where $\mathbf{k}$ is the Bloch vector, and $u_{\mathbf{k}}(\mathbf{r})$ is a function with the same periodicity as the lattice. The energy is a function of $\mathbf{k}$. At the surface, a z plane, the periodic translational symmetry along the z direction is lost. At perfect surfaces, the periodic translational symmetry in the lateral directions, (x, y), are preserved. Any state at the surface should follow the Bloch theorem on the two-dimensional space. In other words, any wavefunction at the surface should have the form

$$\psi_{\mathbf{k}_{\|}} = e^{i\mathbf{k}_{\|}\bullet\mathbf{r}}\, u_{\mathbf{k}_{\|}}(\mathbf{r}), \tag{4.2}$$

where $\mathbf{k}_{\|} = (k_x, k_y)$ is the lateral Bloch vector, and $u_{\mathbf{k}_{\|}}(\mathbf{r})$ is a function having the periodicity of the surface. The energy of this state should be a function of $\mathbf{k}_{\|}$ (that is, $E = E(\mathbf{k}_{\|})$). Similar to the case of three-dimensional space, the concept of the surface Brillouin zone (SBZ) is essential for describing the wavefunctions at surfaces. Details will be presented in Chapter 5.

The vacuum tails of Bloch waves are relatively straightforward to understand. Comparing with other experimental methods, STM is more sensitive to the surface states, both the sample and the tip. Therefore, we will spend more time to explain the concept of surface states, from a theoretical point of view and an experimental point of view.

4.2 The jellium model

The simplest model of a metal surface is the jellium model, which is a Sommerfeld metal with an abrupt boundary. In provides a useful semiquantitative description of the work function and the surface potential (Bardeen, 1936). It validates the independent-electron picture of surface electronic structure: Essentially all the quantum mechanical many-body effects can be represented by the classical image force, which has been discussed briefly in Section 2.2 (Bardeen, 1936; Lang and Kohn, 1970; Appelbaum and Hamann, 1972).

The Sommerfeld theory of free-electron metals

The simplest model of metals is the Sommerfeld theory of free-electron metals (Ashcroft and Mermin 1985, Chapter 2), where a metal is described by a single parameter, the conduction electron density n. A widely used measure of it is the radius of the sphere whose volume is equal to the volume per conduction electron, r_s. By definition,

$$\frac{1}{n} = \frac{4\pi r_s^3}{3} .$$ (4.3)

The value of r_s is always given in the atomic unit of length, bohr (1 bohr = 0.529 Å). For almost all the metals used in STM, $r_s \approx 2$–3 bohr.

According to the Pauli exclusion principle, the conduction electrons occupy the states from the bottom of the conduction band up to an energy level where the metal becomes neutral. This highest energy level occupied by an electron is the Fermi level, E_F. In the Sommerfeld theory of metals, a natural reference point of the energy level is the bottom of the conduction band. The Fermi level with respect to that reference point is

$$E_F \approx \frac{50.1}{r_s^2} \text{ eV.}$$ (4.4)

In STM experiments, we are only interested in the electrons near the Fermi level. An important quantity is the number of electrons per unit volume per unit energy near the Fermi level, or the Fermi-level local density of states (LDOS). Inside a Sommerfeld metal, it is a constant

$$\rho(E_F) = \frac{3}{2} \frac{n}{E_F} \approx \frac{7.15 \times 10^{-3}}{r_s} \text{ eV}^{-1} \text{ bohr}^{-3}.$$ (4.5)

Surface potential in the jellium model

The jellium model for the surface electronic structure of free-electron metals was introduced by Bardeen (1936) for a treatment of the surface potential. In the jellium model, the lattice of positively charged cores is replaced by a uniform positive charge background, which drops abruptly to zero at the surface,

$$n_+(\mathbf{r}) = \begin{cases} \dfrac{3}{4\pi r_s^3}, & x \leq 0, \\ 0, & x > 0. \end{cases}$$ (4.6)

In the bulk, the charge density of electrons n equals in magnitude the charge density of the uniform positive charge background n_+, thus to preserve charge neutrality. The only parameter in the jellium model, r_s, is the same as in the Sommerfeld theory of free-electron metals.

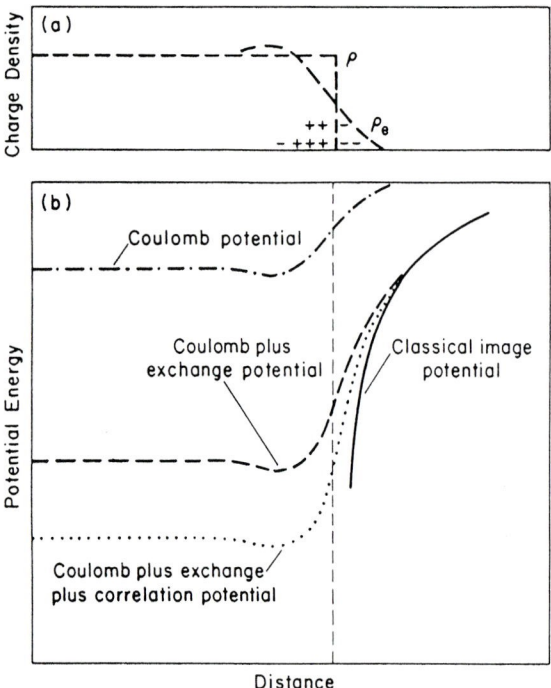

Fig. 4.2. Charge distribution and surface potential in a jellium model. (a) Distribution of the positive charge (a uniform background abruptly drops to zero at the boundary) and the negative charge density, determined by a self-consistent field calculation. (b) Potential energy as seen by an electron. By including all the many-body effects, including the exchange potential and the correlation potential, the classical image potential provides an adequate approximation. (After Bardeen, 1936; see Herring, 1992.)

Bardeen (1936) applied the self-consistent field method (the Hartree–Fock method) to calculate the electron density near the abrupt edge of the uniform positive charge background. The quantum-mechanical many-body effects, including the exchange potential and correlation potential, are taken into account. The results of Bardeen are shown in Fig. 4.2. First, the electrons spill out to the boundary of the uniform positive charge background. This is compensated by the reduced electron density right inside the boundary. It generates a dipole layer, thus an electrostatic potential. The exchange potential and the correlation potential generate additional potential on the electrons. The overall result is that each electron outside the boundary can be

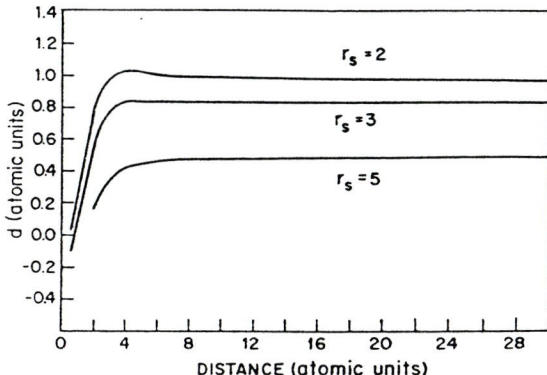

Fig. 4.3. Position of the image plane in the jellium model. The surface potential of an electron in the jellium model is calculated using the local-density approximation. By fitting the numerically calculated surface potential with the classical image potential, Eq. (4.7), the position of the image plane z_0 is obtained as a function of r_s and z. The results show that the classical image potential is accurate down to about 3 bohrs from the boundary of the uniform positive charge background. For metals used in STM, $r_s \approx 2 - 3$ bohr, $z_0 \approx 0.9$ bohr. (Reproduced from Appelbaum and Hamann, 1972, with permission.

considered as individually moving in a potential, described accurately by the classical image potential, Eq. (2.2),

$$U = -\frac{e^2}{4(z - z_0)} , \qquad (4.7)$$

where z_0 is the location of the *image plane,* with respect to the boundary of the uniform positive charge background.   Bardeen's conclusion was verified by Lang and Kohn (1970), and by Appelbaum and Hamann (1972).

The jellium model has only one parameter, r_s.   The distance between the image plane and the boundary of the uniform positive charge background, z_0, should be a function of r_s only.   Appelbaum and Hamann (1972) calculated it for several different values of r_s.   Their results are shown in Fig. 4.3.   The approach of Appelbaum and Hamann is to fit the calculated surface potential with the classical image potential, Eq. (4.7).   The position of the image plane z_0 is obtained as a function of r_s and z.   The results show that the classical image potential is accurate down to about 3 bohrs from the boundary of the uniform positive charge background. For almost all the metals used in STM, $r_s \approx 2 - 3$ bohr.   From Fig. 4.3, one finds $z_0 \approx 0.9$ bohr, or 0.5 Å.   From this Figure, it is clear that for most metals used in STM, the classical image potential is accurate up to 1 Å from the image plane.

Table 4.1. Measured work functions and predictions from the jellium model. First three rows are experimental values for different materials and crystallographic orientations, from *Handbook of Chemistry and Physics,* 69th edition, CRC Press (1988); the last row is the predictions of the jellium model, from Smith (1968)

Surface	Al	Au	Cu	Ir	W
(100)	4.41	5.47	4.59	5.67	4.63
(110)	4.06	5.37	4.48	5.42	5.25
(111)	4.24	5.31	4.94	5.76	4.47
jellium	3.64	3.19	3.32	4.02	3.91

Work function

An early application of the jellium model is to estimate the work function (Bardeen, 1936; Smith, 1968). In the jellium model, there is only one parameter r_s. The work function is then a function of r_s only. In reality, the work function depends not only on the material, but also on the crystallographic orientation of the surface. For most metals used in STM, the work function predicted by the jellium model is 1–2 eV smaller than the experimentally observed values, as shown in Table 4.1.

As shown, the jellium model gives inaccurate predictions for the work functions. The work functions predicted by first-principles calculations (see Section 4.7) are much more accurate.

Surface energy

The surface energy of a crystal is the energy required, per unit area of new surface formed, to split a crystal into two along a plane. In other words, it is the surface tension of a metal surface. The surface energy was an test ground for the validity of the jellium model. Smith (1968) found that although for a few alkali metals, Na, K, Cs, and Rb, fair agreement was obtained, for high-electron-density metals, for example, Al and W, the jellium model predicted a *negative* surface energy. In other words, the jellium model predicts that all the metal parts used in STM experiments would break into tiny pieces spontaneously (Zangwill 1988). Lang and Kohn (1970) performed calculations similar to that of Smith (1968) with great care, and concluded that the jellium model is totally inadequate for describing the surface energy of high-density metals. For comparison, Lang and Kohn (1970) conducted a series of calculations *with crystallographic lattices,* using the pseudopotential method. They found that with the inclusion of crystal lattices, the calculated surface energies show much better agreement with experiments than the oversimplified jellium

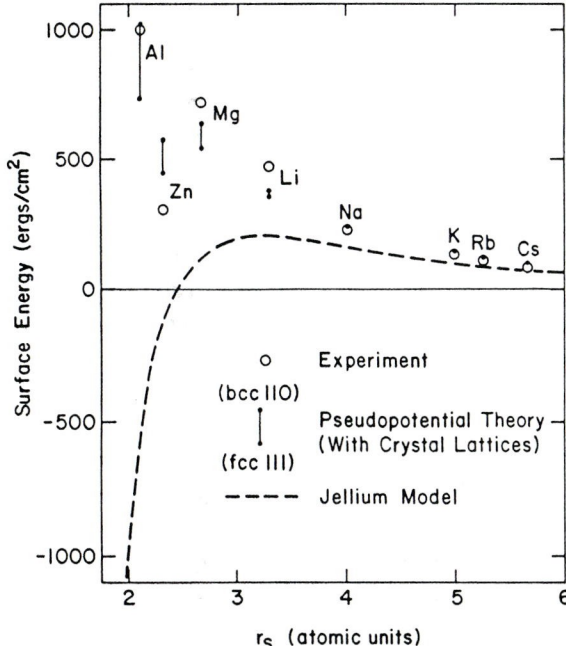

Fig. 4.4. Surface energy predicted by the jellium model. A test case of the accuracy of the jellium model, conducted by Lang and Kohn (1970). Only for four alkali metals, Na, K, Rb, and Cs, are the predictions of surface energy by the jellium model fair. For most metals, with $r_s \leq 2.5$ bohr, the surface energy predicted by the jellium model is negative, contradicting seriously with experimental facts. On the other hand, the calculated values of surface energies with crystal lattices agree much better with the values measured experimentally. (After Lang and Kohn, 1970).

model, thus theoretically rescued all the metal parts used in STM from spontaneously breaking into tiny particles. As expected, different types of lattices resulted in different surface energies (see Fig. 4.4).

Failures of the jellium model

Similar to the failures of the free-electron model of metals (Ashcroft and Mermin, 1985, Chapter 3), the fundamental deficiency of the jellium model consists in its total neglect of the atomic structure of the solids. Furthermore, because the jellium model does not have band structure, it does not support the concept of surface states. Regarding STM, the jellium model predicts the correct surface potential (the image force), and is useful for interpreting the distance dependence of tunneling current. However, it is inapplicable for describing STM images with atomic resolution.

4.3 Concept of surface states

The existence of surface states is a consequence of the atomic structure of solids. In an infinite and uniform periodic potential, Bloch functions exist, which explains the band structures of different solids (Kittel, 1986). On solid surfaces, surface states exist at energy levels in the gap of the energy band (Tamm, 1932; Shockley, 1939; Heine, 1963).

The concept of surface states was proposed by Tamm (1932) using a one-dimensional analytic model. We start with reviewing the proof of the Bloch theorem for a one-dimensional periodic potential $U(x)$ with periodicity a (Kittel, 1986):

$$U(x) = U(x + a). \tag{4.8}$$

The translational symmetry of the potential suggests the wavefunction to have the following property:

$$\psi(x + na) = C^n \, \psi(x), \tag{4.9}$$

where C is a constant, and n is an integer. For an infinite piece of solid, the amplitude of a wavefunction must have a finite value. It follows that the constant C must be a complex number of modulo 1 (otherwise in some directions the wavefunction will become infinity). The Bloch theorem then follows, which states that the wavefunction must have the following form:

$$\psi_k(x) = e^{ikx} \, u_k(x), \tag{4.10}$$

where $u_k(x)$ is a periodic function with periodicity a. At a boundary of the solid, in the direction normal to the surface, the restriction that the constant C must be of modulo 1 can be removed. Especially, it can be a real number other than 1. The wavefunction inside the solid may have the form

$$\psi_\gamma(x) = \pm \, e^{-\gamma x} \, u_\gamma(x), \tag{4.11}$$

where γ is a positive real number, and $u_\gamma(x)$ is again a periodic function with periodicity a. For a semi-infinite solid with a boundary at $x = 0$, it represents a wavefunction exponentially decaying into the solid, that is, a *surface state*. The minus sign represents a state which changes its sign for $x_1 = x + a$.

In the original paper of Tamm (1932), the concept of surface state is demonstrated with a Krönig–Penney potential (Kittel, 1986) with a boundary, as shown in Fig. 4.5. By solving the Schrödinger equation, explicit expressions for the surface states and their energy levels can be obtained. In

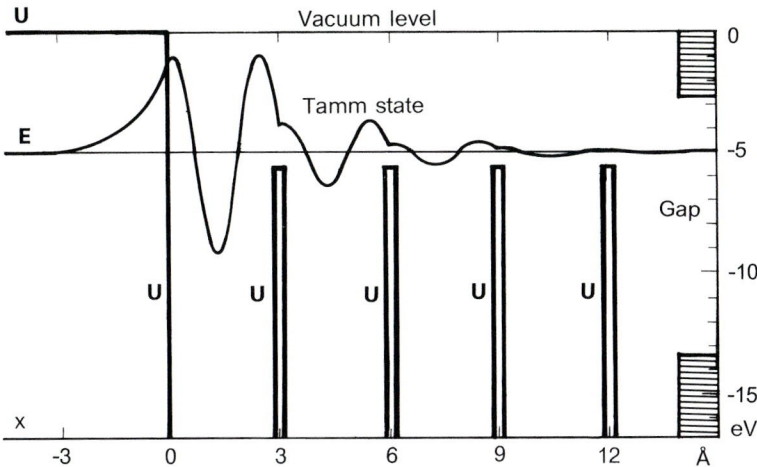

Fig. 4.5. Surface states. By solving the Schrödinger equation for a cut-off Krönig–Penney potential, it is found that in the energy gaps of the corresponding Krönig–Penney solid, there are surface states that decay exponentially into the vacuum *and* into the solid (Tamm, 1932). The explicit wavefunction of a Tamm state with $P = 15$ and $a = 3$ Å at $E = 5$ eV below the vacuum level is shown. The shaded areas represent allowed energy bands in the bulk.

the Krönig–Penney model (Kittel, 1986), the secular equation that determines the energy band is:

$$S(Ka) \equiv \frac{P}{Ka} \sin Ka + \cos Ka = \cos ka, \qquad (4.12)$$

where $K = \sqrt{2mE}/\hbar$, and P is a constant representing the strength of the periodic potential barriers. For a series of intervals on the scale Ka, $|S(Ka)|$ is greater than 1, for which Eq. (4.12) does not have a solution. Those values of Ka correspond to band gaps on the energy scale.

Tamm (1932) proposed that if complex values of k are allowed, for example, by setting $k = i\gamma$ (or $k = i\gamma + \pi/a$), Eq. (4.12) becomes

$$\frac{P}{Ka} \sin Ka + \cos Ka = \pm \cosh \gamma a, \qquad (4.13)$$

which does have solutions in the energy gaps.

The surface states have another boundary condition to be fulfilled. In the vacuum region, $x < 0$, the wavefunction is

$$\psi = \text{const.} \times e^{\kappa x}, \tag{4.14}$$

where $\kappa = \sqrt{2m(U - E)}\,/\hbar$ is the decay constant in the vacuum. The conditions of continuity at $x = 0$ between Eq. (4.14) and the solutions inside the solid result in another equation. Straightforward algebra gives

$$\cos Ka + \frac{\kappa}{K} \sin Ka = \pm\, e^{-\gamma a}. \tag{4.15}$$

This condition has a simple physical explanation. By replacing x for a in Eq. (4.15), the left-hand side is the solution inside the solid with initial value $\psi(0) = 1$, and satisfies the continuity condition for the derivatives with respect to x at $x = 0$. The right-hand side of this equation simply states that at $x = a$, the requirement of an exponentially decaying envelope in the solid, Eq. (4.11), is satisfied.

Combining Equations (4.15) and (4.13), with some algebra, we obtain the equation for the energy levels of surface states:

$$Ka \cot Ka = \frac{q^2 a^2}{2P} - \sqrt{q^2 - K^2}\, a, \tag{4.16}$$

where $q = \sqrt{2mU}\,/\hbar$ is a constant determined by the potential, see Fig. 4.5. The obvious relation $q^2 = \kappa^2 + K^2$ is used to eliminate the variable κ. The explicit wavefunction of a surface state and its energy level are shown in Fig. 4.5. As expected, it decays exponentially into the vacuum. It also decays exponentially into the solid, modulated with a periodic function.

From this elementary model, it is clear that the strength of these surface states depends on the strength of the periodic potential. If $P \ll 1$, there will be no surface states, and the electronic structure of this solid is similar to a free-electron Fermi gas. For large values of P, the energy gaps are wide, and the amplitudes of the surface states at the solid surface are large. Therefore, we expect that for nearly free-electron metals, such as alkali metals and alkaline earth metals, the surface states are very weak (if they ever exist), whereas for transition metals and semiconductors, the surface states are strong. On the other hand, the simple one-dimensional model predicts a discrete energy level of a surface state. In real solids, the energy level is always broadened, because in the x and y directions, the lateral Bloch vector always results in a dispersion.

The surface states play an important role in the STM experiments. On Si and Ge surfaces, near the Fermi level, the surface states usually dominate the surface charge density, which are what STM is probing. The tips are usually made of d band metals, W, Pt, and Ir. Surface states have a strong presence on the tips and often dominate tunneling current.

4.4 Field-emission spectroscopy

After the first theoretical work of Tamm (1932), a series of theoretical papers on surface states were published (for example, Shockley, 1939; Goodwin, 1939; Heine, 1963). However, there has been no experimental evidence of the surface states for more than three decades. In 1966, Swanson and Crouser (1966, 1967) found a substantial deviation of the observed *field-emission spectroscopy* on W(100) and Mo(100) from the theoretical prediction based on the Sommerfeld theory of metals. This experimental discovery has motivated a large amount of theoretical and subsequent experimental work in an attempt to explain its nature. After a few years, it became clear that the observed deviation from free-electron behavior of the W and Mo surfaces is an unambiguous exhibition of the surface states, which were predicted some three decades earlier.

A schematic diagram of the experimental apparatus is shown in Fig. 4.6. A metal tip emits electrons due to the high positive voltage on the anode, usually of the order of 10 kV. The electrons pass the hole of the anode, being decelerated, and reach the collector. The difference between the voltage on the collector and the emitter is adjusted by a potentiometer. The current detected by the collector is amplified and recorded. With improved design and electronics, the energy resolution of such a field-emission spectrometer can be better than ±0.03 eV (Swanson and Crouser, 1967).

The theory of field-emission spectroscopy for free-electron metals was developed by Young (1959). We present here a simplified version of Young's theory, which includes all the essential physics related to the experimental observation of surface states.

According to the Fowler–Nordheim equation, Eq. (1.34), the field-emission current from electrons at energy $|E|$ below the vacuum level is proportional to

$$T \approx \exp\left(-\frac{4\sqrt{2m}\,|E|^{3/2}}{3e\hbar F}\right). \tag{4.17}$$

Consider now energy levels not far from the Fermi level, that is, by a small energy deviation $|\epsilon| \ll \phi$

$$E = -\phi + \epsilon. \tag{4.18}$$

Expand the exponent in Eq. (4.17) into power series, we obtain

$$T = T_0 \exp\left(-\frac{2\kappa\epsilon}{eF}\right), \tag{4.19}$$

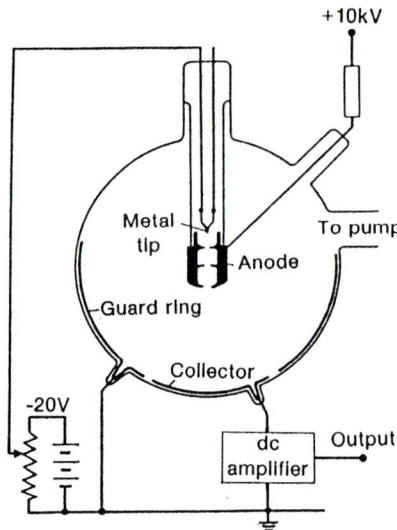

Fig. 4.6. Schematic diagram of a field-emission spectrometer. (After Good and Müller, 1956).

where $\kappa = \sqrt{2m\phi}\,/\hbar$ is the decay constant of the surface wavefunction near the Fermi level [see Equations (1.8) and (1.9)], and T_0 is a constant. If the emitter is a piece of free-electron metal, the density of states near the Fermi level should be a constant, and the electron density at a given temperature is proportional to the Fermi distribution function, $[1 + \exp(\epsilon/k_B T)]^{-1}$. The field-emission current is then

$$I = I_0 \; \frac{\exp\,(-2\kappa\epsilon/eF)}{1 + \exp\,(\epsilon/k_B T)} \;, \qquad (4.20)$$

where I_0 is a normalization constant. Experimentally, the emission current is always expressed in arbitrary units. Eq. (4.20) is shown by the dotted curve in Fig. 4.7.

Equation (4.20) has been verified by a large number of metals at various temperatures, from 77 K to 900 K. Most metals conform to this equation very well. However, Swanson and Crouser (1966) found that the field-emission spectra on W(100) deviates substantially from Eq. (4.20). There is a pronounced peak at about 0.35 eV below the Fermi level (Fig. 4.7). This phenomenon later acquired a nickname, the *Swanson hump*. A series of more extensive experiments were then conducted to investigate its nature (Swanson and Crouser 1967). The main results are: (1) The spectrum depends dramatically on the crystallographic orientation. On W(310), W(211), W(111), and W(611), the measured field-emission spectra agree well with the free-electron

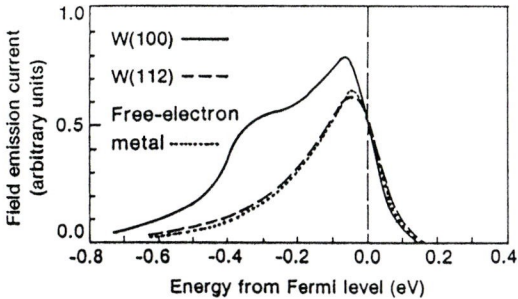

Fig. 4.7. Field emission spectra of W(112) and W(100). Dotted curve: theoretical field emission spectrum for free electron metals. Dashed curve: experimental field emission spectrum for W(112). Solid curve: experimental field emission spectrum for W(100). A substantial deviation from the free electron metal behavior is observed. The deviation, so-called Swanson hump, is due to the dominating role of localized surface states near the Fermi level at W(100) surface in field emission. (After Swanson and Crouser, 1967).

model. Substantial deviation is only observed on W(100) and W(110). (2) The anomaly is independent of temperature, ranging from 77 K to about 900 K. (3) After the surface is coadsorbed with Zr and O, the humps disappeared. Similar peaks in field-emission spectra were found on Mo(100) (Swanson and Crouser, 1967a), see Fig. 4.8.

With a combined experimental and theoretical study, the observed Swanson humps on W(100), W(110), and Mo(100) were soon identified as due to *surface states* (see Plummer and Gadzuk, 1970).

Fig. 4.8. Field-emission spectrum of Mo(100). The quantity displayed, R, is the ratio between the observed field-emission current and the prediction based on a free-electron model, Eq. (4.20). As shown, the field-emission spectrum of Mo(100) near the Fermi level is substantially different from a free-electron-metal behavior. (After Weng, 1977.)

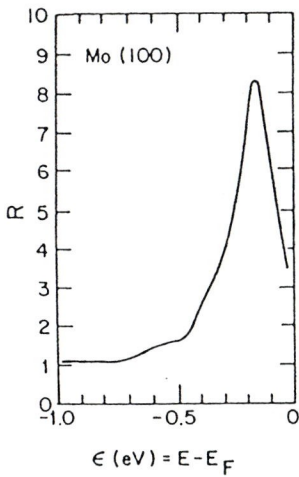

The surface states observed by field-emission spectroscopy have a direct relation to the process in STM. As we have discussed in the Introduction, field emission is a *tunneling phenomenon*. The Bardeen theory of tunneling (1960) is also applicable (Penn and Plummer, 1974). Because the outgoing wave is a structureless plane wave, as a direct consequence of the Bardeen theory, the tunneling current is proportional to the density of states near the emitter surface. The observed enhancement factor on W(100), W(110), and Mo(100) over the free-electron Fermi-gas behavior implies that at those surfaces, near the Fermi level, the LDOS at the surface is dominated by surface states. In other words, most of the surface densities of states are from the surface states rather than from the bulk wavefunctions. This point is further verified by photoemission experiments and first-principles calculations of the electronic structure of these surfaces.

4.5 Photoemission studies

Photoemission spectroscopy is based on the photoelectric effect, discovered more than a century ago by Hertz, in 1887. Einstein received his Nobel prize because of his interpretation of this effect. After rapid development in the 1960s and 1970s, photoemission has become a powerful tool for studying surface phenomena (see, for example, Feuerbacher et al., 1978; Cardona and Ley, 1978; Ley and Cardona, 1979; Himpsel, 1983). A schematic diagram of the photoemission experiment is shown in Fig. 4.9. As shown, monochromatic light with photon energy $\hbar\omega$ falls on the sample surface. Electrons emit from the surface as a result of the photoelectric effect. According to Einstein's theory of the photoelectric effect, for an electron in the sample with energy ϵ below the Fermi level, the emitted electron has a kinetic energy

$$E_K = \hbar\omega - \phi - \epsilon, \tag{4.21}$$

where ϕ is the work function of the material. The energy distribution of the emitted electrons contains information on the energy states in the solid near the surface. There are several methods to separate the contributions from bulk states and from surface states (Feuerbach et al., 1978; Himpsel, 1983). One of the methods is to choose a photon energy to generate electrons having the shortest escape depth from the solid, which can be as short as 4 Å. In this case, the emitted electrons come primarily from the surface and carry information about the wavefunctions at the surface.

The differential cross section for the photoelectron can be calculated using first-order time-dependent perturbation theory (see Cardona and Ley, 1978). The incident light is treated as a perturbation,

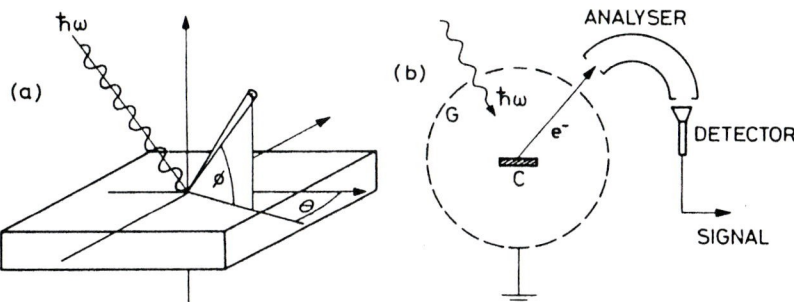

Fig. 4.9. Schematic of photoemission experiments. A beam of incident photons with energy $\hbar\omega$ induces electrons to emit from the sample. The photoelectrons are collected by the velocity analyzer and the electron detector at angles θ and ϕ with respect to the solid surface. (Afrer Feuerbacher et al., 1978.)

$$H_{\hbar\omega} = \frac{e}{2mc}\,(\mathbf{p} \bullet \mathbf{A} + \mathbf{A} \bullet \mathbf{p}), \tag{4.22}$$

where $\mathbf{p} = -i\hbar\nabla$ is the momentum operator, and $\mathbf{A}$ is the vector potential of the electromagnetic field, which carries information about the frequency, wave vector, polarization, and amplitude of the incident photons. The photoemission differential cross section follows the Fermi golden rule (Landau and Lifshitz, 1977),

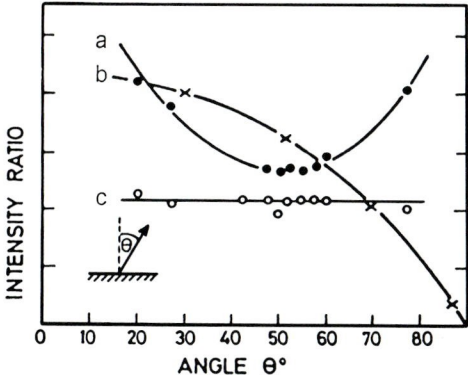

Fig. 4.10. Angle dependence of photoelectrons from different states. The polar-angle dependence of the photoelectrons reflects the orbital characteristics of the wavefunctions at surfaces. (a) d_z peak on MoS_2. (b) p_z peak on GaSe. (c) An s like state on MoS_2. (Reproduced from Cardona and Ley, 1978, with permission.)

$$\frac{d\sigma}{d\Omega} \propto \sum_i \left| \langle e^{i\mathbf{k}\cdot\mathbf{R}} | H_{\hbar\omega} | \Psi_i \rangle \right|^2 \delta(E_f - E_i - \hbar\omega). \tag{4.23}$$

The delta function corresponds to Einstein's equation, which says that the kinetic energy of the emitted electron E_f equals the difference of the photon energy $\hbar\omega$ and the energy level of the initial state of the sample, E_i. The final state $e^{i\mathbf{k}\cdot\mathbf{R}}$ is a plane wave with wave vector $\mathbf{k}$, which represents the electrons emitted in the direction of $\mathbf{k}$. Apparently, the dependence of the matrix element $\langle e^{i\mathbf{k}\cdot\mathbf{R}} | H_{\hbar\omega} | \Psi_i \rangle$ on the direction of the exit electron, $\mathbf{k}$, contains information about the angular distribution of the initial state on the sample. For semiconductors and d band metals, the surface states are linear combinations of atomic orbitals. By expressing the atomic orbital in terms of spherical harmonics (Appendix A),

$$\psi = R(r)Y_{lm}(\theta, \phi), \tag{4.24}$$

the transition matrix element can be factorized. Thus, the transition matrix elements reflect the angular dependence of the atomiclike orbital. The angular dependence of the emitted electron is proportional to

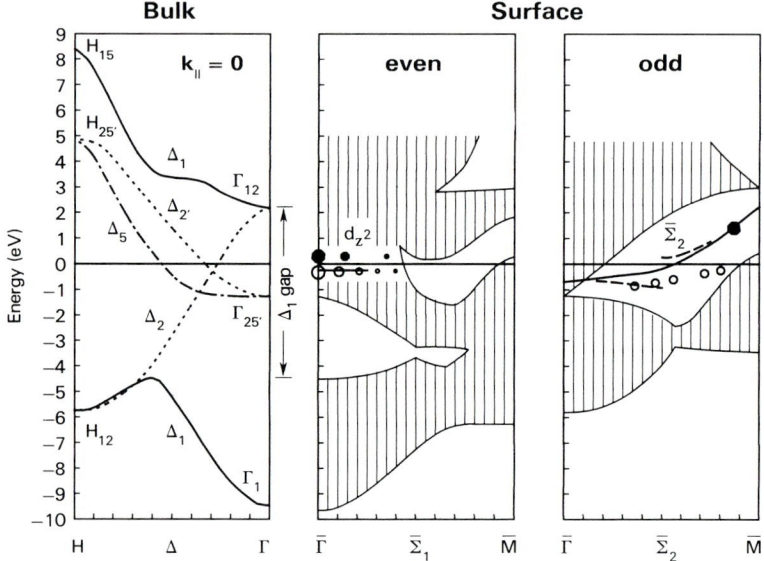

Fig. 4.11. Photoemission and inverse photoemission studies of W(001). The results of photoemission are shown as open circles, and those of inverse photoemission as dots. The surface states, both below and above the Fermi level, are of a d_{z^2} character. Reproduced from Drube et al. (1986) with permission.

$$\frac{d\sigma}{d\Omega} \propto |Y_{lm}(\theta, \phi)|^2. \tag{4.25}$$

For example, for an s state, the angular distribution should be a constant; for a p_z state, the dependence should like $\cos^2 \theta$; and for a d_{z^2} state, a dependence of $(3 \cos^2 \theta - 1)^2$ should be observed. This model has been verified by experiments (Smith, 1978). Figure 4.10 shows the polar-angle dependence of certain peaks in the spectra, which match what is expected from the knowledge of the orbital characteristics of the wavefunctions at these surfaces.

Angle-resolved photoemission can provide information for both energy distribution and orbital characteristics of wavefunctions at surfaces. With additional care, the signals from bulk states and from surface states can be distinguished. First, the surface state is very sensitive to adsorbates. Therefore, if a feature observed with photoemission can be eliminated by a fraction of a monolayer of certain adsorbates, it is due to a surface state. For example, the surface state on W(100) was observed by Waclawski and Plummer (1972) as well as by Feuerbacher and Fitton (1972) using this method. The position of the surface state fits well with the measurement by field-emission spectroscopy. By exposing the surface to a fraction of a monolayer of CO or H_2, the very pronounced peak at about 0.4 eV below the Fermi level disappears (Feuerbacher, et al. 1978). Second, the escape length of the photoelectron depends sharply on the wavelength of the incident photon. At about 70 eV, the escape length of the photoelectrons reaches a minimum, and the intensity of the surface state exhibits a sharp peak. By varying the wavelength of the incident light, the contributions from the surface and from the bulk can be distinguished.

Using *inverse photoemission,* the unoccupied electronic states of solid surfaces are being studied. Here, instead of injecting an UV light onto the surface and analyzing the emitted electrons, an electron beam is injected onto the surface and the spectrum of the emitted photons is analyzed. Fig. 4.11 shows a summary of the results of photoemission and inverse photoemission of one of the most exhaustively studied surfaces, W(001) [Drube et al. (1986)]. As shown, strong surface states immediately below and above the Fermi level are observed. Both are of a d_{z^2} character.

4.6 Atom-beam diffraction

Diffraction, by X-rays or neutrons, has been the standard method for determining the structures of crystals. The mean free path of X-rays and neutrons is very long, and thus is not sensitive to surfaces. To probe the structures of surfaces, the probing particles must have a very short mean free path in solids. Two methods are extensively used for determining surface structures: low-energy electron diffraction (LEED) and atomic-beam diffraction. A helium atom beam is the dominating probe used in atomic beam diffraction, which is often called helium atom diffraction.

The de Broglie relation, $\mathbf{p} = \hbar\mathbf{k}$, is valid for any particle. An beam of particles with mass M and kinetic energy E is associated with a wavelength

$$\lambda = \frac{2\pi\hbar}{\sqrt{2ME}} . \tag{4.26}$$

For electrons,

$$\lambda \approx \frac{12.3}{\sqrt{E(\text{eV})}} \text{ Å}. \tag{4.27}$$

An electron beam with energy 150 eV has a wavelength of 1 Å, which is suitable to probe crystallographic structures. The typical electron energy used in LEED is 20–500 eV. The mean free path of such electrons in metals is 5–10 Å. The electrons scatter elastically from the surface. LEED has been discussed in many textbooks (Ashcroft and Mermin, 1985; Zangwill, 1988).

Using a nozzle, He beams with narrow kinetic energy distribution can be obtained. The de Broglie wavelength of He^4 with kinetic energy E is

$$\lambda \approx \frac{0.144}{\sqrt{E(\text{eV})}} \text{ Å}. \tag{4.28}$$

The typical kinetic energy of a He beam generated by a nozzle is between 20 and 200 meV, corresponding to wavelength ranges between 0.1 and 1.0 Å. By directing an atomic beam to a solid surface, diffraction occurs.

Atomic-beam diffraction was first demonstrated in 1930, as a verification of the concept of the de Broglie wave (Estermann and Stern, 1930). In the 1970s, it was developed into an extremely informative method for determining topography and atomic structure of solid surfaces (Steele, 1974; Goodman and Wachman, 1976).

A typical setup for an atomic-beam diffraction experiment is shown in Fig. 4.12. The incident atomic helium beam is generated by a nozzle. The diffracted beam from the solid surface is detected by a movable detector. From the angle distribution of the diffracted He atoms, the entire He–surface interaction potential function $V(\mathbf{r})$ can be determined. (Further, by analyzing the energy distribution of the diffracted He beam, by studying the *inelastic atomic-beam scattering,* information about the surface phonon spectra can be obtained. Here, concerning STM, we will discuss elastic He scattering only.)

In general, the He–surface interaction potential has two regimes. At relatively large distances (6 Å and up), the van der Waals attraction dominates.

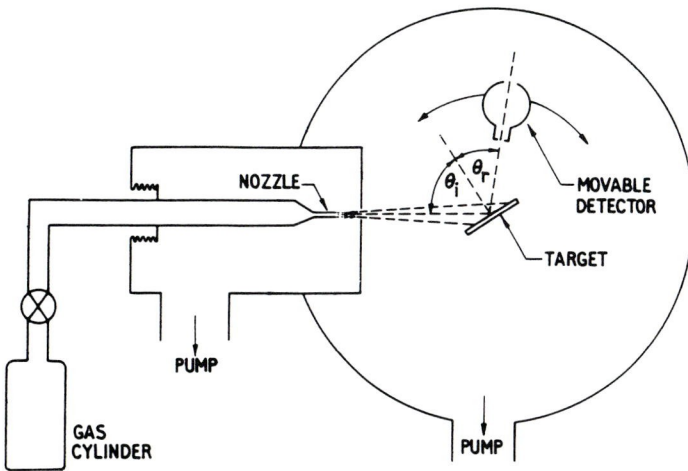

Fig. 4.12. Atomic-beam diffraction. A nearly monochromatic beam of helium, generated by a nozzle, falls on the solid surface with an angle of incidence. The diffracted beam is collected at an outgoing angle. The angular distribution of the diffracted helium beam contains the information about the topography and structure of the surface.

At shorter distances, repulsive force dominates. For typical He kinetic energies, the van der Waals energy, typically a few meV, has very little effect. The diffraction is essentially determined by the repulsive interactions.

4.6.1 The Esbjerg–Nørskov approximation

Based on the first-principles study of helium adsorption on metals (Zaremba and Kohn, 1977), Esbjerg and Nørskov (1980) made an important observation. Because the He atom is very tight (with a radius about 1 Å), the surface electron density of the sample does not vary much within the volume of the He atom. Therefore, the interaction energy should be determined by the electron density of the sample at the location of the He nucleus. A calculation of the interaction of a He atom with a homogeneous electron distribution results in an explicit relation between the He scattering potential $V(\mathbf{r})$ and the local electron density $\rho(\mathbf{r})$. For He atoms with kinetic energy smaller than 0.1 eV, Esbjerg and Nørskov (1980) obtained

$$
\begin{aligned}
V(\mathbf{r}) &\approx 110\, \rho(\mathbf{r}) \; (\text{eV Å}^{-3}) \\
&\approx 750\, \rho(\mathbf{r}) \; (\text{eV au}^{-3}).
\end{aligned}
\tag{4.29}
$$

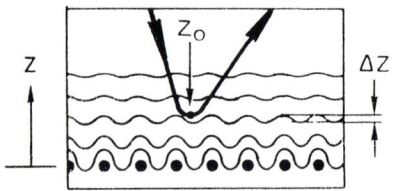

Fig. 4.13. Classical turning point in atomic-beam scattering. When the repulsive potential on a He atom at the sample surface equals the kinetic energy of the He atom, as a classical particle, the He atom is turned back.

This relation is found to agree with results from experiments and configuration interaction calculations (Esbjerg and Nørskov, 1980).

The detailed data from He-scattering experiments provide information about the electron density distribution on crystalline solid surfaces. Especially, it provides direct information on the corrugation amplitude of the surface charge density at the classical turning point of the incident He atom, as shown in Fig. 4.13. As a classical particle, an incident He atom can reach a point at the solid surface where its vertical kinetic energy equals the repulsive energy at that point. The corrugation amplitude of the surface electron density on that plane determines the intensity of the diffracted atomic beam.

The electron density distribution of a known surface structure can be calculated from first-principles. Thus, the He diffraction data can be compared with theoretical results, in particular, to verify different structural models. Hamann (1981) performed first-principles calculations of the charge-density distributions of the GaAs(110) surface, for both relaxed and unrelaxed configurations. The He diffraction data are in excellent agreement with the calculated charge-density distributions of the relaxed GaAs(110) surface, and are clearly distinguished from the unrelaxed ones (Hamann, 1981).

4.6.2 The method of Harris and Liebsch

It is rather cumbersome to use the results of first-principles calculations directly for interpreting the atom-beam scattering experiments. To simplify calculation and to gain physical insight, Harris and Liebsch (1982, 1982a) developed an approximate method for evaluating corrugations of charge-density distributions from the calculated electronic structure of the surface. The spirit of this method is as follows. The helium scattering potential of the surface is related to the electron density distribution function, weighed by an energy-dependent factor,

$$V(\mathbf{r}) = \int_{-\infty}^{E_F} g(E)\,\rho(\mathbf{r}, E)\,dE. \tag{4.30}$$

where $g(E)$ is a weighting factor. By expanding the surface wavefunctions in terms of a Fourier integral and analyzing their z dependence, they observed that most of the terms decay very quickly with z. For a crystalline surface of a one-dimensional periodicity with a reflection symmetry at the origin, by retaining only the slowest-decaying terms, the surface charge density should follow a universal exponential form

$$V(x, z) = V_0 e^{-\alpha z}\left[1 + \tfrac{1}{2}\alpha h_0 e^{-\beta(z-z_0)}\cos(g_1 x)\right]. \tag{4.31}$$

Here, g_1 is the length of the primitive reciprocal lattice vector. The constants α, β, and h_0 are determined either by considering leading Bloch waves or by fitting with first-principles calculations. The method of Harris and Liebsch is used extensively in the treatment of atom scattering data. With some modifications, the method of Harris and Liebsch is also applicable to calculate STM images. We will discuss it in detail in Chapter 5.

4.6.3 Atom charge superposition

Another method for interpreting atom beam scattering data was proposed Haneman and Haycock (1982), who suggested the use of superposition of atomic charge densities as a suitable approach for estimating surface charge densities. By superposing atomic charge densities calculated by Herman and Skillman (1963), using $[s^2p^3]$ orbitals for the As atom, and $[s^2p]$ orbitals for the Ga atom, they reproduced the results of first-principles calculations on the GaAs(110) surface (Hamann, 1981) with adequate accuracy.

A crude estimation of the charge-density distribution on simple metal surfaces can be made by assuming that the electron charge for each atom is spherical. Especially, as shown by Cabrera and Goodman (1972), by representing the atomic charge distribution with a Yukawa function,

$$\rho(r) = \frac{\beta}{r}\,e^{-\alpha r}, \tag{4.32}$$

where r is the distance from the nucleus and β and α are constants, then the charge-density function of periodic surfaces has a *finite analytic form*. This makes the estimation of surface electron charge density extremely simple. However, only for very simple surfaces, such as close-packed simple-metal surfaces, is this simple method adequate. For most solid surfaces, it is not accurate. Sakai, Cardillo, and Hamann (1986) showed that for Si(100), the simple spherical charge superposition is incapable of generating a scattering potential for interpreting the observed He diffraction data. To account for the experimental data, Si atoms in different positions have to be treated differently, and the charge anisotropy due to the p orbitals must be explicitly included (Sakai, Cardillo, and Hamann, 1986).

After some modification, the method of atomic charge superposition of Haneman and Heydock (1982), together with the analytic summation of Cabrera and Goodman (1972), can be applied to calculate STM and AFM images. We will discuss it in Chapter 6.

4.7 First-principles theoretical studies

After the discovery of the relativistic wave equation for the electron by Dirac in 1928, it seems that all the problems in condensed-matter physics become a matter of mathematics. However, the theoretical calculations for surfaces were not practical until the discovery of the density-functional formalism by Hohenberg and Kohn (1964). Although it is already simpler than the Hartree-Fock formalism, the form of the exchange and correlation interactions in it is still too complicated for practical problems. Kohn and Sham (1965) then proposed the *local density approximation*, which assumes that the exchange and correlation interaction at a point is a universal function of the total electron density at the same point, and uses a semiempirical analytical formula to represent such universal interactions. The resulting equations, the Kohn–Sham equations, are much easier to handle, especially by using modern computers. This method has been the standard approach for first-principles calculations for solid surfaces.

4.7.1 The density-functional formalism

The density-functional formalism is based on a theorem by Hohenberg and Kohn (1964), which states that for any many-electron system, the *total electron density distribution function $n(\mathbf{r})$ of its ground state contains all the information about that system*. This extremely powerful theorem enables a substantial reduction of the numerical procedures for determining the ground state of a many-electron system and makes first-principles calculations of complex many-electron systems practical. There are many review articles and books available (for example, Kohn and Vashishta, 1983). In this section, we present the basic concepts of this method, aiming at a conceptual understanding of the kind of information it provides, regarding the understanding of STM.

The Born–Oppenheimer approximation

Any condensed-matter system includes nuclei and electrons. The nuclei are about 10^5 times heavier than the electrons. Therefore, in calculating the structure of matter, the first step is to assume that the positions of the nuclei are fixed. The potential of the fixed nuclei is the external potential in which the electrons are moving about. To find out the most stable nuclei configuration, different nucleus configurations are assumed, and the corresponding total energies are calculated through their electronic structures. The configuration with the lowest total energy is then the most stable configuration. The vibration of the nuclei can be taken into account by starting with stationary nucleus configurations and calculate the potential energy with respect to nucleus positions. This approach is called the Born–Oppenheimer approximation, or the adiabatic approximation (see Landau and Lifshitz, 1977; Ashcroft and Mermin, 1976).

The Hohenberg–Kohn theorem

Consider a system of N electrons moving in an *external potential* $v(\mathbf{r})$, which can be the field of all the fixed nuclei in the system in the Born–Oppenhiemer approximation. The external potential $v(\mathbf{r})$ can also include the potential of an electrical field applied to this system. The potential energy of the system is the sum of the potential energy between the field $v(\mathbf{r})$ and the electrons and the interaction potential among the electrons. Once the external potential $v(\mathbf{r})$ is given, the entire system is uniquely determined through the Schrödinger equation. The electron density distribution of its ground state, $n(\mathbf{r})$, as a property of the system, is also completely determined by the external potential.

The spirit of the Hohenberg–Kohn theorem is that the inverse statement is also true: The external potential $v(\mathbf{r})$ is uniquely determined by the ground-state electron density distribution, $n(\mathbf{r})$. In other words, for two different external potentials $v_1(\mathbf{r})$ and $v_2(\mathbf{r})$ (except a trivial overall constant), the electron density distributions $n_1(\mathbf{r})$ and $n_2(\mathbf{r})$ must not be equal. Consequently, *all aspects of the electronic structure of the system are functionals of $n(\mathbf{r})$, that is, completely determined by the function $n(\mathbf{r})$.*

The total energy E of the system is also a functional of the density distribution, $E = E[n(\mathbf{r})]$. Therefore, if the form of this functional is known, the ground-state electron density distribution $n(\mathbf{r})$ can be determined by its Euler–Lagrange equation. However, except for the electron gas of almost constant density, the form of the functional $E[n(\mathbf{r})]$ cannot be determined a priori.

Self-consistent equations

To make practical calculations using the density-functional formalism, Kohn and Sham (1965) show that the condition of minimizing the energy is equivalent to a set of ordinary differential equations that can be solved by a self-consistent procedure. In other words, by assuming an initial effective potential $v_{\text{eff}}(\mathbf{r})$, one solves a set of single-particle Schrödinger equations

$$\left[-\frac{1}{2} \nabla^2 + v_{\text{eff}}(\mathbf{r}) \right] \psi_j(\mathbf{r}) = \epsilon_j \psi_j(\mathbf{r}).$$ (4.33)

The N solutions with the lowest energy values are used to calculate the electron density:

$$n(\mathbf{r}) = \sum_{j=1}^{N} |\psi_j(\mathbf{r})|^2.$$ (4.34)

The effective potential in Eq. (4.33) is then calculated from this charge-density function as a sum of the classical Coulomb energy and the exchange-correlation energy,

$$v_{\text{eff}}(\mathbf{r}) = v(\mathbf{r}) + \int \frac{n(\mathbf{r}')}{|\mathbf{r} - \mathbf{r}'|} \, d\mathbf{r}' + v_{\text{xc}}(\mathbf{r}).$$ (4.35)

Substituting the new $v_{\text{eff}}(\mathbf{r})$ for Eq. (4.33), and obtaining a new set of single-electron wavefunctions, $v(\mathbf{r})$ is recomputed and the process is iterated to convergence. The final $n(\mathbf{r})$ then is the correct ground-state electron distribution function.

Local-density approximation

The exchange-correlation energy $v_{\text{xc}}(\mathbf{r})$ in Eq. (4.35) is still a complicated matter. In the great majority of all practical calculations, it is assumed that $v_{\text{xc}}(\mathbf{r})$ is a universal function of the charge density at a point $\mathbf{r}$,

$$v_{\text{xc}}(\mathbf{r}) \approx f(n),$$ (4.36)

where $n(\mathbf{r})$ is the electron density at a point $\mathbf{r}$. The explicit form of the function $f(n)$ can be adapted from the case of an almost uniform electron gas. Several different forms of such functions have been used in actual calculations. The Wigner interpolation form is a widely used one (see Kohn and Vashishta, 1983),

$$f(n) = -\frac{0.458}{r_s} - \frac{0.44}{r_s + 7.8},$$ (4.37)

where r_s is the radius of a unit-charge sphere, defined in Section 4.2.1. Strictly speaking, those forms are *additional assumptions in the numerical calculation,* and these calculations are not strictly ab initio. As the theoretical justification of the local-density approximation and the explicit functional form of the exchange-correlation potential are still under investigation (Kohn and Vashishta, 1983), a practical justification is that by comparing with experimental results, the theoretical results are often surprisingly accurate.

4.7.2 Electronic structures of bulk solids

Surface states on d band metals and semiconductors are important examples of surface wavefunctions, which may dominate the tunneling current. On many metal surfaces, the tails of the bulk states dominate. For example, on the surfaces of Pt and Ir, the tails of the bulk states dominate -he wavefunctions at surfaces, and can be represented with reasonable accuracy as linear combinations of atomic states (LCAO).

The band structure and Bloch functions of metals have been extensively published. In particular, the results are compiled as standard tables. The book *Calculated Electronic Properties of Metals* by Moruzzi, Janak, and Williams (1978) is still a standard source, and a revised edition is to be published soon. Papaconstantopoulos's *Handbook of the Band Structure of Elemental Solids* (1986) listed the band structure and related information for 53 elements. In Fig. 4.14, the electronic structure of Pt is reproduced from Papaconstantopoulos's book. Near the Fermi level, the DOS of s and p states are much less than 1%. The d states are listed according to their symmetry properties in the cubic lattice (see Kittel, 1963). Type t_{2g} includes atomic orbitals with basis functions $\{xy, yz, xz\}$, and type e_g includes $\{(3z^2 - r^2), (x^2 - y^2)\}$. The DOS from d orbitals comprises 98% of the total DOS at the Fermi level.

Table 4.2 is a summary of the Fermi-level LDOS for three commonly used tip materials: Pt, Ir, and W. At the Fermi level, the d states dominate the bulk DOS. Therefore, on surfaces of Pt, Ir, and W, the wavefunctions as tails of bulk states are also dominated by d states.

Table 4.2. Fermi-level DOS of common tip materials

Material	Pt	Ir	W
s state	0.77%	0.94%	3.1%
p state	1.2%	3.1%	12%
d state	98%	96%	85%

PLATINUM

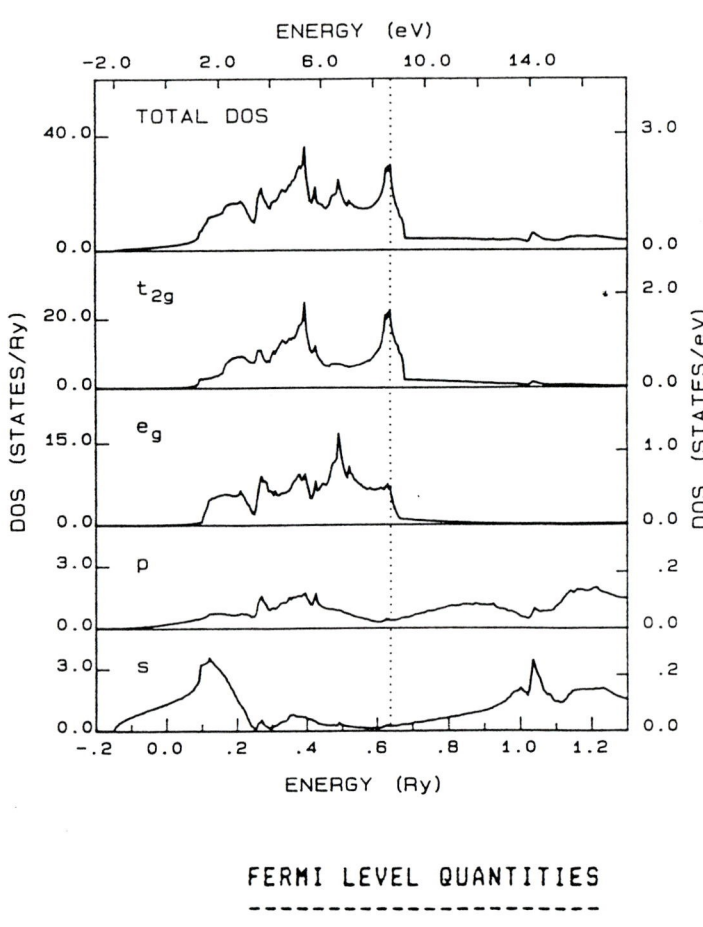

Fig. 4.14. Calculated electronic structure of Pt. The total DOS and the decomposition into atomic orbitals are shown. The d states are listed under two symmetry types. Type t_{2g} includes $\{xy,\ yz,\ xz\}$, and type e_g includes $\{(3z^2 - r^2),\ (x^2 - y^2)\}$. The Fermi-level quantities are also listed. The d states comprise about 98% of the Fermi-level DOS. (Reproduced from Papaconstantopoulos, 1986, with permission.)

4.7.3 Surface electronic structures

The problem of first-principles calculations of the electronic structure of solid surface is usually formatted as a problem of *slabs,* that is, consisting of a few layers of atoms. The translational and two-dimensional point group symmetry further reduce the degrees of freedom. Using modern supercomputers, such first-principles calculations for the electronic structure of solid surfaces have produced remarkably reproducible and accurate results as compared with many experimental measurements, especially angle-resolved photoemission and inverse photoemission.

The first successful first-principle theoretical studies of the electronic structure of solid surfaces were conducted by Appelbaum and Hamann on Na (1972) and Al (1973). Within a few years, first-principles calculations for a number of important materials, from nearly free-electron metals to d-band metals and semiconductors, were published, as summarized in the first review article by Appelbaum and Hamann (1976). Extensive reviews of the first-principles calculations for metal surfaces (Inglesfeld, 1982) and semiconductors (Lieske, 1984) are published. A current interest is the reconstruction of surfaces. Because of the refinement of the calculation of total energy of surfaces, tiny differences of the energies of different reconstructions can be assessed accurately. As examples, there are the study of bonding and reconstruction of the W(001) surface by Singh and Krakauer (1988), and the study of the surface reconstruction of Ag(110) by Fu and Ho (1989).

Another method for calculating electronic structures of complex surfaces is the *cluster calculation.* As the electronic state of an atom is mostly affected by the nearest and second-nearest neighbors (Heine, 1980), the results of cluster calculations provide a reasonably accurate account of the electronic states of the top atoms on a surface. Fig. 4.17 is the result of a calculation of W clusters by Ohnishi and Tsukada (1989).

As we have mentioned, the results of first-principles calculations are usually presented in the form of charge-density contours. The following are the commonly used forms of presenting results of first-principles calculations.

Total charge density

This is the most common form of data. The units of charge density, however, differ from paper to paper. Although the number of electrons per Å^3 and per a.u.^{3} are commonly used, several authors use density units of electrons per unit cell or per atomic volume.

Charge density for a particular state

This form of information is particularly common for the surface states. The units used are similar to that of the total charge density.

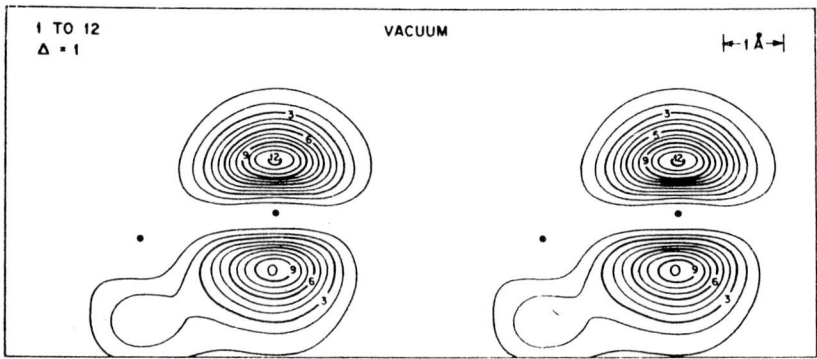

SILICON TOP SURFACE STATE CHARGE DENSITY

Fig. 4.15. Contours of constant charge density for Si(111). The occupied portion of the dangling-bond surface state on Si(111) is shown. Dots locate nuclei of surface atoms, the vacuum is above, and the charge density is in a.u.$\times 10^3$. (Reproduced from Appelbaum and Hamann, 1976, with permission.)

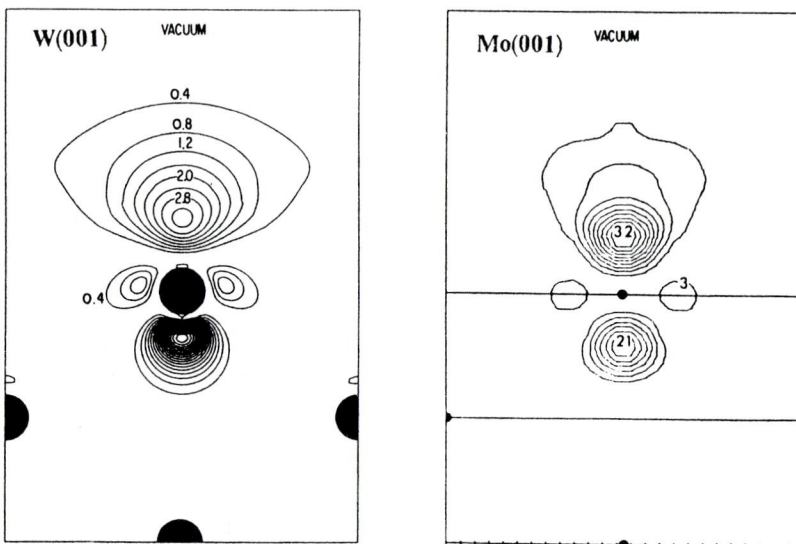

Fig. 4.16. Charge density of surface states on W(001) and Mo(001). (a) Charge-density contours of a localized surface state on W(001), located 0.3 eV below the Fermi level. Contours are in units of electrons per unit cell. (Reproduced from Posternak et al., 1980, with permission.) (b) A d_z localized surface state on Mo(001) 0.2 eV below the Fermi level. The contour plane is perpendicular to the surface. Dots indicate nuclei of surface atoms. The charge density is in units of electrons per unit cell. (Reproduced from Kerker et al., 1978, with permission.)

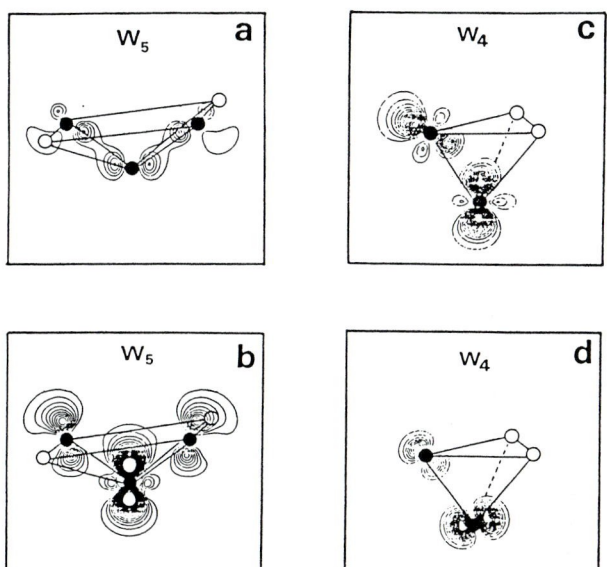

Fig. 4.17. Electronic states of W clusters near the Fermi level. Charge-density contours for several states on W clusters. (a) and (c) The highest occupied molecular orbitals (HOMO) of W_4 and W_5. (b) and (d) The eigenstates just below HOMO. (Reproduced from Ohnishi and Tsukada, 1989, with permission.)

Local density of states

A commonly used quantity to present the information obtained from a first-principles calculation based on the density-functional method is the local density of states (LDOS) at every energy value below the Fermi level at zero absolute temperature. Because every state has an energy eigenvalue, the information with both spatial and energetic distributions is important for many experiments involving energy information. The LDOS $\rho(\mathbf{r}, E)$ at a point $\mathbf{r}$ and at an energy level E is defined as

$$\rho(\mathbf{r}, E) \equiv \lim_{\epsilon \to 0} \frac{1}{\epsilon} \sum_{E_\mu = E}^{E+\epsilon} |\psi_\mu(\mathbf{r})|^2$$

$$= \sum_\mu |\psi_\mu(\mathbf{r})|^2 \delta(E - E_\mu), \tag{4.38}$$

where E_μ is the energy eigenvalue of the state $\psi_\mu(\mathbf{r})$. The total charge density is related to the LDOS by

$$\rho(\mathbf{r}) = \int_{-\infty}^{E_F} \rho(\mathbf{r}, E)\, dE. \qquad (4.39)$$

If available, the LDOS at different energy levels, for the tip and the sample, is very useful information for predicting STM images. Several examples of surface electronic structures from first-principles calculations are reproduced as illustrations.

Many, if not most, of the perfect surfaces studied by STM have also been studied by first-principles calculations with adequate accuracy. A fast growing field in theoretical surface physics is in first-principles calculations of the surfaces with adsorbates. A recent review of this field in given by Feibelman (1990). As the STM experiments are moving rapidly to the study of adsorbates as well, a direct comparison between the experimental observations and the theoretical predictions becomes practical and desirable.

CHAPTER 5
IMAGING CRYSTALLINE SURFACES

5.1 Types of STM images

Mathematically, an image is a matrix, or a two-dimensional array of numbers, $v(x, y)$. The variables x and y, in any practical case, are discrete. Therefore, they can be represented as natural numbers. Typically, the range of x and y is $100 \sim 1000$. The step size in the $\mathbf{x} \equiv (x, y)$ plane represents the lateral resolution of the data recording device. For example, if a 100×100 matrix is assigned to a 100×100 Å^2 image, the lateral resolution is limited to 1 Å. A point on the $\mathbf{x}$ plane with a value $v(\mathbf{x})$ assigned is called a *pixel* in computer jargon, which means a "picture element." The variable v is a physical quantity, usually considered as a real number. Because of the limited accuracy of the data and the cost of storing it in the computer memory, it is often an integer ($-32768 \sim +32767$), sometimes even a byte ($-128 \sim +127$).

For a given bias, the basic physical quantity measured by the STM is the tunneling current, which is a function of $\mathbf{x} \equiv (x, y)$ and z:

$$I = I(\mathbf{x}, z). \tag{5.1}$$

If z is perpendicular to a nearly perfect surface, the tunneling current can be decomposed into a constant (that is, independent of $\mathbf{x}$), and a small variable component that represents the features, or *corrugation* of the surface,

$$I(\mathbf{x}, z) = I_0(z) + \Delta I(\mathbf{x}, z), \tag{5.2}$$

with the condition

$$|\Delta I(\mathbf{x}, z)| \ll |I_0(z)|. \tag{5.3}$$

By taking the current I as the variable v, a *constant-z current image* is defined. Sometimes, for convenience, the relative variation of the tunneling current is taken as the image variable:

$$\delta(\mathbf{x}) = \frac{\Delta I(\mathbf{x}, z)}{I_0(z)} .$$

(5.4)

The *constant-current topographic image* can be derived from the current images as follows. Making an *Ansatz,*

$$z(\mathbf{x}) = z_0 + \Delta z(\mathbf{x})$$

(5.5)

and substituting it into Eq. (5.2),

$$I = I_0(z_0) + \left(\frac{dI_0(z)}{dz} \right) \Delta z(\mathbf{x}) + \Delta I(\mathbf{x}, z).$$

(5.6)

Due to the smallness of ΔI, the variation of the second term in Eq. (5.2) is neglected. The topographic image is then defined by the condition of constant current:

$$\Delta z(\mathbf{x}) = - \frac{\Delta I(\mathbf{x})}{\left(\dfrac{dI_0(z)}{dz} \right)} .$$

(5.7)

The variation of the tunneling current versus z is related to the apparent barrier height ϕ through Eq. (1.13). Due to the (x, y) dependence of the tunneling current, the apparent barrier height also exhibits an (x, y) dependence. The quantity $(\partial I/\partial z)$ can be taken as the image variable. The image variable can also be the apparent barrier height, defined by Eq. (1.13). In this case, straightforward algebra gives the approximate formula for the barrier-height image:

$$\phi(\mathbf{x}) = \phi_0 + 1.95 \sqrt{\phi_0} \ \frac{\partial}{\partial z} \left(\frac{\Delta I(\mathbf{x}, z)}{I_0(z)} \right),$$

(5.8)

in units of Å and eV, and the constant term is

$$\phi_0 = 0.95 \left(\frac{d \ln I_0(z)}{dz} \right)^2 .$$

(5.9)

By varying the bias V, a new dimension is opened to the images. The tunneling current is then $I(\mathbf{x}, z, V)$. We will discuss this case in Chapter 14.

In this chapter, we discuss the images of perfect crystalline surfaces. First, we present the analytic method for handling surface wavefunctions —

the leading-Bloch-waves approximation, which is a special case of the method of Harris and Liebsch (1982) in the theory of atom-beam diffraction. Within this approximation, explicit analytic forms of STM images with various tip states are obtained, which can be directly used to compare with experimental STM images.

5.2 Surfaces with one-dimensional corrugation

We start with the simplest case of a one-dimensional metal surface. Many of the basic concepts are demonstrated with this case. Then, we discuss several other types of surfaces that are frequently imaged by STM.

5.2.1 Leading-Bloch-waves approximation

With modern computational methods, the surface wavefunctions are routinely obtained from first-principles calculations. However, the data from first-principles calculations are overwhelming. Approximate methods that can yield analytic results are thus highly desirable. In Chapter 4, we discussed a method proposed by Harris and Liebsch (1982, 1982a) for interpreting atom-beam scattering data. The spirit of the Harris–Liebsch method is to fit the z dependence of both the uncorrugated and the corrugated components of the surface charge by an exponential function. In this section, we discuss the simplest case: a metal surface of a one-dimensional periodicity a with a reflection symmetry at $x = 0$ (see Fig. 5.1). The general formula for the electron charge density distribution is (Harris and Liebsch, 1982)

$$\begin{aligned}
\rho(x, z) &= C_0 e^{-\alpha z} + C_1 e^{-\gamma z} \cos^2(qx) \\
&= C_0 e^{-\alpha z}\left[1 + (C_1/C_0)e^{-\beta z}\cos^2(qx)\right],
\end{aligned} \tag{5.10}$$

where $\beta \equiv \gamma - \alpha$, and

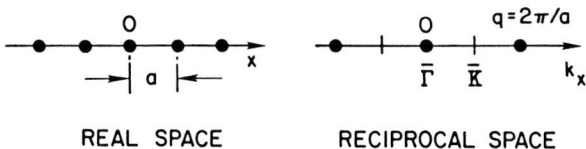

REAL SPACE   RECIPROCAL SPACE

Fig. 5.1. A metal surface with one-dimensional periodicity. The lowest Fourier components of the charge-density distribution are determined by the Bloch functions at the $\bar{\Gamma}$ and the $\bar{K}$ points in reciprocal space.

$$q = \frac{1}{2} g = \frac{\pi}{a}, \tag{5.11}$$

and g is the primitive reciprocal lattice vector. The constants C_0, C_1, α, and γ are determined by fitting with results from first-principles calculations. The second term in Eq. (5.10), the corrugated term, is much smaller than the first term, the constant term. The factor $\cos gx$ in Eq. (4.31) is rewritten as $\cos^2(qx)$ for convenience. (Actually, for $\cos^2(qx)$, the corrugation amplitude is 1 rather than 2 as for $\cos(gx)$, and the origin of the corrugation from the relevant Bloch wave becomes more explicit, as will be shown.) In STM, only the electron states near the Fermi level are involved. The constants α and γ in Eq. (5.10) can be obtained from general considerations of the surface wavefunctions.

Near the Fermi level, the surface wavefunctions in the vacuum region satisfy the Schrödinger equation in the vacuum:

$$(\nabla^2 - \kappa^2)\psi = 0, \tag{5.12}$$

where, as usual, $\kappa = (2m\phi)^{1/2}/\hbar$. As shown in Fig. 5.1, the first term in Eq. (5.10) originates from the constant term in the Bloch functions, that is, the Bloch functions near $\overline{\Gamma}$. From Eq. (5.12), the lowest Fourier component is:

$$\psi_{\overline{\Gamma}} = \text{const.} \times e^{-\kappa z}, \tag{5.13}$$

which makes the first term $C_0 e^{-2\kappa z}$. Therefore, we identified the first decay constant in Eq. (5.10),

$$\alpha = 2\kappa. \tag{5.14}$$

The Bloch functions near the $\overline{K}$ points have a long decay length and contribute to the second term of Eq. (5.10). Following Eq. (5.12), in general, a surface Bloch function at that point has the form:

$$\psi_{\overline{K}} = e^{iqx} \sum_{n=-\infty}^{\infty} c_n e^{-\sqrt{\kappa^2 + (q + 2nq)^2}\, z} e^{2inqx}. \tag{5.15}$$

In addition to the term with $n = 0$, the term with $n=-1$ have the same decay length, and thus have the same magnitude. Also, the Bloch function that generates the symmetric charge density must also be symmetric. The lowest-order symmetric Fourier sum of the Bloch function near $\overline{K}$ is:

$$\psi_{\overline{K}} \propto e^{-\sqrt{\kappa^2 + q^2}\, z} \cos qx. \tag{5.16}$$

The charge density is proportional to $|\psi_K|^2$. We then find the second constant in Eq. (5.10),

$$\gamma = 2\sqrt{\kappa^2 + q^2} \ . \tag{5.17}$$

Therefore, an approximate expression for the Fermi-level LDOS is

$$\rho(\mathbf{r}, E_F) = \frac{1}{\epsilon} \sum_{E_\mu = E_F}^{E_F + \epsilon} |\psi_\mu(\mathbf{r})|^2 \tag{5.18}$$

$$= C_0 \, e^{-2\kappa z} + C_1 \, e^{-2\sqrt{\kappa^2 + q^2}\, z} \cos^2 qx.$$

The corrugation amplitude of the Fermi-level LDOS for a metal surface with one-dimensional corrugation can be obtained using Equations (5.7) and (5.18),

$$\Delta z = \frac{C_1}{2\kappa C_0} \, e^{-\beta z}, \tag{5.19}$$

where

$$\beta \equiv \gamma - \alpha = 2\sqrt{\kappa^2 + q^2} - 2\kappa. \tag{5.20}$$

The expression of γ for metals with one-dimensional periodicity, Eq. (5.17), was first obtained by Tersoff and Hamann (1985). The ratio C_1/C_0 can be obtained from first-principles calculations. The correctness of the expression of the constant γ, Eq. (5.17), was verified by comparing with the scattering-theoretical calculations of Garcia et al. (1983), Stoll et al. (1984), and Stoll (1984), as we have discussed in Section 1.5.1 (see Eq. (1.25)).

5.2.2 Topographic images

Using the expressions of the tunneling matrix elements derived in Chapter 3, theoretical STM images can be calculated. In this section, we discuss the theoretical STM images of the simple metal surface we presented in the previous subsection.

s-wave tip state

The tunneling matrix element for an *s*-wave tip state is proportional to the amplitude of the sample wavefunction at the nucleus of the apex atom. The tunneling current is:

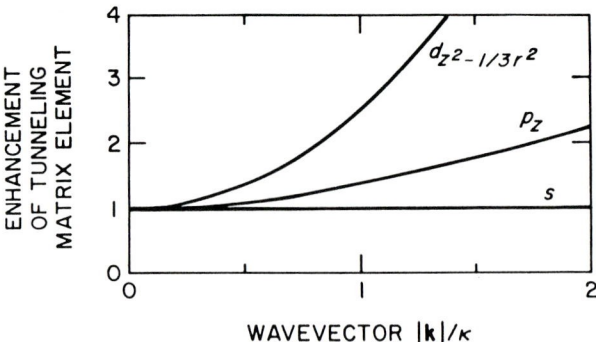

Fig. 5.2. Corrugation enhancement arising from different tip states. Solid curves, enhancement of tunneling matrix elements arising from different tip states. The tunneling current is proportional to the square of the tunneling matrix element. Therefore, the enhancement factor for the corrugation amplitude is the square of the enhancement factor for the tunneling matrix element, dotted curves. (Reproduced from Chen, 1990b, with permission.)

$$I = \text{const.} \times \sum_{E_F - eV}^{E_F} |\psi(\mathbf{r}_0)|^2$$

$$\propto V\left(C_0\, e^{-2\kappa z} + C_1\, e^{-2\sqrt{\kappa^2 + q^2}\, z}\, \cos^2 qx\right). \tag{5.21}$$

For free-electron metals, the local density of states near the Fermi level is proportional to the total valence-electron charge density. Therefore, up to an overall constant depending on the bias V, the tunneling current is proportional to the charge density at the nucleus of the apex atom:

$$I = \text{const.} \times \rho(x, z). \tag{5.22}$$

The topographic image can be calculated from Eq. (5.7). It is

$$\Delta z(x) = \frac{C_1}{2\kappa C_0}\, e^{-\beta z} \cos^2 gx. \tag{5.23}$$

The corrugation of the charge density on metal surfaces can be obtained from first-principles calculations or helium scattering experiments. The theory and the experiments match very well. A helium atom can reach to about 2.5–3 Å from the top-layer nuclei. At that distance, the repulsive force between the helium atom and the surface is already strong. The corrugation at that distance is about 0.03 Å, from both theory and experiments. For STM, with a pure s-wave tip state, if the nucleus of the apex atom reaches as deep as a helium atom can reach into the electron cloud at the surface, a corrugation of about 0.03 Å would be observed. At that distance, a strong repulsive force is present. However, experimentally, corrugations as high as 0.3 Å are routinely observed, where the force is still attractive. Therefore, the observed atomic resolution must be due to other tip states.

p_z tip state

According to the derivative rule, the tunneling matrix element for surface wavefunction at $\overline{\Gamma}$ from a p_z tip state is identical to that from a spherical tip state. However, for a surface wavefunction at $\overline{K}$, the tunneling matrix element from a p_z tip state is:

$$M_{\overline{K}} \propto \left(1 + \frac{q^2}{\kappa^2}\right)^{1/2} \psi_{\overline{K}}, \qquad (5.24)$$

and the topographic image arising from a p_z tip state is:

$$\Delta z(x) = \left(1 + \frac{q^2}{\kappa^2}\right) \frac{C_1}{2\kappa C_0} e^{-\beta z} \cos^2 qx. \qquad (5.25)$$

Therefore, the corrugation amplitude arising from a p_z tip state gains a factor of $[1 + (q^2/\kappa^2)]$ over that of the charge-density contour: see Fig. 5.2. This is the quantitative explanation of the resolution enhancement due to p-like localized tip states, as proposed by Demuth et al. (1989).

d_z tip state

Using the expression for the transmission matrix element of a d_z tip state, for a sample wavefunction at $\overline{\Gamma}$, it picks up a factor of 2/3, whereas for a sample wavefunction at $\overline{K}$ it picks up a factor of $[(2/3) + (q^2/\kappa^2)]$. Similar to the case of p_z tip state, we find the topographic image to be:

$$\Delta z(x) = \left(1 + \frac{3q^2}{2\kappa^2}\right)^2 \frac{C_1}{2\kappa C_0} e^{-\beta z} \cos^2 qx. \qquad (5.26)$$

The enhancement for the tunneling matrix element is shown in Fig. 5.2. The enhancement factor for the corrugation amplitude, $[1 + (3q^2/2\kappa^2)]^2$, could be substantial. For example, on most close-packed metal surfaces, $a \approx 2.5$ Å, which implies $q \approx 1.25$ Å^{-1}. An enhancement of 11.2 is expected. Most of the commonly used tip materials are d-band metals, for example, W, Pt, and Ir. As we have shown in Chapter 4, localized d_z states often occur on the surfaces. These states will enhance the corrugation amplitude by more than one order of magnitude.

p_x tip state

For a pure p_x tip state, the tunneling matrix element for the sample wavefunction at $\overline{\Gamma}$ is negligible, whereas for sample wavefunctions at $\overline{K}$, a phase shift of 90° in the x direction is obtained. The tunneling current is:

$$I = \text{const.} \times \frac{q^2}{\kappa^2} e^{-\beta z} \sin^2 qx. \tag{5.27}$$

The meaning of this result was discussed by Lawunmi and Payne (1990) in conjunction with the theory of anomalous corrugation (Tersoff, 1986; see Section 5.7). According to the theory of anomalous corrugation, the relevant surface wavefunction may occur only at the edge of the first Brillouin zone, i. e., the $\overline{K}$ point. In one-dimensional models, for the lowest Fourier component, a single sinusoidal wave may dominate the surface wavefunction. By considering the s-wave tip state only, the nodal structure of this state could give rise to an infinitely large corrugation. However, such an enhancement is extremely unstable. With the presence of a minute amount of other tip states, the corrugation enhancement is completely suppressed (Tersoff and Lang, 1990).

d_{xz} tip state

A straightforward calculation using the tunneling matrix elements listed in Table 3.2 shows that the d_{xz} state results in a large but inverted corrugation amplitude on metal surfaces, because the tunneling matrix element for the sample wavefunction at the $\overline{\Gamma}$ point vanishes. The role of this state and the d_{x^2} state in the inverted corrugation will be discussed in Section 5.5.

5.3 Surfaces with tetragonal symmetry

The one-dimensional case we presented in the previous section is directly applicable to surfaces with rectangular lattices, with lattice constants a and b, respectively. The corrugations in the x and y directions can be treated separately. Many of the (110) surfaces of cubic crystals fall into this category. The (001) surfaces of cubic crystals, where $a = b$, are also commonly encountered. In this section, we discuss this case as an extension of the one-dimensional case.

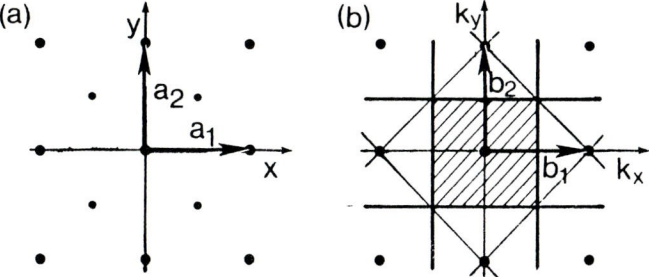

Fig. 5.3. Surface with tetragonal symmetry.. The example of Cu(001) is shown. The lattice in real space. The large dots represent the nuclei in the top layer. The small dots represent the nuclei in the second layer. The reciprocal space is also shown.

An example of surfaces with tetragonal symmetry is the Cu(001) surface, as shown in Fig. 5.3. The top-layer nuclei form a two-dimensional square lattice on the x,y plane with lattice constant a. The origin of the coordinate system is chosen to be at one of the top-layer nuclei. The $+z$ direction is defined as pointing into the vacuum. The reciprocal lattice is also shown in Fig. 5.3, with a lattice constant of

$$k = \frac{2\pi}{a}. \tag{5.28}$$

The first Brillouin zone, formed by the bisector lines between the center and the nearest lattice points in reciprocal space, is a square bounded with lines $k_x = \pm\,\pi/a$ and $k_y = \pm\,\pi/a$. The lowest Fourier components of the sum of the local density of states (LDOS) over a range of energy ΔE should have the form:

$$\rho(\mathbf{x}, z) = \sum_{E_F - \Delta E}^{E_F} |\psi|^2 = a_0(z) + a_1(z)\,\phi^{(4)}(k\mathbf{x}), \tag{5.29}$$

where a square cosine function is defined for convenience,

$$\phi^{(4)}(\mathbf{X}) \equiv \frac{1}{2} + \frac{1}{4}\,(\cos X + \cos Y). \tag{5.30}$$

The function $\phi^{(4)}(k\mathbf{x})$ has a value 1 at each lattice point in real space, and 0 at the center of four neighboring lattice points. Similar to the one-dimensional case, the $a_0(z)$ term in Eq. (5.29) comes mainly from the Bloch functions near

Γ, which has the form $a_0(z) \propto \exp(-2\kappa z)$. By making an analogy to the one-dimensional case, the second term comes from four $\overline{M}$ points in the Brillouin zone. The Bloch functions should satisfy the Schrödinger equation, Eq. (5.12). At one of these points, the lowest Fourier component of a Bloch wave with Bloch vector $\mathbf{q}_n$ (see Fig. 5.3) is:

$$\psi_n = c_n e^{i\mathbf{q}_n \bullet \mathbf{x}} e^{-\frac{1}{2}\gamma z},$$

(5.31)

where

$$\gamma = 2\sqrt{\kappa^2 + |\mathbf{q}|^2}.$$

(5.32)

A single plane wave such as Eq. (5.31) is not a good Bloch function because it does not satisfy the symmetry of these points, $2mm$ (see Appendix E). The appropriately symmetrized Bloch functions are

$$\Psi_1 = \cos qx\, e^{-\frac{1}{2}\gamma z},$$

(5.33)

and

$$\Psi_2 = \cos qy\, e^{-\frac{1}{2}\gamma z}.$$

(5.34)

The two wavefunctions, Ψ_1 and Ψ_2, should have the same amplitude because of the fourfold symmetry. A direct calculation gives the charge density:

$$
\rho(\mathbf{x}, z)_{\Delta E} = \sum_{E_F - \Delta E}^{E_F} |\psi(\mathbf{r})|^2
$$

$$
= \Delta E \left[C_0 e^{-2\kappa z} + C_1 e^{-\gamma z} \phi^{(4)}(k\mathbf{x}) \right],
$$

(5.35)

where C_0 and C_1 are constants characterizing the uniform term and the corrugated term of the LDOS near the Fermi level, which can be obtained from first-principles calculations. If the constants do not change appreciably over a few Å below the Fermi level, then these constants can be inferred from the calculated *total* charge densities. In this case, the total charge density has the following asymptotic form:

$$\rho(\mathbf{r}) = \frac{C_0\,\phi}{\kappa z} e^{-2\kappa z} \left[1 + \frac{\gamma C_1}{2\kappa C_0} e^{-\beta z} \phi^{(4)}(k\mathbf{x}) \right],$$

(5.36)

where

$$\beta = \gamma - 2\kappa. \tag{5.37}$$

Therefore, the constants C_0 and C_1 can be obtained from the charge-density contours of first-principles calculations.

In the following, we derive the STM images for the $m = 0$ tip states. For an s-wave tip state, the image is the contour of the LDOS near the Fermi level. Following Eq. (5.7), the topographic image is

$$\Delta z(\mathbf{x}) = \frac{C_1}{2\kappa C_0} \, e^{-\beta z} \phi^{(4)}(k\mathbf{x}). \tag{5.38}$$

Following the procedure for the one-dimensional case, the STM image for the p_z tip state is

$$\Delta z(\mathbf{x}) = \frac{C_1}{2\kappa C_0} \left(1 + \frac{q^2}{\kappa^2} \right) e^{-\beta z} \phi^{(4)}(k\mathbf{x}), \tag{5.39}$$

and the STM image for the d_{z^2} tip state is

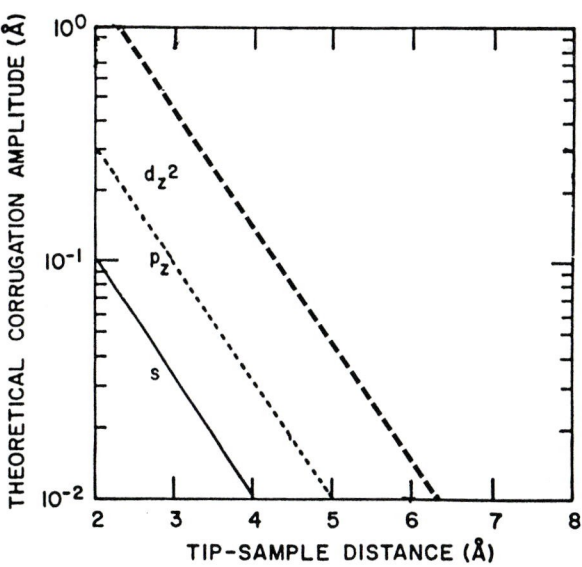

Fig. 5.4. **The corrugation amplitude of the STM images for Cu(001).** Calculated for different $m = 0$ tip states and different tip–sample distances. The corrugation amplitude of charge density contours is obtained from Gay et al. (1977).

$$\Delta z(\mathbf{x}) = \frac{C_1}{2\kappa C_0} \left(1 + \frac{3q^2}{2\kappa^2}\right)^2 e^{-\beta z} \phi^{(4)}(k\mathbf{x}). \qquad (5.40)$$

Figure 5.4 shows the calculated corrugation amplitudes of STM images of Cu(001) with different tip states. The s-wave tip state does not provide atomic resolution, as expected. The large corrugation amplitude observed on Cu(001) (Samsavar et al., 1990) is probably due to an d_z tip state.

5.4 Surfaces with hexagonal or trigonal symmetry

Probably the most commonly encountered surfaces in STM experiments are of trigonal symmetry. The close-packed metal surfaces and most cleaved surfaces of layered materials belong to this category. In Fig. 5.5, the structure of a close-packed metal surface is shown. The large dots represents the atoms in the top layer. The circles represent the atoms in the second layer. The small dots are those in the third layer. However, experimentally, it was found that only the atoms in the first layer are observed. Therefore, the surface has an approximate hexagonal symmetry, $p6mm$, which is the highest symmetry in all plane groups (see Appendix E). The high symmetry makes the treatment much simpler, since the basic features of the images with the lowest nontrivial Fourier components are determined by symmetry only. In this case, the charge density should have a hexagonal symmetry, i. e., invariant with respect to plane group $p6mm$ (see Fig. 5.5). Up to the lowest nontrivial Fourier components, the most general form of surface charge density with hexagonal symmetry is:

$$\rho(\mathbf{r}) = \sum_{E_F - \Delta E}^{E_F} |\psi(\mathbf{r})|^2 \approx a_0(z) + a_1(z)\, \phi^{(6)}(k\mathbf{x}), \qquad (5.41)$$

where $\mathbf{x} = (x, y)$, and $k = 4\pi/\sqrt{3}\,a$ is the length of a primitive reciprocal lattice vector. A hexagonal cosine function is defined for convenience,

$$\phi^{(6)}(\mathbf{X}) \equiv \frac{1}{3} + \frac{2}{9} \sum_{n=0}^{2} \cos \omega_n \bullet \mathbf{X}, \qquad (5.42)$$

where $\omega_0 = (0, 1)$, $\omega_1 = (-\tfrac{1}{2}\sqrt{3}, -\tfrac{1}{2})$, and $\omega_2 = (\tfrac{1}{2}\sqrt{3}, -\tfrac{1}{2})$, respectively. By plotting it directly, it is clear that the function $\phi^{(6)}(k\mathbf{x})$ has a maximum value 1 at each atomic site, and nearly 0 in the space between atoms. The function $[1 - \phi^{(6)}(k\mathbf{x})]$ has a minimum value 0 at each atomic

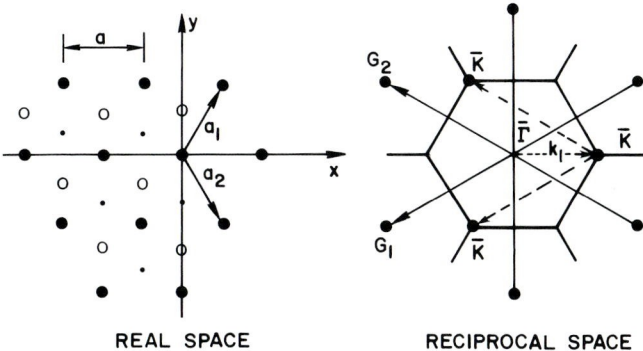

REAL SPACE   RECIPROCAL SPACE

Fig. 5.5. Geometrical structure of a close-packed metal surface. Left, the second-layer atoms (circles) and third-layer atoms (small dots) have little influence on the surface charge density, which is dominated by the top-layer atoms (large dots). The top layer exhibits sixfold symmetry, which is invariant with respect to the plane group *p6mm* (that is, point group C_{6v} together with the translational symmetry.). Right, the corresponding surface Brillouin zone. The lowest nontrivial Fourier components of the LDOS arise from Bloch functions near the $\overline{\Gamma}$ and $\overline{K}$ points. (The symbols for plane groups are explained in Appendix E.)

site, and nearly 1 in the space between atoms, which describes an inverted corrugation (see Fig. 5.6).

The problem of the electron charge-density distribution of a surface with hexagonal symmetry has been treated by Liebsch, Harris, and Weinert (1984). Similar to previous cases, the $a_0(z)$ term in Eq. (5.41) comes mainly from the Bloch functions near $\overline{\Gamma}$, whose lowest Fourier component is:

$$a_0(z) \propto e^{-2\kappa z}. \tag{5.43}$$

The Bloch functions near the $\overline{K}$ points have the longest decay length, which are the dominating contribution to the second term in Eq. (5.41). In general, a surface Bloch function at that point has the form:

$$\psi_{\overline{K}} = \sum_{\mathbf{G}} a_{\mathbf{G}} e^{-\sqrt{\kappa^2 + |\mathbf{k}_1 + \mathbf{G}|^2}\, z}\, e^{i(\mathbf{k} + \mathbf{G}) \cdot \mathbf{x}}, \tag{5.44}$$

with $|\mathbf{k}_1| \equiv q = k/\sqrt{3}$. By inspecting Eq. (5.44) and Fig. 5.5, one finds that the only slow-decaying symmetric Fourier sums of the Bloch functions near $\overline{K}$ are:

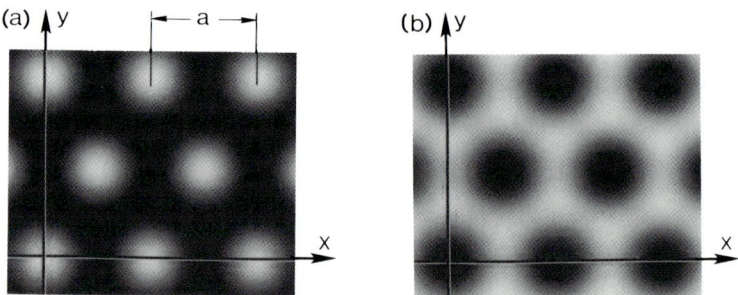

Fig. 5.6. The hexagonal cosine function and its complementary function. (a) The hexagonal cosine function defined by Eq. (5.42), $\phi^{(6)}(k\mathbf{x})$, has a maximum value 1 at each atomic site, and nearly 0 in the space between atoms. The function $[1 - \phi^{(6)}(k\mathbf{x})]$ has a minimum value 0 at each atomic site, and nearly 1 in the space between atoms, which describes an inverted corrugation. (Reproduced from Chen, 1992c, with permission.)

$$\psi_1 = B\, e^{-\frac{1}{2}\gamma z} \sum_{n=0}^{2} \cos(q\boldsymbol{\tau}_n \bullet \mathbf{x}), \qquad (5.45)$$

and

$$\psi_2 = B\, e^{-\frac{1}{2}\gamma z} \sum_{n=0}^{2} \sin(q\boldsymbol{\tau}_n \bullet \mathbf{x}), \qquad (5.46)$$

where $\boldsymbol{\tau}_0 = (1, 0)$, $\boldsymbol{\tau}_1 = (-\frac{1}{2}, \frac{1}{2}\sqrt{3})$, $\boldsymbol{\tau}_2 = (-\frac{1}{2}, -\frac{1}{2}\sqrt{3})$; B is a real constant; and $\gamma = 2(\kappa^2 + q^2)^{1/2}$ is the corresponding decay constant. The symmetric wavefunctions at the $\overline{K}$ point, Equations (5.45) and (5.46), are invariant under a rotation $2m\pi/3$, individually. Under a rotation of π, ψ_1 is invariant, whereas ψ_2 changes sign. These two wavefunctions are degenerate because, while shifting the origin by one lattice constant, these two wavefunctions mix with each other by the matrix

$$\begin{bmatrix} \dfrac{1}{2}, & \pm\dfrac{\sqrt{3}}{2} \\[2ex] \mp\dfrac{\sqrt{3}}{2}, & \dfrac{1}{2} \end{bmatrix}. \qquad (5.47)$$

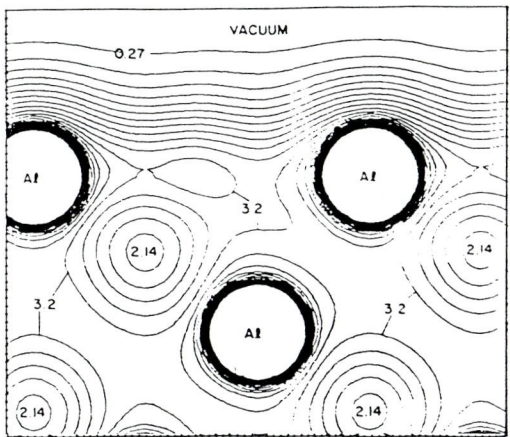

Fig. 5.7. Charge-density contour plot of Al(111) film. Results of a first-principles calculation of a nine-layer Al(111) film. The contours are in steps of 0.27 electrons per atom. (Reproduced from Wang et al., 1981, with permission.)

This can be easily shown by direct calculation. The charge density is the sum of Eq. (5.43) and the sum of the charge density proportional to $|\psi_1^*\psi_1| + |\psi_2^*\psi_2|$. A straightforward calculation gives

$$\rho(\mathbf{r}) \propto \sum_{E_F - \Delta E}^{E_F} |\psi(\mathbf{r})|^2$$
$$= \Delta E \left[C_0 e^{-2\kappa z} + C_1 e^{-\gamma z} \phi^{(6)}(k\mathbf{x}) \right], \tag{5.48}$$

where C_0, C_1 are constants. The corrugation charge-density contour, Δz, as a function of z, can be obtained from Eq. (5.48)

$$\Delta z(x) = \frac{C_1}{2\kappa C_0} e^{-\beta z} \phi^{(6)}(k\mathbf{x}). \tag{5.49}$$

Similarly, $\beta = \gamma - 2\kappa$. The ratio (C_1/C_0) can be determined by comparing Eq. (5.49) with the corrugation amplitudes of the charge-density contours obtained from first-principles calculations. For example, from Fig. 5.7, averaged from five contours ranging from three contours of thinnest densities, we find $(C_1/C_0) \approx 5.7 \pm 1.0$. Following the procedure for the one-dimensional case, the STM image for the p_z tip state is

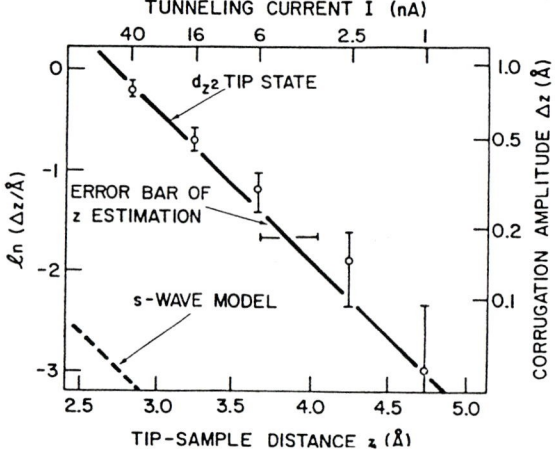

Fig. 5.8. Interpretation of the STM corrugation observed on Al(111). The predicted corrugation amplitude with a d_z tip state (solid curve) agrees well with the experimental data from Wintterlin et al. (1989) (circles with error bars). The parameters of the theoretical curve are taken from Fig. 5.7. The corrugation from an s-wave tip state (dashed curve), that is, the corrugation of Fermi-level LDOS contour, is included for comparison. (Reproduced from Chen, 1990, with permission.)

$$\Delta z(\mathbf{x}) = \frac{C_1}{2\kappa C_0} \left(1 + \frac{q^2}{\kappa^2}\right) e^{-\beta z} \phi^{(6)}(k\mathbf{x}), \tag{5.50}$$

and the STM image for the d_z tip state is

$$\Delta z(\mathbf{x}) = \frac{C_1}{2\kappa C_0} \left(1 + \frac{3q^2}{2\kappa^2}\right)^2 e^{-\beta z} \phi^{(6)}(k\mathbf{x}). \tag{5.51}$$

A comparison of the theory with experiments is shown in Fig. 5.8. For Al(111), $a = 2.88$ Å, $\phi = 3.5$ eV, it follows that $\kappa = 0.96$ Å^{-1}, $\gamma = 3.48$ Å^{-1}. The slope of the $\ln \Delta z \sim z$ curve from Equations (5.49) through (5.51) fits well with experimental data. The absolute tip–sample distance is obtained from curve fitting, which gives the shortest average tip–sample distance at $I = 40$ nA (with bias 50 mV) to be about 2.9 Å, consistent with the measured tip–sample distance (Dürig et al, 1986, 1988), about 1 Å before mechanical contact.

The approximate sixfold symmetry does not apply to many surfaces such as graphite, MoS_2, and most of the layered materials. In this case, we have to consider the irreducible representations of the point group $3mm$ or C_{3v}. Actually, this can be achieved by considering the irreducible representations of the plane group $p3m1$. This group is not Abelian. Therefore, at least one of its irreducible representations is not one-dimensional. In other words, the "good" wavefunctions are not necessarily invariant under a $2\pi/3$ rotation. If a pair of wavefunctions mixes with each other under such a rotation, they are good wavefunctions as well. Actually, there are other two pairs of wavefunctions that mix with each other through the matrix Eq. (5.47) not only by translation, but also by a rotation of $2n\pi/3$:

$$\psi_3 = A\, e^{-\frac{1}{2}\gamma z} \sum_{n=0}^{2} \cos\left(q\, \boldsymbol{\tau}_n \bullet \mathbf{x} + \frac{2n\pi}{3}\right), \tag{5.52}$$

$$\psi_4 = A\, e^{-\frac{1}{2}\gamma z} \sum_{n=0}^{2} \sin\left(q\, \boldsymbol{\tau}_n \bullet \mathbf{x} + \frac{2n\pi}{3}\right), \tag{5.53}$$

$$\psi_5 = H\, e^{-\frac{1}{2}\gamma z} \sum_{n=0}^{2} \cos\left(q\, \boldsymbol{\tau}_n \bullet \mathbf{x} - \frac{2n\pi}{3}\right), \tag{5.54}$$

$$\psi_6 = H\, e^{-\frac{1}{2}\gamma z} \sum_{n=0}^{2} \sin\left(q\, \boldsymbol{\tau}_n \bullet \mathbf{x} - \frac{2n\pi}{3}\right), \tag{5.55}$$

where A and H are constants. The charge-density distribution produced by ψ_3 and ψ_4, together with that at the $\overline{\Gamma}$ point, is:

$$\rho_A(\mathbf{r}) \propto C_{A0}\, e^{-2\kappa z} + C_{A1}\, e^{-\gamma z}\, \phi^{(6)}[k(\mathbf{x} - \mathbf{x}_A)]. \tag{5.56}$$

Similarly, for ψ_5 and ψ_6,

$$\rho_H(\mathbf{r}) \propto C_{H0}\, e^{-2\kappa z} + C_{H1}\, e^{-\gamma z}\, \phi^{(6)}[k(\mathbf{x} - \mathbf{x}_H)], \tag{5.57}$$

respectively.

5.5 Corrugation inversion

In this section, we discuss the effect of $m \neq 0$ tip states. Those tip states will create inverted images, where the sites of surface atoms are minima rather than maxima in the topographic images (Barth et al. 1990).

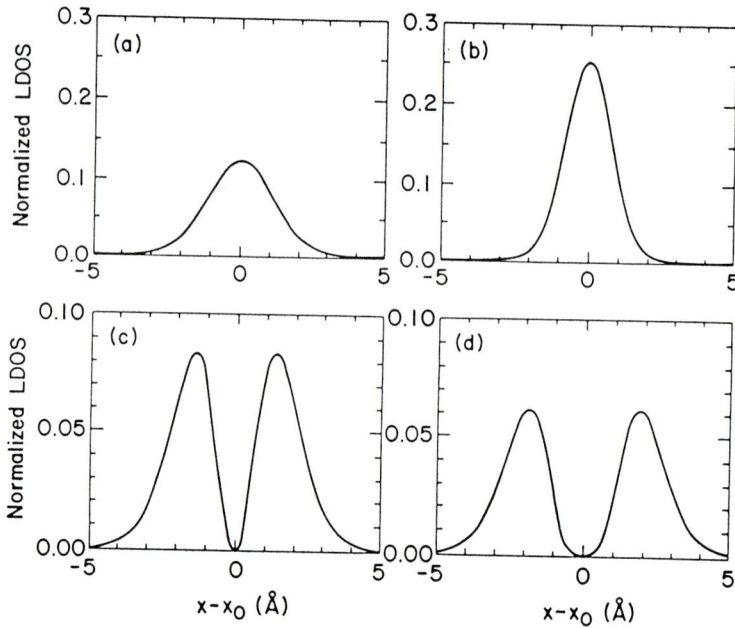

Fig. 5.9. LDOS of several tip electronic states. Evaluated and normalized on a plane $z_0 = 3$ Å from the nucleus of the apex atom. Axial symmetry is assumed. (a) s state. (b) $l = 2$, $m = 0$ state ($d_{3z^2-r^2}$). (c) $l = 2$, $m = 1$ states (d_{xz} and d_{yz}). (d) $l = 2$, $m = 2$ states ($d_{x^2-y^2}$ and d_{xy}). (Reproduced from Chen, 1992c, with permission.)

Qualitatively, the explanation of the effect of $m \neq 0$ tip states is as follows. For simplicity, we assume that the tip has axial symmetry (actually, a trigonal symmetry suffices). In other words, the two $m=1$ states, xz and yz, are degenerate. Similarly, the two $m=2$ states, xy and $x^2 - y^2$, are also degenerate. The LDOS of those tip states are shown in Fig. 5.9. Consider a simple metal, for example, Au(111). Each Au atom has only s-wave states near the Fermi level, and the tunneling current from each sample atom is additive. Thus, the tunneling current distribution between a single Au atom and the tip is the tip-state LDOS, measured at the center of that Au atom. For a d_{z^2} tip state, it has a sharp peak centered at the atom site. The total current distribution is the sum of tunneling current for all the Au atoms at the surface. The sharpness of the tunneling current distribution for the d_{z^2} tip state, compared with that of the s-wave tip state (Fig. 5.9), again illustrates why the d_{z^2} tip state enhances image corrugation. The $m=1$ and 2 tip states exhibit a ring-shaped LDOS, as shown in Fig. 5.9. The tunneling current distribution for a single Au atom should be proportional to the tip LDOS, which is ringshaped (Fig. 5.9). The total current distribution is the sum of the tunneling current for all the Au atoms at the surface. Therefore, with an $m \neq 0$ tip state, an inverted STM image should be expected. In other words, with an $m \neq 0$ tip state, every Au atom site at the surface should appear as a depression rather than a protrusion in the STM image.

The general expression for the tunneling current can be obtained using the explicit forms of tunneling matrix elements listed in Table 3.2. To put the five d states on an equal footing, normalized spherical harmonics, as listed in Appendix A, are used. The wavefunctions and the tunneling matrix elements are listed in Table 5.1.

Up to a constant, the tunneling current is

$$
\begin{aligned}
I = {} & 4 \left| D_0 B_0 \right|^2 e^{-2\kappa z} \\
& + 9 \left| D_0 B_1 \right|^2 e^{-2\kappa_1 z} \left[3(\kappa_1/\kappa)^2 - 1 \right]^2 \phi^{(6)}(k\mathbf{x}) \\
& + 54 \left| D_1 B_1 \right|^2 e^{-2\kappa_1 z} \left(q\kappa_1/\kappa^2 \right)^2 \left[1 - \phi^{(6)}(k\mathbf{x}) \right] \\
& + (27/2) \left| D_2 B_1 \right|^2 e^{-2\kappa_1 z} \left(q/\kappa \right)^4 \left[1 - \phi^{(6)}(k\mathbf{x}) \right].
\end{aligned}
\tag{5.58}
$$

Table 5.1. Wavefunctions and tunneling matrix elements for different d-type tip states. The tip is assumed to have an axial symmetry. For brevity, the common factor in the normalization constant of the spherical harmonics and a common factor $(2\pi\hbar^2/\kappa m_e)$ in the expressions for the tunneling matrices is omitted.

State	Tip wavefunction	Tunneling matrix element
$3z^2 - r^2$	$D_0\, k_2(\kappa r)\, (3\cos^2\theta - 1)$	$D_0 \left[\dfrac{3}{\kappa^2} \dfrac{\partial^2}{\partial z^2} - 1 \right] \psi(\mathbf{r}_0)$
xz	$D_1\, k_2(\kappa r)\sqrt{3}\ \sin 2\theta \cos\phi$	$D_1 \left[\dfrac{2\sqrt{3}}{\kappa^2} \dfrac{\partial^2}{\partial x \partial z} \right] \psi(\mathbf{r}_0)$
yz	$D_1\, k_2(\kappa r)\sqrt{3}\ \sin 2\theta \sin\phi$	$D_1 \left[\dfrac{2\sqrt{3}}{\kappa^2} \dfrac{\partial^2}{\partial y \partial z} \right] \psi(\mathbf{r}_0)$
$x^2 - y^2$	$D_2\, k_2(\kappa r)\sqrt{3}\ \sin^2\theta \cos 2\phi$	$D_2 \left[\dfrac{\sqrt{3}}{\kappa^2} \left(\dfrac{\partial^2}{\partial x^2} - \dfrac{\partial^2}{\partial y^2} \right) \right] \psi(\mathbf{r}_0)$
xy	$D_2\, k_2(\kappa r)\sqrt{3}\ \sin^2\theta \sin 2\phi$	$D_2 \left[\dfrac{2\sqrt{3}}{\kappa^2} \dfrac{\partial^2}{\partial x \partial y} \right] \psi(\mathbf{r}_0)$

The first term in Eq. (5.58) represents the uncorrugated tunneling current, which decays much more slowly than the corrugated terms. Therefore, if D_0 is not too small, the corrugation of the topographic image is

$$\Delta z = \left[\left(\frac{3\kappa_1^2}{2\kappa^2} - \frac{1}{2} \right)^2 - \frac{3}{2} \left| \frac{D_1}{D_0} \right|^2 \left(\frac{q\kappa_1}{\kappa^2} \right)^2 \right.$$
$$\left. - \frac{3}{8} \left| \frac{D_2}{D_0} \right|^2 \left(\frac{q}{\kappa} \right)^4 \right] \Delta z_0, \tag{5.59}$$

where

$$\Delta z_0 = \frac{9}{2\kappa} \left| \frac{B_1}{B_0} \right|^2 e^{-2(\kappa_1 - \kappa)z} \phi^{(6)}(k\mathbf{x}) \tag{5.60}$$

is the corrugation of the Fermi-level LDOS of the sample. The ratio $|B_1/B_0|$ can be determined independently by first-principles calculations or independent experimental measurements, such as helium atom scattering. For Au(111), $a = 2.87$ Å, $q = 1.46$ Å^{-1}, $\kappa = 0.96$ Å^{-1}, and $\kappa_1 = 1.74$ Å^{-1}. From Eq. (5.59), we obtain

$$\Delta z = \left(19.6 - 11.4 \left| \frac{D_1}{D_0} \right|^2 - 2.0 \left| \frac{D_2}{D_0} \right|^2 \right) \Delta z_0. \tag{5.61}$$

The enhancement factor E, that is, the quantity in the parenthesis in this equation, is displayed in Fig. 5.10. Because the corrugation amplitude depends only on the relative intensities of different components, we normalize it through

$$|D_0|^2 + |D_1|^2 + |D_2|^2 = 1. \tag{5.62}$$

Naturally, the results can be represented by a diagram similar to a three-component phase diagram, as shown in Fig. 5.10. Several interesting features are worth noting. First, when the $m = 0$ or d_{z^2} state dominates, a large, positive enhancement is expected. The condition for a substantial enhancement is quite broad. For example, when the condition $|D_0|^2 > 1.2 |D_1|^2 + 0.2 |D_2|^2$ is satisfied, the positive enhancement should be greater than 10, or a full order of magnitude. It is about 15% of the total phase space. To have an enhancement of more than 5, one third of the total phase space is available. Therefore, the

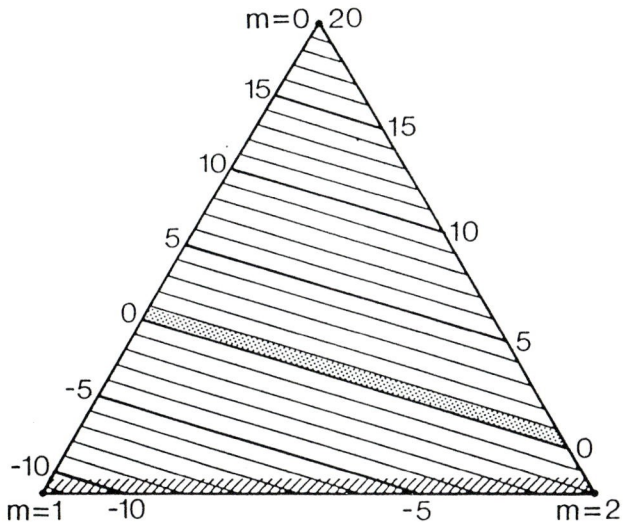

Fig. 5.10. Enhancement factor for different tip states. The shaded area near $E=0$ is the area where the corrugation amplitude is within the limit of the Fermi-level LDOS contours. In the hatched area near the bottom, the theoretical amplitude of the negative corrugation shows a spurious divergence. (Reproduced from Chen, 1992c, with permission.)

experimental observation of large positive corrugation enhancement should be frequent. Second, when $m \neq 0$ states dominate, an inverted corrugation should be observed. Again, the probability for a negative image to occur is large. Actually, when the condition $|D_0|^2 < 0.58|D_1|^2 + 0.1|D_2|^2$ is fulfilled, the image corrugation is inverted. This is about 43% of the total phase space. To have negative corrugations with an enhancement factor of 5 or more, 14% of the total phase space is available. Third, from Eq. (5.61) and Fig. 5.10, it is apparent that the effect of $m=1$ states in generating inverted corrugation is much stronger than that of $m=2$ states. This is expected from Fig. 5.9. The $m=1$ states have a much sharper rim than the $m=2$ states. Finally, there is a small region in which an almost complete cancellation of the positive enhancement and the negative enhancement can occur, as indicated by the shaded area near zero corrugation. In this case, the image is similar to the prediction of the s-wave model. The observed image corrugation in this case should be equal to or smaller than the corrugation of the Fermi-level LDOS. From Eq. (5.61) and Fig. 5.10, the available phase space is about 2.8% of the total phase space. Therefore, the probability is small. Practically, when this situation occurs, an almost flat image is observed. The experimentalist explains it as a bad tip. A

Fig. 5.11. Au(111) topograph taken while a change in the tip state reversed the corrugation. The upper part exhibits a positive corrugation, whereas the lower part exhibits a negative corrugation. Individual Au atoms on both parts are clearly resolved. (Reproduced from Barth et al., 1990, with permission.)

tip-sharpening procedure is then conducted until a large corrugation is observed, which is explained as having a good tip.

An experimental observation of the corrugation reversal during a scan is shown in Fig. 5.11. Owing to a sudden change of the tip state, the image switched from positive corrugation to negative corrugation. An interpretation is that before the tip restructuring, an $m=0$ tip state dominates; and after the tip restructuring, an $m \neq 0$ tip state dominates.

The corrugation inversion due to $m \neq 0$ tip states is a universal phenomenon in the STM imaging of low-Miller-index metal surfaces. For most metals (except several alkali and alkaline earth metals, which have rarely been imaged by STM), the nearest-neighbor atomic distance $a \approx 3$ Å. Consequently, the numerical coefficients on Eq. (5.61) are very close to those for Au(111).

5.6 The s-wave-tip model

The STM images of large superstructures on metal surfaces exhibit a very simple form. As shown first time by Tersoff and Hamann (1983, 1985), at the low-bias limit, the STM images of large superstructures on metal surfaces are *independent of tip electronic states,* and an STM image is simply a contour of an important quantity of the *sample* surface only: *the Fermi-level local density of states (LDOS), taken at the center of curvature of the tip.* An attempt was also made to interpret the observed atom-resolved images of semiconductors and semimetals (such as Si(111)-2×1 and graphite) using the *s*-wave-tip model, via a theory of *anomalous corrugation* (Tersoff 1986). However, according to a recent analysis of Tersoff and Lang (1990), the *s*-wave-tip model breaks down for atomic-size features, especially for graphite. In contrast to the large structures on metal surfaces, where the STM image is roughly independent of tip electronic structure, Tersoff and Lang (1990) showed clearly and unambiguously that the tip electronic states have dramatic effects on the STM images of atomic-size features, especially the STM images of graphite.

The Tersoff–Hamann approximation

For profiles of surface reconstructions, the periodicity a can be much larger than the interatomic distance on metal surfaces, which is 2.5–3 Å. Tersoff and Hamann (1983, 1985) showed that under the condition

$$\left(1 + q^2/\kappa^2\right)^{l/2} \approx 1, \tag{5.63}$$

the STM images are independent of tip electronic states. In this equation, $q = \pi/a$ is the relevant wave vector at the edge of the Brillouin zone. Therefore, for sufficiently large a, $(\kappa_1/\kappa) \rightarrow 1$, and $(q/\kappa) \rightarrow 0$. The STM images of the profiles of reconstructions gradually become independent of tip states. The validity of the Tersoff-Hamann approximation for the profiles of large superstructures is demonstrated clearly in the case of the $22 \times \sqrt{3}$ reconstruction of Au(111) (Barth et al. 1990): Although the atomic corrugation depends dramatically on tip states, the average contour of the reconstruction in the $[1\bar{1}0]$ direction (with periodicity 63 Å) is essentially independent of tip states and independent of tunneling conditions. Consider its second harmonics, which represent the details of the reconstruction contour. The charge density is proportional to $|\psi|^2$, whose 31.5 Å periodicity corrugation is determined by the fundamental Fourier component of a Bloch wave of periodicity 63 Å. The relevant parameters are $q \approx 2\pi/63$ Å ≈ 0.1 Å^{-1}, and $\kappa_1 = (0.96^2 + 0.1^2)^{1/2} \approx 0.965$ Å^{-1}. Substituting these numbers into Eq. (5.59), we obtain

$$\Delta z = \left(1.03 - 0.016 \left|\frac{D_1}{D_0}\right|^2 - 4.4 \times 10^{-5} \left|\frac{D_2}{D_0}\right|^2\right) \times Ce^{-0.01z}\cos^2(0.1x), \tag{5.64}$$

where C is the peak-to-peak amplitude of the second harmonics in the reconstruction contour.

From Eq. (5.64), it is clear that in the STM image of the Au(111)-$22 \times \sqrt{3}$ reconstruction, the average contour (with the atomic corrugation neglected) is almost exactly the Fermi-level LDOS contour. In fact, for the contour of the reconstruction, the d_{z^2} tip state behaves almost exactly like an s state, and the effects of the $m \neq 0$ tip states are negligible.

In the original papers of Tersoff and Hamann (1983, 1985), the result of the s-wave approximation was compared with the STM images of the super-structure profiles of the reconstructed Au(110) surfaces. By assuming the center of the s-wave is approximately 15 Å from the top-layer nuclei of the gold surface, the Fermi-level LDOS contour approximately matches the meas-ured image of Au(110)-2×1 and Au(110)-3×1 surfaces. For Au(110)-3×1, $q \approx 0.25$ Å^{-1}, and $q^2/\kappa^2 \approx 0.1$. For small l, the Tersoff–Hamann condition, Eq. (5.63), is approximately satisfied, and the STM images are approximately independent of tip electronic states.

Furthermore, in the case of Au(110)-3×1, the topographic images are approximately independent of the tip–sample distance, or independent of tunneling conditions. Actually, here $q \approx 0.25$ Å^{-1}, $\kappa \approx 1$ Å^{-1}. Thus,

$$\Delta z = \text{const.} \times e^{-0.063z}. \tag{5.65}$$

By moving the tip 1 Å in the z direction, the corrugation amplitude changes by about 6%. In other words, by moving the tip a few Å outwards or inwards, there should be no experimentally detectable difference in corrugation. For the same reason, because the lateral variation of the surface structure under consideration is slow, the exact lateral location of the center of the s-wave is not critical. Therefore, for the cases where the s-wave model is valid, the exact location of the geometrical point, which represents the center of curvature of the tip in the s-wave model, is not critical. Furthermore, since the tip-sample distance is closely related to tunneling conditions, the STM images of the contours of large reconstructions should be approximately inde-pendent of tunneling conditions. This was actually observed for the profiles of the $22 \times \sqrt{3}$ superstructure of Au(111) (Barth et al. 1990). Hence, for features of which the Tersoff–Hamann condition [Eq. (5.63)] is well satisfied, the atomic structure and the electronic states of the tip can be completely ignored, and the STM images should be approximately independent of tunneling condi-tions.

The s-wave-tip model is applicable to another case: the smoothed profile of a large atom (such as Na, Xe) adsorbed on metal surfaces, if the atomic structure of the metal surface is completely neglected (Lang, 1986; Eigler et al, 1991). Actually, if the tip-sample distance is large enough (approximately 10 Å), the atomic structure of the metal surface is not observable. Only the smoothed profile of the large absorbate is observable. In both cases (Lang 1987, Eigler et al. 1991), the size of the image of the smoothed profile of the absorbed atom is 7–9 Å. The smoothed profile is determined by the long wave length Fourier components, which is relatively insensitive to tip electronic states. At shorter tip-sample distances, the underlying atomic structure of the metal substrate becomes observable. The STM images of the metal surfaces depend dramatically on tip conditions and tunneling conditions (Zeppenfeld et al. 1992).

Anomalous corrugation theory

Around 1986, atomic resolution was observed on a few semiconductor and semimetal surfaces, including graphite and Si(111)2 × 1, which well exceeds the resolution limit predicted by the *s*-wave-tip model. A theory of *anomalous corrugation* was proposed to interpret it (Tersoff 1986). The theory was based on a singularity in the Fermi-level LDOS of the *sample*. For example, graphite, a semimetal, is a well-known example of this nature. The Fermi surface shrinks to six points at the six corners of the Brillouin zone. If the *s*-wave-tip theory is correct, there should be an anomaly in the STM image of graphite, as illustrated by the following one-dimensional model.

If at the Fermi level, the only surface Bloch wave of the material is a sinusoidal function with Bloch vector q,

$$\Psi_q = \sin(qx)\, e^{-\frac{1}{2}\gamma z}, \tag{5.66}$$

where $\gamma = 2(\kappa^2 + q^2)^{1/2}$, and $\kappa = (2m\phi)^{1/2}/\hbar$; then according to the *s*-wave model, the tunneling conductance is

$$\sigma = A_0 \sin^2(qx)\, e^{-\gamma z}, \tag{5.67}$$

where x and z specify the center of curvature of the tip, and A_0 is a constant. The topographic STM image is a contour of constant conductance, for example, $\sigma = \sigma_0$. From Eq. (5.67), one obtains immediately

$$z = z_0 + \frac{1}{\gamma} \ln \sin^2(qx). \tag{5.68}$$

where the reference point of z is $z_0 = -\ln(\sigma_0/A_0)/\gamma$. Because of the logarithmic singularity in Eq. (5.68), *the theoretical corrugation amplitude is infinity.*

There are two consequences of the *s*-wave model on graphite that can be tested experimentally. First, from Eq. (5.67), the average tunneling conductance varies with z as

$$\bar{\sigma} = \frac{A_0}{2}\, e^{-\gamma z}. \tag{5.69}$$

The typical lattice constant of a close-packed surface (metal or graphite) is $a = 2.5$ Å, and $q = \pi/a \approx 1.26$ Å^{-1}, which makes $\gamma = 2(1.26^2 + 0.95^2)^{1/2}$

≈ 3.2 Å^{-1}. Therefore, the tunneling conductance should decay about 25 times per Å instead of 8 times per Å as in metals. In other words, the apparent barrier height, calculated from Eq. (1.13), should be 9.5 eV instead of the normal value, 4–5 eV. (In the two-dimensional model, the tunneling conductance should decay about 40 times per Å. In other words, an apparent barrier height of about 25 eV should be observed. See Tersoff, 1986.) Second, according to Eq. (5.68), the corrugation amplitude should be independent of z. Experimentally, however, the apparent barrier height observed on graphite is much less than 1 eV. Furthermore, the corrugation amplitudes depend dramatically on tunneling conductance, thus on tip–sample distance, as shown by Binnig et al. (1986a), Soler et al. (1986), and Mamin et al. (1986). These experimental facts were explained by a deformation of the sample surface due to the tip–sample interaction, as proposed by Soler et al. (1986).

Breakdown of the s-wave-tip model

As shown by Lawunmi and Payne (1990), the anomalous corrugation due to the singularity of sample LDOS disappears completely with a minute amount of $m \neq 0$ tip state; thus the theory of anomalous corrugation is extremely unstable. Using various atoms adsorbed on jellium surface as the model of the tip, Tersoff and Lang (1990) reached the same conclusion: The image of a single sinusoidal Bloch wave, as described in Equations (5.66) through (5.68), is extremely sensitive to the choice of the tip atom, indicating a breakdown of the s-wave-tip model in this case. For different tip atoms, the degree of deviation from the s-wave-tip model is very different. For a Na atom, 97% of the tunneling current is associated with an $m = 0$ state, accounting for the close agreement with the s-wave-tip model. For Ca, the deviation increases substantially. For Si, the $m = 1$ contribution is almost as large as the $m = 0$, which includes both s and p_z. And for Mo, while the p contribution is reduced, this is more than compensated by the large contribution from states of d like symmetry. In both cases, the theoretical images are substantially different from the predictions of the s-wave-tip model, as shown in Table 5.2.

It is natural to consider the case of a carbon-atom tip, which may be picked up by the tip end from the sample surface. According to the calculation of Tersoff and Lang (1990), as listed in Table 5.2, the image of graphite from a carbon-atom tip (adsorbed on jellium) is dominated by $m = 1$ states. The corrugation is *reversed,* as shown in the right column of Table 5.2. Tersoff and Lang (1990) show that this situation is a coincidence, which depends dramatically on the surface lattice constant, and depends on the actual environment in which the carbon atom is. Following the derivative rule (see Section 5.2), it is clear that the smaller the lateral wave vector q, the smaller the contribution from the $m = 1$ tip state. By increasing the lattice constant by a factor of 2 (that is, by increasing the lattice constant of graphite from the actual value 2.46 Å to a hypothetical value 4.92 Å), the image becomes positive. By further increasing the lattice constant to 9.84 Å, the predicted image comes very close to that from an s-wave tip state. Similarly, for a C atom in a very different environment (for example, bonded to a transition-metal tip or a carbon cluster instead of to jellium), such a cancellation seems unlikely. Thus, Tersoff and Lang (1990) concluded that for atomic-scale features, specifically for graphite, tip electronic states have dramatic effects on STM images.

Table 5.2. Contributions of various tip electronic states to tunneling current, in percent. Calculated for five different tip atoms. The corrugations for graphite, for five different tip atoms, are shown in the right-most column. For the s-wave-tip theory, the corrugation is infinity. After Tersoff and Lang (1990).

Tip atom	m=0	m=1	m=2	Corrugation (Å)
Na	97	2	1	0.9
Ca	88	9	3	0.5
Si	59	41	0	0.2
Mo	46	28	26	0.1
C	1	99	0	−0.2

CHAPTER 6

IMAGING ATOMIC STATES

The capability of imaging perfect crystalline surfaces down to the level of single atoms is not the only merit of STM. The unique capability of STM is to image local phenomena on an atomic scale, such as a single-atom defect or an adsorbed atom or molecule, and to image every single atom at a step, a dislocation, or a twin boundary. This is far beyond the capability of diffraction methods, such as low-energy electron diffraction (LEED), atom-beam diffraction, and x-ray diffraction. The direct information obtained by STM is the local electronic structure (Binnig and Rohrer, 1982), rather than the positions of the atomic nuclei. In this chapter, we discuss the process of STM imaging of individual atomic states at solid surfaces.

6.1 Slater atomic wavefunctions

A widely used approximate method of describing atomic states is the Slater atomic wavefunctions (Zener, 1930; Slater, 1930). In this section, we show that regarding STM, the Slater wavefunction is a convenient tool for describing localized atomic states at surfaces.

The understanding of electronic states in atoms is to a great extent based on Schrödinger's solution of the hydrogen-atom problem. These wavefunctions have the general form (Landau and Lifshitz, 1977):

$$\psi_{nlm}(r, \theta, \phi) = C e^{-\zeta r} r^l L_{n+l}^{2l+1}(2\zeta r) Y_{lm}(\theta, \phi), \tag{6.1}$$

where L_{n+l}^{2l+1} is a polynomial of order $(n - l - 1)$, and ζ is a constant. In other words, the radial factor of a hydrogen wavefunction is an exponential factor times a polynomial of order $(n - 1)$.

In treating complex atoms and molecules, these hydrogenlike wavefunctions are still too complicated in most applications. In treating low-energy (up to a few eV) problems, the behavior of the wavefunctions near the nuclei of the atoms is not important. Therefore, in the hydrogen wavefunctions, only the term with the highest power of r is significant. Based on this observation, Zener (1930) and Slater (1930) proposed a simplified form of the atomic wavefunctions:

To determine the image, the first step is to determine the distribution of tunneling current as a function of the position of the apex atom. We set the center of the coordinate system at the nucleus of the sample atom. The tunneling matrix element as a function of the position $\mathbf{r}$ of the nucleus of the apex atom can be evaluated by applying the derivative rule to the Slater wavefunctions. The tunneling conductance as a function of $\mathbf{r}$, $g(\mathbf{r})$, is proportional to the square of the tunneling matrix element:

$$g(\mathbf{r}) \propto |M|^2. \tag{6.12}$$

Because the equal-conductance contour is independent of an overall constant, we neglect it for convenience. In Table 6.1, the results for three types of $m=0$ tip states and three types of $m=0$ sample states are listed.

6.2.1 Apparent radius and apparent curvature

A convenient quantity for comparing with the STM image profile for a single atomic state is the apparent radius of the image near its peak, R, or the apparent curvature, $K = 1/R$. Considering the $m = 0$ state only, these quantities are related to the tunneling conductance distribution by

$$K \equiv \frac{1}{R} = \left(\frac{\partial g(\mathbf{r})}{\partial z} \right)^{-1} \frac{\partial^2 g(\mathbf{r})}{\partial x^2}, \tag{6.13}$$

which is evaluated at $(0, 0, z)$. To illustrate the meaning of Eq. (6.13), we make an explicit calculation for a spherical tunneling-conductance distribution,

$$g(\mathbf{r}) \equiv g(r), \tag{6.14}$$

where $r^2 = x^2 + y^2 + z^2$. Using the following relations,

$$\left(\frac{\partial g}{\partial z} \right)_{z=r} = \left(g' \frac{z}{r} \right)_{z=r} = g', \tag{6.15}$$

$$\left(\frac{\partial^2 g}{\partial x^2} \right)_{x=0} = \left(g'' \frac{x^2}{r^2} + \frac{g'}{r} - g' \frac{x^2}{3r^3} \right)_{x=0} = \frac{g'}{r}, \tag{6.16}$$

Fig. 6.1. Apparent radius for a spherical conductance distribution. For a spherical tunneling conductance distribution, the apparent radius equals the nominal tip–sample distance, regardless of the specific functional form of the distribution.

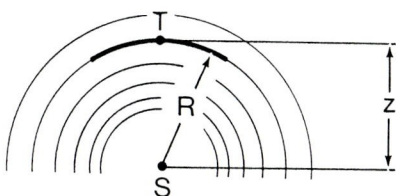

we find that the definition of the radius, Eq. (6.13), coincides with the nominal distance between the two nuclei, as shown in Fig. 6.1.

Intuitively, we will expect that for p_z or d_z states on the tip and on the sample, the images should be sharper; that is, the apparent radius should be smaller. This is indeed true. For example, for a d_z tip state and an s-wave sample state, the tunneling conductance distribution is

$$
g(\mathbf{r}) = \left(\frac{3}{2} \frac{z^2}{r^2} - \frac{1}{2} \right)^2 e^{-2\kappa r}.
\tag{6.17}
$$

Using Eq. (6.13), we find

$$
K \equiv \frac{1}{R} = \frac{1}{z}\left(1 + \frac{3}{\kappa z}\right).
\tag{6.18}
$$

In general, the apparent curvature is

$$
K \equiv \frac{1}{R} = \frac{1}{z}\left(1 + \frac{h_S + h_T}{\kappa z}\right),
\tag{6.19}
$$

where h_S is determined by the atomic state at the sample surface under probing: for an s state, $h_S = 0$; for a p_z state, $h_S = 1$; and for a d_z state, $h_S = 3$. The same holds for the tip states. The results are listed in Table 6.1, and are illustrated in Fig. 6.2. Clearly, the apparent radius is reduced substantially for p and d states on the tip as well as on the sample. In other words, with p and d states, the images of individual atoms at surfaces look much sharper than those with s states, which is expected. The images for mixed states, for example, sp^3 states, can be treated using the same method. For states with $n > 1$, the enhancement is more pronounced. As we have shown, the apparent radius for a spherical distribution does not depend on the specific functional form of the distribution. Therefore, the algebraic factor in the radial Slater functions results in a small correction to the apparent radius for $l \neq 0$

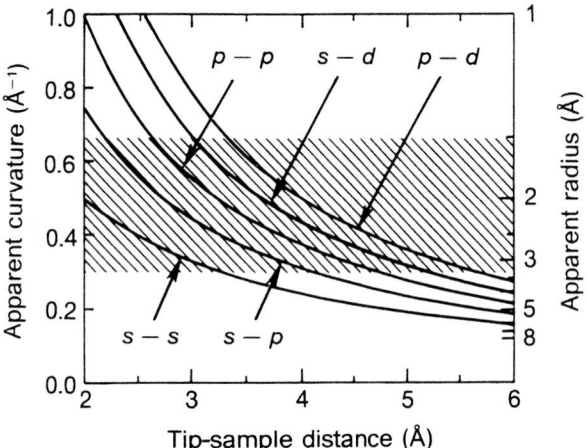

Fig. 6.2. Apparent radius as a function of tip–sample distance. For different combinations of tip states and sample states, the apparent radii are given by Eq. (6.19). The shaded area indicates the condition for achieving atomic resolution (that is, having an apparent radius comparable to the actual radius of an atom). The minimum tip–sample distance is limited by the mechanical contact (that is, about 2.5 Å). Therefore, most of the images with true atomic resolution are obtained with p_z or d_z tip states.

states. It is easy to show that asymptotically the correction term is proportional to $n - 1$ and to z^{-3}.

A glance in Table 6.1 reveals an interesting fact. Not only does the radius for s states depend on the tip–sample distance z, but also the *differences* among different types of states. At large distances, for which $\kappa z \gg 1$, the apparent sizes of images with different types of atomic states differ by a constant. The relative difference is reduced with increasing distance. At short distances, where $\kappa z \approx 1$, the apparent sizes for different types of atomic states can differ by a factor of 2, 4, or even more. This is in accordance with the result in Chapter 5 that at large tip-sample distances, that is, $\kappa z \gg 1$, and for large features, where $\kappa a \gg 1$, the difference between different tip states is reduced.

Naturally, we observed the *reciprocity principle*: By interchanging the tip state and the sample state, the conductance distribution, and consequently, the apparent radius of the image, are unchanged.

Table 6.1. Conductance distribution and apparent curvature of the STM images of individual atomic states

Tip	Sample	Conductance distribution	Apparent curvature
s	s	$e^{-2\kappa r}$	$\dfrac{1}{z}$
s	p_z	$(z/r)^2 \, e^{-2\kappa r}$	$\dfrac{1}{z}\left(1 + \dfrac{1}{\kappa z}\right)$
s	d_{z^2}	$\left[3(z/r)^2 - 1\right]^2 e^{-2\kappa r}$	$\dfrac{1}{z}\left(1 + \dfrac{3}{\kappa z}\right)$
p_z	s	$(z/r)^2 \, e^{-2\kappa r}$	$\dfrac{1}{z}\left(1 + \dfrac{1}{\kappa z}\right)$
p_z	p_z	$(z/r)^4 \, e^{-2\kappa r}$	$\dfrac{1}{z}\left(1 + \dfrac{2}{\kappa z}\right)$
p_z	d_{z^2}	$\left[3(z/r)^2 - 1\right]^2 (z/r)^2 \, e^{-2\kappa r}$	$\dfrac{1}{z}\left(1 + \dfrac{4}{\kappa z}\right)$
d_{z^2}	s	$\left[3(z/r)^2 - 1\right]^2 e^{-2\kappa r}$	$\dfrac{1}{z}\left(1 + \dfrac{3}{\kappa z}\right)$
d_{z^2}	p_z	$\left[3(z/r)^2 - 1\right]^2 (z/r)^2 \, e^{-2\kappa r}$	$\dfrac{1}{z}\left(1 + \dfrac{4}{\kappa z}\right)$
d_{z^2}	d_{z^2}	$\left[3(z/r)^2 - 1\right]^4 e^{-2\kappa r}$	$\dfrac{1}{z}\left(1 + \dfrac{6}{\kappa z}\right)$

6.2.2 Comparison with experiments

The results for the apparent radius of STM images for individual states can be used to interpret experimental images directly. For surfaces with complex periodic structures, such as Si(111)-7 × 7 and Si(111)-5 × 5, the concept of imaging individual atomic states is a much better description than surface Bloch functions. For adatoms and defects, the individual state description is the only possible one.

The apparent radii of atomic states in an experimental STM image can be readily obtained from a trace. The profile of an atomic state near its peak can always be approximated as a parabola. By measuring the width $2b$ at a distance h from the peak, the radius at the peak is

$$r = \frac{b^2}{2h} . \tag{6.20}$$

As an example, Fig. 6.3 shows experimental profiles of adatoms on the Si(111)-7 × 7 surface, as collected from four publications by Demuth et al. (1988). The apparent radius, as obtained from direct measurements from these experimental profiles, ranges from 2.8 to 4.0 Å. The normal tip–sample dis-

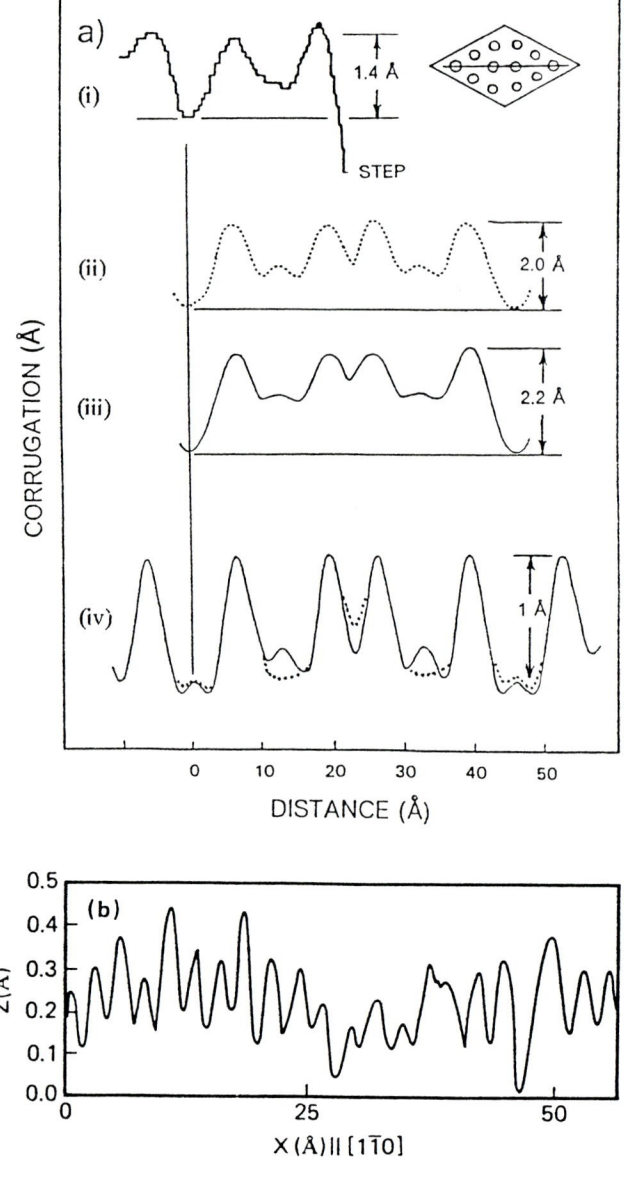

Fig. 6.3. Experimental profile of STM images. The apparent radii of the atomic states are determined from the measured profile measurements. (a) Image profiles of the Si(111)-7 × 7 surface: (i) Becker et al. (1985); (ii) Tromp et al. (1986); (iii) Berghaus et al. (1988); and (iv) Demuth et al. (1988. (From Demuth et al., 1988.) The apparent radius is $r \approx 2.8 - 4.0$ Å. (b) Image profile of Au(111), from Hallmark et al. (1988). The apparent radius obtained from direct measurement is $r \approx 1.5 - 2.0$ Å.

tance in imaging Si surfaces is about 5 Å (Chen and Hamers, 1991a). Considering that the dangling-bond state at each adatom is essentially a p_z state, the tip state should be either another p_z state or a d_z state. By approximating the tip state and the electronic state at the top-layer Si atom by s-wave states, even at a very short tip–sample distance, 4 Å, the apparent radius of the image of the Si atom must be greater than 4 Å.

Another example in Fig. 6.3 is the image of the Au(111) surface with a step, as observed by Wöll et al. (1989). Direct measurement gives an apparent radius of 1.5–2.0 Å. At the Fermi level, the electronic states of a gold surface are dominated by s atomic states. The closest tip–sample distance that is possible experimentally is 3 Å. Therefore, the tip state is likely to be a d state.

6.3 The Na-atom-tip model

Lang (1985, 1986, 1987) developed a theory of STM by modeling the tip as a Na atom weakly adsorbed on a jellium surface. As shown by Tersoff and Lang (1990), the electronic state of the adsorbed Na atom has 97% $m=0$ state near the Fermi level. Therefore, such a tip should reproduce the predictions of the s-wave model (Tersoff and Hamann, 1983) precisely. In the calculation of Lang (1986), the distance between the tip (a Na atom) and the sample (also a Na atom) is 12 Å. The tunneling current is calculated with Bardeen's original perturbation approach. The current from the bare jellium is subtracted from the total current. By moving the two atoms relative to each other while keeping the tunneling current constant, the topographic image of the STM is generated numerically. The result is shown in Fig. 6.4. The apparent radius of Lang's simulated STM image can be measured directly from this figure, which gives $r \approx 12$ Å.

Lang's result from the numerical theory is in excellent agreement with the general theory in the previous section. If both tip state and sample state are s-wave states, the tunneling conductance should have a spherical distribution, and the apparent radius should equal the nominal distance between the two nuclei. The radius obtained from the figure fits well with this expectation.

Similar to the s-wave model, the Na-atom-tip model predicts a poor resolution. The agreement of the Na-atom-tip model with the s-wave-tip model does not mean that the s-wave-tip model describes the actual experimental condition in STM. According to the analysis of Tersoff and Lang (1990), real tips are neither Na or Ca, but rather transition metals, probably contaminated with atoms from the surface (for example, Si and C are common sample materials). For a Si-atom tip, the p state dominates the Fermi-level LDOS of the tip. For a Mo-atom tip, while the p contribution is reduced, this is more than compensated by the large contribution from states of d like symmetry. The STM images from a Si, C, or Mo tip, as predicted by Tersoff and

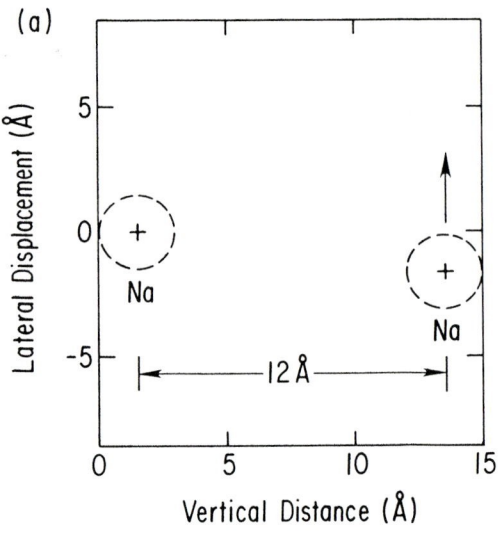

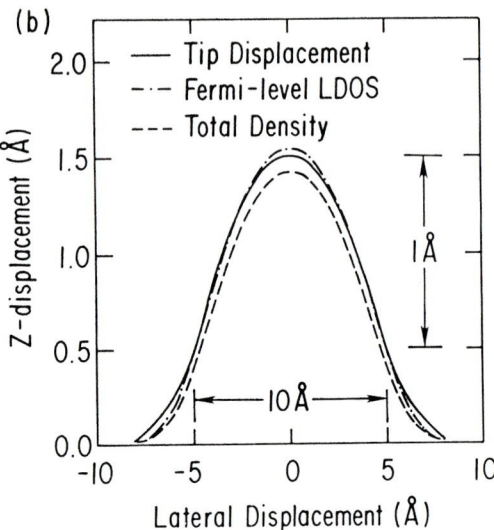

Fig. 6.4. Image profile with a Na-atom tip. (a) Geometry of the simulation. Two flat and structureless jellium surfaces, each with an extra Na atom adsorbed on it, represent the tip and the sample, respectively. The net current from these Na atoms is kept constant while moving the Na atoms across each other. The path is generated numerically. (b) The simulated image (solid curve) is in good agreement with the contour of the Fermi-level LDOS and the total-charge-density contour. The apparent radius, as determined from the curve, is about 12 Å. (After Lang, 1986.)

Lang (1990), are dramatically different from the images predicted by the *s*-wave-tip model.

6.4 Images of surfaces: Independent-orbital approximation

In the previous chapter, we have presented a method for calculating the STM images of simple surfaces using the method of leading Bloch functions. However, this method can only treat simple surfaces. For more complex surfaces, a few special points on the surface Brillouin zone are not sufficient. The choice of special points on the surface Brillouin zone becomes ambiguous. We have mentioned in Chapter 4 that in the theory of gas–surface interactions (Steele, 1974; Goodman and Wachman, 1976), the summation of pairwise interactions has been a powerful method for treating interactions of a single atom with surfaces. In this section, we present an alternative method for predicting STM images for crystalline surfaces utilizing the pairwise summation concept. It is based on the assumption that the total tunneling current can be approximated as the sum of the tunneling currents from individual atomic states on the sample surface. In other words, the atomic orbitals are assumed to be independent. This assumption is not always valid. For example, it does not work for solids exhibiting charge-density waves. However, for many metal and semiconductor surfaces, it provides an adequate description. We start with discussing the general method using the example of a crystalline surface with square lattice and tetragonal symmetry, that is, belonging to the plane group *p4mm*. General expressions for the Fourier coefficients are given. We then discuss the application of this method to the images of Si(111)2 × 1 and close-packed metal surfaces.

6.4.1 General expression for the Fourier coefficients

A tetragonal lattice is the combination of two perpendicular one-dimensional lattices with the same lattice constant and having a reflection symmetry, as shown in Fig. 6.5. Therefore, we start with a one-dimensional case. The tunneling conductance from the *n*th atom is $g(x - na, z)$. The total tunneling conductance from all the atoms is

$$G(x, z) = \sum_{n = -\infty}^{\infty} g(x - na, z). \tag{6.21}$$

Apparently, it is a periodic function with periodicity a. Therefore, it can be expressed as a Fourier series,

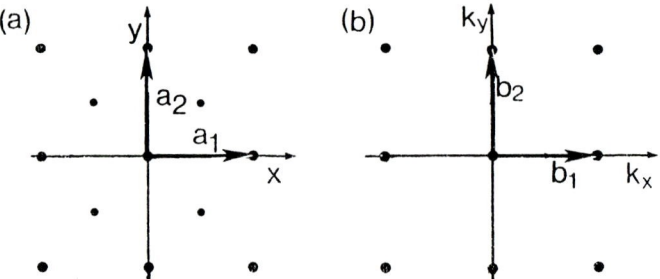

Fig. 6.5. Close-packed surface with tetragonal symmetry. (a) The square lattice in real space. There is an atom on each lattice point. (b) The reciprocal space.

$$G(x,z) = \sum_{j=-\infty}^{\infty} \tilde{G}_j(z)\, e^{ijbx}, \tag{6.22}$$

where $b = 2\pi/a$ is the length of the primitive vector of the reciprocal lattice. The Fourier coefficient is

$$\begin{aligned}
\tilde{G}_j(z) &= \frac{1}{a} \int_0^a dx\, e^{-ijbx} \sum_{n=-\infty}^{\infty} g(x - na, z) \\
&= \frac{1}{a} \int_{-\infty}^{\infty} dx\, e^{-ijbx} g(x, z).
\end{aligned} \tag{6.23}$$

As shown, due to the infinite sum in Eq. (6.21), the integration extends over the entire x axis.

To construct an image including the lowest nontrivial Fourier components, only three terms in the Fourier series are significant. Those are: $\tilde{G}_0(z)$, $\tilde{G}_{-1}(z)$, and $\tilde{G}_1(z)$. Because of the reflection symmetry of the conductance function $g(x,z)$, the last two Fourier coefficients are equal, and are denoted as $\tilde{G}_1(z)$. Up to this term,

$$G(x,z) = \tilde{G}_0(z) + 2\tilde{G}_1(z)\, \cos(bx). \tag{6.24}$$

Following the general relation between current images and topographic images, Eq. (5.7), the topographic image is

$$\Delta z(x) = -\frac{2\tilde{G}_1(z_0)}{\left(\dfrac{d\tilde{G}_0(z_0)}{dz_0}\right)} \cos bx. \tag{6.25}$$

Therefore, the problem of calculating the STM images reduces to the problem of evaluating the Fourier coefficients for the tunneling conductance distribution of a single atomic state, Eq. (6.23).

This treatment can be extended immediately to the case of surfaces with tetragonal symmetry. The Fourier coefficients for the tunneling conductance of a tetragonal lattice is

$$\tilde{G}_{nm}(z) = \frac{1}{a^2} \int \int dx\, dy\, e^{-i(nx+my)} g(\mathbf{r}). \tag{6.26}$$

Because of the axial symmetry of $g(\mathbf{r})$, $\tilde{G}_{1,0} = \tilde{G}_{-1,0} = \tilde{G}_{0,1} = \tilde{G}_{0,-1}$. We denote them as $\tilde{G}_1$. The tunneling current distribution, up to the lowest non-trivial Fourier components, is

$$G(\mathbf{r}) = \tilde{G}_0(z) + 2\tilde{G}_1(z)\, (\cos bx + \cos by). \tag{6.27}$$

The maximum value of the function $(\cos bx + \cos by)$ is 2, and the minimum value of it is −2. Because $\tilde{G}_0$ is much larger than $\tilde{G}_1$, it is convenient to use the function defined by Eq. (5.30) and rewrite Eq. (6.27) as

$$G(\mathbf{r}) = \tilde{G}_0(z) + 8\tilde{G}_1(z)\, \phi^{(4)}(b\mathbf{x}). \tag{6.28}$$

The function $\phi^{(4)}(\mathbf{X})$ has a maximum value of 1 and minimum value of 0. The topographic image is

$$\Delta z(\mathbf{x}) = -\frac{8\tilde{G}_1(z_0)}{\left(\dfrac{d\tilde{G}_0(z_0)}{dz_0}\right)} \phi^{(4)}(b\mathbf{x}). \tag{6.29}$$

In order to obtain the final form of the theoretical images, we need to calculate the Fourier coefficients for the functions

$$f(\mathbf{r}) = r^{n-1} \cos^m \theta\, e^{-2\kappa r}. \tag{6.30}$$

We start with the mathematical identity:

$$I \equiv \int \int \frac{dx\,dy}{r} e^{-2\kappa r + ipx + iqy} = \frac{2\pi}{\gamma} e^{-\gamma z}, \qquad (6.31)$$

where $r^2 = x^2 + y^2 + z^2$ and $\gamma^2 = 4\kappa^2 + p^2 + q^2$. A proof of this mathematical identity can be found in Goodman (1976). In some cases, the value of γ calculated from the individual-state approximation coincides with the result of the leading-Bloch-wave approximation, Eq. (5.17). In many cases, the values are different. For example, for hexagonal surfaces, the value of γ calculated from the independent-state approximation is slightly higher than that calculated from the leading-Bloch-wave approximation.

By taking $\partial/\partial\kappa$ of both sides of Eq. (6.31), we obtain:

$$\int \int dx\,dy\, r^{n-1} e^{-2\kappa r + ipx + iqy}$$
$$= 2\pi \left(-\frac{\partial}{2\partial\kappa} \right)^n \frac{e^{-\gamma z}}{\gamma}. \qquad (6.32)$$

By taking derivatives with respect to z, a factor $x/r = \cos\theta$ is generated. Within the same approximation of the Slater atomic wavefunctions, the differentiation acts on the exponential factor only. Therefore,

$$\int \int dx\,dy\, r^{n-1} \cos^m \theta\, e^{-2\kappa r + ipx + iqy}$$
$$= 2\pi \left(-\frac{\partial}{2\partial\kappa} \right)^n \left(-\frac{\partial}{2\kappa\partial z} \right)^m \frac{e^{-\gamma z}}{\gamma}. \qquad (6.33)$$

Using these equations and the conductance distribution functions listed in Table 6.1, the corrugation amplitudes for a tetragonal close-packed surface with different tip states and sample states can be obtained. For example, for a $1s$ state, using Eq. (6.32), we have

$$\tilde{G}_0(z) = \frac{\pi z}{\kappa} e^{-2\kappa z}, \qquad (6.34)$$

and

$$\tilde{G}_1(z) = \frac{4\pi\kappa z}{\gamma^2} e^{-\gamma z}. \qquad (6.35)$$

Table 6.2. Independent-state model: corrugation amplitudes for surfaces with tetragonal symmetry

Tip	Sample	Corrugation amplitude
s	s	$\dfrac{16\kappa}{\gamma^2} e^{-\beta z}$
s	p_z	$\left(\dfrac{\gamma}{2\kappa}\right)^2 \dfrac{16\kappa}{\gamma^2} e^{-\beta z}$
s	d_{z^2}	$\left(\dfrac{3}{2}\dfrac{\gamma^2}{4\kappa^2} - \dfrac{1}{2}\right)^2 \dfrac{16\kappa}{\gamma^2} e^{-\beta z}$
p_z	s	$\left(\dfrac{\gamma}{2\kappa}\right)^2 \dfrac{16\kappa}{\gamma^2} e^{-\beta z}$
p_z	p_z	$\left(\dfrac{\gamma}{2\kappa}\right)^4 \dfrac{16\kappa}{\gamma^2} e^{-\beta z}$
p_z	d_{z^2}	$\left(\dfrac{\gamma}{2\kappa}\right)^2 \left(\dfrac{3}{2}\dfrac{\gamma^2}{4\kappa^2} - \dfrac{1}{2}\right)^2 \dfrac{16\kappa}{\gamma^2} e^{-\beta z}$
d_{z^2}	s	$\left(\dfrac{3}{2}\dfrac{\gamma^2}{4\kappa^2} - \dfrac{1}{2}\right)^2 \dfrac{16\kappa}{\gamma^2} e^{-\beta z}$
d_{z^2}	p_z	$\left(\dfrac{\gamma}{2\kappa}\right)^2 \left(\dfrac{3}{2}\dfrac{\gamma^2}{4\kappa^2} - \dfrac{1}{2}\right)^2 \dfrac{16\kappa}{\gamma^2} e^{-\beta z}$
d_{z^2}	d_{z^2}	$\left(\dfrac{3}{2}\dfrac{\gamma^2}{4\kappa^2} - \dfrac{1}{2}\right)^4 \dfrac{16\kappa}{\gamma^2} e^{-\beta z}$

From Eq. (6.29), the corrugation amplitude is

$$\Delta z \equiv - \frac{8\widetilde{G}_1(z_0)}{\left(\dfrac{d\widetilde{G}_0(z_0)}{dz_0}\right)} = \frac{16\kappa}{\gamma^2} e^{-\beta z}. \tag{6.36}$$

The definition of β is identical to that of Harris and Liebsch (1982, 1982a), as is given by Eq. (5.20): $\beta = \gamma - 2\kappa$. For p and d states, similar results can be obtained using Eq. (6.33). A list of corrugation amplitudes for $n=1$ states is shown in Table 6.2. Clearly, with p_z and d_{z^2} states on the tip as well as on the sample, substantial corrugation enhancements should be observed. Using similar method, the corrugations for $n>1$ and $m>0$ cases as well as mixed states, such as sp^3 states, can be obtained.

6.4.2 Application to the images of Si(111)2×1

The STM images of the Si(111)-2×1 were first obtained by Feenstra and Stroscio (1987). The structure of the Si(111)-2×1 surface is discussed in Chapter 1; see Fig. 1.11. The unit cell of Si(111)-2×1 is shown in Fig. 6.6. The unit cell dimensions are $a_1 = 6.65$ Å and $a_2 = 3.84$ Å, in the $[\bar{2}11]$ and $[0\bar{1}1]$ directions, respectively. The lengths of the primitive reciprocal lattice vectors are

$$b_1 = \frac{2\pi}{a_1} \approx 0.945 \text{ Å}^{-1},$$ (6.37)

in the $[\bar{2}11]$ direction and

$$b_2 = \frac{2\pi}{a_2} \approx 1.636 \text{ Å}^{-1},$$ (6.38)

in the $[0\bar{1}1]$ direction, respectively.

We now present a simple model for the Si(111)-2×1 surface. At a positive polarity, the electrons tunnel into the empty dangling bonds on the A atoms. On each A atom, there is a $3p_z$ state. The origin of the coordinate system is set at one of the A atoms. Three tip states are considered: the s state, the p_z state, and the d_z state. By keeping the leading exponential term only, simple explicit expressions for the Fourier coefficients are obtained. The corrugation functions are then derived.

For convenience, the following parameters are introduced. The decay constant κ of the average (uncorrugated) tunneling current depends on the average energy level of the empty dangling-bond states. If it is higher than the Fermi level by ΔE, the decay constant is

$$\kappa = \frac{\sqrt{2m_e(\phi - \Delta E)}}{\hbar}.$$ (6.39)

In the following calculation, the value $\kappa \approx 0.96$ Å^{-1} is taken. In the $[0\bar{1}1]$ direction and the $[\bar{2}11]$ direction, the decay constants γ of the corrugation component of the tunneling current are different. We denote them as

$$\gamma_1 = \sqrt{4\kappa^2 + b_1^2},$$ (6.40)

and

$$\gamma_2 = \sqrt{4\kappa^2 + b_2^2}.$$ (6.41)

For a Yukawa distribution

$$g(r) = \frac{1}{r} e^{-2\kappa r}, \tag{6.42}$$

the first three Fourier components are

$$\tilde{G}_{00}(z) = \frac{\pi}{a_1 a_2 \kappa} e^{-2\kappa z}, \tag{6.43}$$

$$\tilde{G}_{10}(z) = \frac{2\pi}{a_1 a_2 \gamma_1} e^{-2\gamma_1 z}, \tag{6.44}$$

and

$$\tilde{G}_{01}(z) = \frac{2\pi}{a_1 a_2 \gamma_2} e^{-2\gamma_2 z}. \tag{6.45}$$

The rest of the calculation closely follows the previous subsection. For an s-wave tip state, the corrugation is

$$\Delta z(x, y) = \frac{16\kappa^2}{\gamma_1^3} e^{-(\gamma_1 - 2\kappa)z} \cos^2 \frac{b_1 x}{2}$$
$$+ \frac{16\kappa^2}{\gamma_2^3} e^{-(\gamma_2 - 2\kappa)z} \cos^2 \frac{b_2 y}{2}. \tag{6.46}$$

For a p_z tip state,

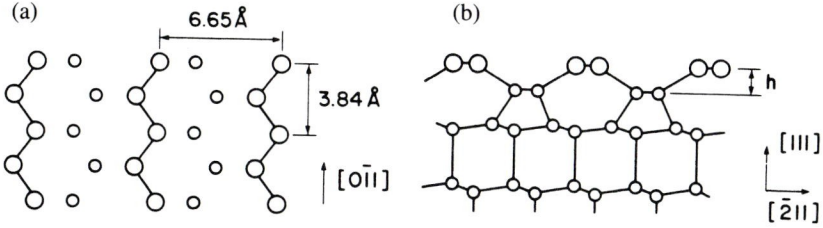

Fig. 6.6. Unit cell of Si(111)-2×1. (a) top view, (b) side view. (Reproduced from Feenstra et al., 1986, with permission.)

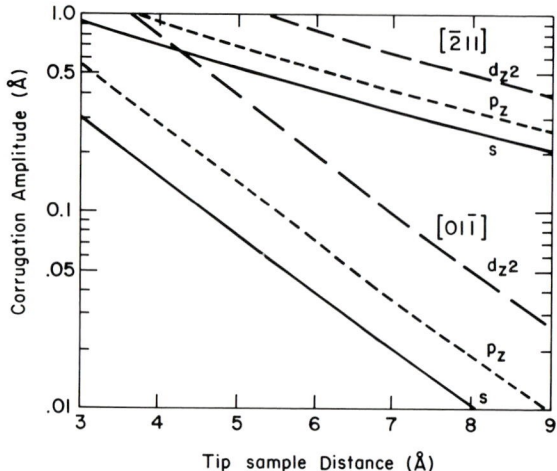

Fig. 6.7. Corrugation amplitudes of the STM images of the Si(111)-2×1 surface. The dependence of corrugation amplitudes on tip–sample distances are calculated using the independent-state model. The corrugation in the $[\bar{2}11]$ direction is much easier and much less dependent on tip electronic states than the corrugations in the $[0\bar{1}1]$ direction.

$$\Delta z(x, y) = \frac{4}{\gamma_1} e^{-(\gamma_1 - 2\kappa)z} \cos^2 \frac{b_1 x}{2}$$
$$+ \frac{4}{\gamma_2} e^{-(\gamma_2 - 2\kappa)z} \cos^2 \frac{b_2 y}{2} . \tag{6.47}$$

And for a d_z tip state,

$$\Delta z(x, y) = \frac{16\kappa^2}{\gamma_1^3} \left[\frac{3}{2} \left(\frac{\gamma_1}{2\kappa} \right)^2 - \frac{1}{2} \right]^2 e^{-(\gamma_1 - 2\kappa)z} \cos^2 \frac{b_1 x}{2}$$
$$+ \frac{16\kappa^2}{\gamma_2^3} \left[\frac{3}{2} \left(\frac{\gamma_2}{2\kappa} \right)^2 - \frac{1}{2} \right]^2 e^{-(\gamma_2 - 2\kappa)z} \cos^2 \frac{b_2 y}{2} . \tag{6.48}$$

The corrugation amplitudes in the $[0\bar{1}1]$ direction and in the $[\bar{2}11]$ direction are displayed in Fig. 6.7. Some general features are worth noting: First, the corrugation in the $[\bar{2}11]$ direction is much easier to observe and much less dependent on tip states than the corrugation in the $[0\bar{1}1]$ direction. Second, the decay constant for the corrugations in the two directions are quite different. Third, those decay constants are independent of the tip state.

6.4.3 Application to close-packed metal surfaces

In general, for an oblique lattice in two dimensions with primitive vectors $\mathbf{a}_1$ and $\mathbf{a}_2$, the total conductance G can be expanded into a two-dimensional Fourier series,

$$
\begin{aligned}
G(\mathbf{r}) &= \sum_{n,m = -\infty}^{\infty} g(\mathbf{r} + n\,\mathbf{a}_1 + m\,\mathbf{a}_2) \\
&= \sum_{j,k = -\infty}^{\infty} \tilde{G}_{jk}(z)\, e^{i(j\,\mathbf{b}_1 + k\,\mathbf{b}_2)\,\bullet\,\mathbf{x}}.
\end{aligned}
\tag{6.49}
$$

The Fourier coefficients are

$$
\tilde{G}_{jk}(z) = \frac{1}{|\mathbf{a}_1 \times \mathbf{a}_2|} \int d^2\mathbf{x}\, G(\mathbf{r}) e^{-i(j\,\mathbf{b}_1 + k\,\mathbf{b}_2)\,\bullet\,\mathbf{x}},
\tag{6.50}
$$

where $\mathbf{b}_1 = 2\pi(\mathbf{a}_2 \times \mathbf{e}_z)/|\mathbf{a}_1 \times \mathbf{a}_2|$ and $\mathbf{b}_2 = 2\pi(\mathbf{e}_z \times \mathbf{a}_1)/|\mathbf{a}_1 \times \mathbf{a}_2|$ are primitive vectors of the reciprocal lattice, and the integration extends over the entire surface. $\mathbf{e}_z$ is a vector in the z direction.

For a close-packed surface with hexagonal symmetry, up to the lowest nontrivial Fourier components, only the Fourier coefficients at $\mathbf{k} = 0$ and the six equivalent points are significant. Because of the axial symmetry of the conductance function $g(\mathbf{r})$ on each site, the Fourier coefficients at these six points, [$\tilde{G}_{1,0}(z)$, $\tilde{G}_{-1,0}(z)$, $\tilde{G}_{0,1}(z)$, $\tilde{G}_{0,-1}(z)$, $\tilde{G}_{1,1}(z)$, and $\tilde{G}_{-1,-1}(z)$], are equal. We denote it as $\tilde{G}_1(z)$. Up to this term,

$$
G(\mathbf{r}) = \tilde{G}_0(z) + \frac{9}{2}\,\tilde{G}_1(z)\,\phi^{(6)}(b\mathbf{x}),
\tag{6.51}
$$

where b is the magnitude of the primitive reciprocal vectors, $\mathbf{b}_1$ and $\mathbf{b}_2$. An inspection to Fig. 6.8 gives $b = (4\pi/\sqrt{3}\,a)$. The hexagonal cosine function is defined by Eq. (5.42).

The theoretical expression of the topological STM image can be written in the form:

$$
z = z_0 + \Delta z\,\phi^{(6)}(b\mathbf{x}),
\tag{6.52}
$$

with

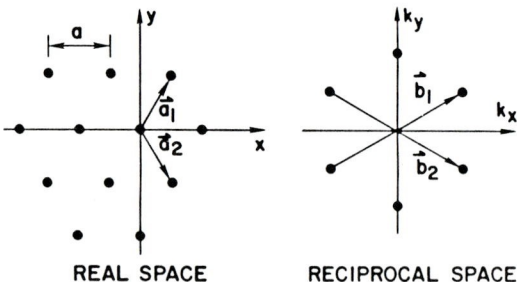

REAL SPACE   RECIPROCAL SPACE

Fig. 6.8. Hexagonal close-packed surface. (a) Real space. (b) Reciprocal space. The length of the primitive reciprocal lattice vector is $b = (4\pi/\sqrt{3}\ a)$.

$$\Delta z(z_0) = -\frac{9}{2} \frac{\tilde{G}_1(z_0)}{\left(\dfrac{d\tilde{G}_0(z_0)}{dz_0}\right)} . \tag{6.53}$$

By comparing this with the results for surfaces with tetragonal symmetry, it is clear that the only difference is the factor of 8 in Eq. (6.29) is replaced by 9/2. With the same lattice constant, the corrugation amplitude of a surface with hexagonal symmetry is smaller than that for a surface with tetragonal symmetry by a factor of 9/16=0.5625. The decay constant of the corrugation is

$$\gamma = \sqrt{4\kappa^2 + b^2} . \tag{6.54}$$

In the last column of Table 6-2, the ratios of the corrugation amplitudes with respect to the s-wave case are given for a surface with atomic distance a=2.88 Å and work function ϕ=3.5 eV. The relevant quantities are: $\kappa = 0.96$ Å^{-1}, $b = 2.52$ Å^{-1}, and $\gamma = 3.17$ Å^{-1}. For most metals, the enhancements can be greater then one order of magnitude. In Table 6-2, only cases with n=1 and m=0 are listed. It is easy to extend the results for n>1 and m>0 cases, as well as mixed states, such as sp^3 states.

Figure 6.9 is a comparison of the results discussed previously with the first-principles calculation of the Al(111) surface as well as the experimental results of STM images on Al(111) by Wintterlin et al. (1989). A very simple model of the Al(111) surface is used: On each surface Al atom, there is an independent 1s state near the Fermi level. The charge density contour (i.e., the image with an s-wave tip state) agrees with the extrapolated corrugation amplitudes of the first-principles calculation (Mednick and Kleinman, 1980;

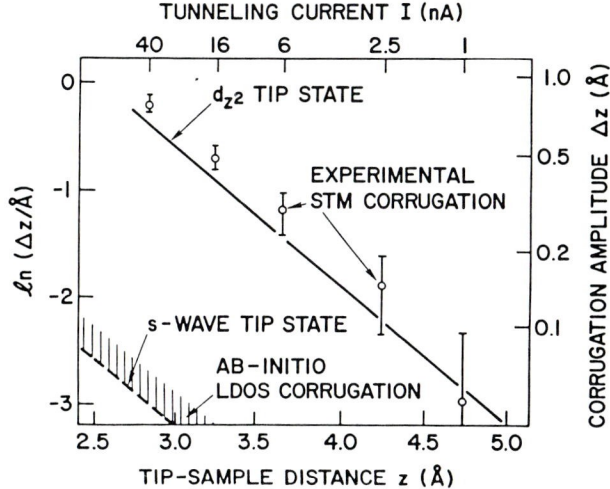

Fig. 6.9. Corrugation amplitudes of a hexagonal close-packed surface. Solid curve, theoretical corrugation amplitude for an s and a d_z tip state, on a close-packed metal surface with a=2.88 Å and ϕ=3.5 eV. The orbitals on each metal atom on the sample is assumed to be 1 s-type. Measured STM corrugation amplitudes are from the data of Wintterlin et al. (1989). The first-principle calculation of Al(111) is taken from Mednick and Kleinman (1980). The corrugation amplitude for a s-wave tip state is more than one order of magnitude smaller then the experimental corrugation. (Reproduced from Chen, 1991, with permission.)

Wang et al., 1981). The theoretical STM image with a d_z tip state agrees well with the "best" observed STM image on Al(111) (Wintterlin et al., 1989). The tip–sample distance is taken from the measured tip–sample distance, 1 Å, from a mechanical contact (Dürig et al., 1986, 1988), and the calculated distance of a mechanical contact of an Al–Al system (Ciraci et al., 1990a), 2.3 Å.

On surfaces of some d band metals, the d_z states dominated the surface Fermi-level LDOS. Therefore, the corrugation of charge density near the Fermi level is much higher than that of free-electron metals. This fact has been verified by helium-beam diffraction experiments and theoretical calculations (Drakova, Doyen, and Trentini, 1985). If the tip state is also a d state, the corrugation amplitude can be two orders of magnitude greater than the predictions of the s-wave tip theory, Eq. (1.27) (Tersoff and Hamann, 1985). The maximum enhancement factor, when both the surface and the tip have d_z states, can be calculated from the last row of Table 6.2. For Pt(111), the lattice constant is 2.79 Å, and $b = 2.60$ Å^{-1}. The value of the work function is $\phi \approx 4$ eV, and $\kappa \approx 1.02$ Å^{-1}. From Eq. (6.54), $\gamma \approx 3.31$ Å^{-1}. The enhancement factor is

$$E = \left(\frac{3}{2} \frac{\gamma^2}{4\kappa^2} - \frac{1}{2} \right)^4 \approx 140. \qquad (6.55)$$

Experimental observation of atomic-resolution STM images of Pt(111) was reported by Zeppenfeld et al. (1992).

CHAPTER 7

ATOMIC FORCES AND TUNNELING

7.1 Effect of atomic force in STM

As we have discussed in Chapter 1, the importance of forces in vacuum tunneling was first realized by Teague (1978), a few years before the invention of the STM. The attractive and repulsive forces between the two gold spheres cause minute deformations near the tunneling gap, which is enough to produce a substantial effect. Coombs and Pethica (1985) reported that the apparent barrier height in STM could be much lower than the work function. Such an anomalously low apparent barrier height is due to a mechanical contact between the tip and the sample, which makes the z reading from the piezo voltage different from the actual gap displacement. Those original ideas have played an important role in the development of the concept of the STM imaging mechanism. Soler et al. (1986) explained the observed giant corrugation on graphite and other layered materials in terms of the repulsive force and the deformation of the sample surface. The existence and importance of the force and deformation in the STM imaging mechanism was further verified by Mamin et al. (1986). The observation of the role of force in STM also inspired the invention of the atomic force microscope (AFM).

The AFM was invented by Binnig et al. (1986) using an instrumentation concept very similar to the STM. By measuring the displacement of weak springs to an accuracy of $10^{-4}\,\text{Å}$, the AFM is sensitive enough to measure the force of individual chemical bonds, that is, 10^{-10} N. With a combined AFM–STM experiment on metals, Dürig et al. (1988, 1990) have shown that the observed attractive force obeys an exponential law with respect to tip–sample separation over a range of 3 Å. The range of the exponential attractive force overlaps that of normal STM operation. The exponential dependence indicates that the nature of the force cannot be van der Waals. The magnitude of the measured force is much larger than the theoretical value of the van der Waals force. In summary, experimental evidence shows that in the range of normal operation of STM, atomic forces always accompany tunneling. Therefore, an understanding of the relation between atomic force and tunneling conductance is important for the understanding of STM as well as AFM.

The problem of the relation between atomic forces and tunneling phenomena is closely connected to the theory of the chemical bond of Pauling, which is based on the concept of resonance (Heisenberg, 1926; Pauling, 1977; Pauling and Wilson, 1935; Feynman, 1967). The most illustrative application of this idea, as described by Pauling (1960), is the problem of the hydrogen molecular ion. Suppose at time $t = 0$, the electron is in the vicinity of proton A. If proton B is sitting nearby, then the electron has a finite probability to migrate to the vicinity of proton B, and from B back to A. The resonance of the electron back and forth between the two protons, with a frequency v, gives rise to a resonance energy hv, which forms the chemical bond. Pauling's theory was criticized because the starting states, such as the state in the hydrogen molecular ion, with an electron staying in the vicinity of one proton only, is not an eigenstate of the system, the choice of which is arbitrary (see Pauling, 1977). With the advancements in STM and AFM, the arbitrariness of the resonance theory can be removed, and the independent measurement of the energy lowering (through force) and the resonance frequency (through tunneling conductance) becomes practical. As we have discussed in Chapter 2, an STM can be considered as a giant molecule consisting of two component molecules with a controllable intermolecular distance. The binding energy as a function of distance, and the tunneling conductance as a function of distance, can be measured independently. The resonance theory of the chemical bond simply means that there is a correlation between the two measurable physical quantities.

We have shown in Chapter 2 that the tunneling conductivity can be evaluated using Bardeen's surface integral (from properly modified wavefunctions). In this chapter, using a time-dependent Schrödinger equation, we show that for two atomic systems weakly coupled, Bardeen's tunneling matrix element and Heisenberg's resonance energy are equal. By applying this fundamental equality to the problem of the hydrogen molecular ion, we show that in the entire range of attractive force, the coupling energy can be accurately represented as a sum of the van der Waals energy and the resonance energy. By evaluating the resonance energy using Bardeen's expression for the tunneling matrix element, a simple and accurate analytic expression for the potential curve of the hydrogen molecular ion is derived. The accuracy of this formula is a proof of the accuracy of the perturbation theory we presented in Chapter 2.

This fundamental relation can be extended to the many-body case, and a correlation between the interatomic force in the attractive-force regime and the tunneling conductance can be established. For metals, an explicit equation between two sets of *measurable quantities* is derived. Of course, the simple relation between the measured force and measured tunneling conductance is not valid throughout the entire distance range. First, the total force has three components, namely, the van der Waals force, the resonance force, and the repulsive force. Second, the actual measurement of the force in STM and

AFM is further complicated by the deformation of the tip and sample near the gap region (Teague, 1978; Coombs and Pethica, 1985; Chen and Hamers, 1991a). At very short tip–sample distances, the reading from the z piezo no more represents the gap displacement. Nevertheless, in the normal working distance range of STM, the dominant force is the resonance force, and the simple relation is valid.

7.2 Attractive atomic force as a tunneling phenomenon

In this section, we present the perturbation theory for the resonance energy, and consequently the attractive atomic force, exemplified by the hydrogen molecular ion. The hydrogen molecular ion is the simplest system containing more than one atom. It is one of a handful of problems in quantum mechanics for which exact analytic solutions exist. Therefore, it is the cornerstone for the understanding of interatomic forces and even all of condensed-matter physics (Slater, 1963). Throughout the 20th century, the hydrogen molecular ion problem has attracted the attention of many prominent physicists and chemists. Using the old quantum theory established by Bohr and Sommerfeld, Wolfgang Pauli (1922) did his thesis on the hydrogen molecule ion problem. He concluded that within the framework of the Bohr–Sommerfeld quantum theory, no stable states can be explained. Yet, a stable H_2^+ with a binding energy of 2.7 eV was observed experimentally. After the discovery of quantum mechanics in 1925–1926, the H_2^+ problem became one of the first test grounds for the validity of the nascent quantum mechanics (the other one was the helium atom). Burrou (1927) showed that the Schrödinger equation for the H_2^+ problem is separable. Using numerical integration of the ordinary differential equation, he obtained a stable ground state, with a bonding energy consistent with experimental observations. Eduard Teller (1930) did his thesis on the systematic analytic solution of the H_2^+ problem (not the hydrogen bomb!), and worked out the potential curves of a series of excited states. This problem is still of considerable current interest, as evidenced by continuing publications up to recent years (Bates et al., 1953; Holstein, 1955; Herring, 1962; Damburg and Propin, 1968; Landau and Lifshitz, 1977; Cizek et al., 1986; Tang, Toennies, and Yiu, 1991; Kais, Morgan, and Herschbach, 1991).

We will analyze the hydrogen molecular ion problem from a completely different perspective: the intimate relation between attractive atomic force and electron tunneling. The goal is to achieve a deeper understanding of STM, and to establish a relation between STM and AFM. Using time-dependent perturbation theory, we show that the resonance energy of Heisenberg and Pauling is exactly equal to the modified Bardeen integral we introduced in Chapter 2. We then establish a quantitative relation between attractive atomic force and tunneling, as well as a unified view of STM and AFM.

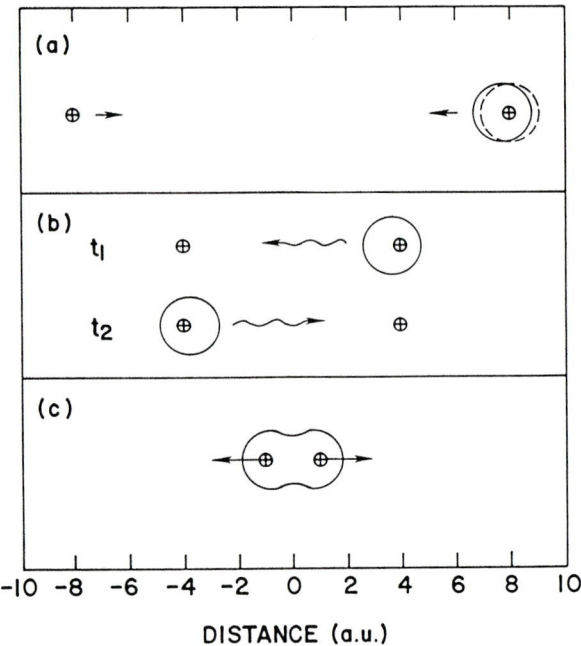

Fig. 7.1. Three regimes of interaction in the hydrogen molecular ion. (a) At large distances, $R>16$ a.u., the system can be considered as a neutral hydrogen atom plus a proton. The polarization of the hydrogen atom due to the field of the proton generates a van der Waals force. (b) At intermediate distances, $16>R>4$ a.u. the electron can tunnel to the vicinity of another proton, and vice versa. A resonance force is generated, which is either attractive or repulsive. (c) At short distances, $R<4$ a.u., proton–proton repulsion becomes important. (Reproduced from Chen, 1991c, with permission.)

For convenience, in this section, atomic units are used, where three constants are set to unity: $e = \hbar = m = 1$. The atomic unit of length is the bohr (0.529 Å), and the atomic unit of energy is the hartree (27.21 eV).

7.2.1 Three regimes of interaction

Figure 7.1 shows three regimes of interaction in a hydrogen molecular ion. At large distances, the system can be considered as a neutral hydrogen atom plus a proton. The electrical field of the proton polarizes the hydrogen atom. As a result, a van der Waals force is induced. The van der Waals force dominates as the proton–proton distance $R>8$ Å. At a distance $R<8$ Å, the $1s$ electron in the vicinity of the right proton has an appreciable probability of tunneling into the $1s$ state of the left proton, and vice versa. This tunneling phenomenon gives rise to a *resonance* and results in a lowering of the total energy (Pauling, 1977). The resonance gives rise to a bonding state, with a lower total energy (attractive force), and an antibonding state, with a higher total energy (repulsive force). At even shorter distances (namely, $R < 2.5$ Å), the repulsive force between the protons becomes important, and the net force becomes repulsive regardless of the type of the state.

In the following, we discuss the three regimes individually.

7.2.2 van der Waals force

As shown in Fig. 7.1, at large distances, the system can be considered as a neutral hydrogen atom plus an isolated proton. The field of the proton polarizes the hydrogen atom to induce a dipole. The interaction between the proton and the induced dipole generates a van der Waals force. The van der Waals force can be treated as a classical phenomenon by introducing a phenomenological polarizability α:

$$\mathbf{p} = \alpha \mathbf{E}, \tag{7.1}$$

where $\mathbf{p}$ is the induced dipole of the neutral hydrogen atom, and $\mathbf{E}$ is the electrical field of another proton. The coupling energy E between the proton and the neutral hydrogen atom is:

$$E = -\frac{\alpha}{2} R^{-4}. \tag{7.2}$$

The polarizability of the hydrogen atom can be calculated accurately using perturbation theory (Pauling and Wilson, 1935). The result is

$$\alpha = \frac{9}{2}. \tag{7.3}$$

The mechanism can be described as follows: A p_z component is induced by the external electrical field, which in turn generates a shift of the center of negative charge from the position of the proton. This p_z component is of the same nature as the tip induced local states (TILS) in the theory of STM introduced by Ciraci et al. (1990). From Eq. (7.2) and Eq. (7.3), the energy lowering due to van der Waals force is:

$$E = -\frac{9}{4} R^{-4}. \tag{7.4}$$

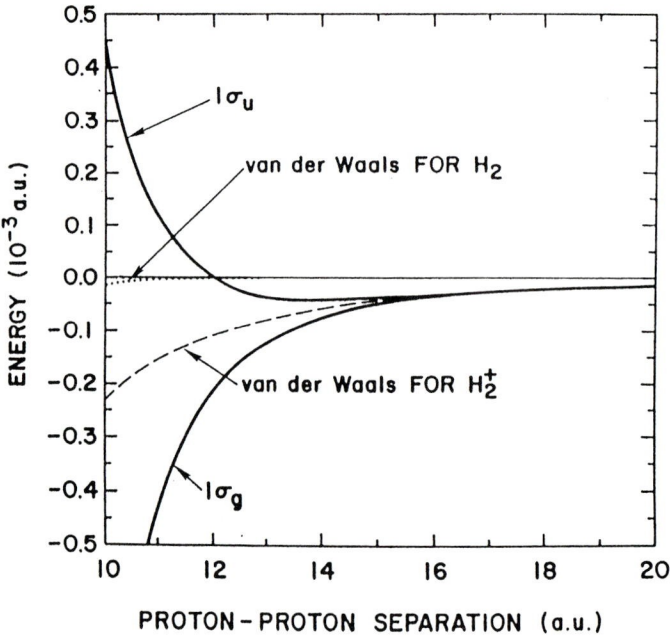

PROTON – PROTON SEPARATION (a.u.)

Fig. 7.2. Potential curve for the hydrogen molecular ion at large distances. At R >16 a.u. (8 Å), the van der Waals force dominates. For R <16 a.u., the resonance energy becomes important. The van der Waals energy for a pair of neutral hydrogen atoms is also shown for comparison. (Reproduced from Chen, 1991c, with permission.)

The van der Waals force occurring in STM and AFM is much smaller than the van der Waals force between a neutral hydrogen atom and a proton. Actually, in STM and AFM experiments on conducting materials, the atoms near the gap are nearly neutral, which is similar to the situation of a pair of neutral hydrogen atoms rather than a proton with a neutral hydrogen atom. The van der Waals force between a pair of neutral hydrogen atoms is also a well-studied problem. The exact result at large distances is (Landau and Lifshitz, 1977):

$$E = -\frac{6.50}{R^6}. \tag{7.5}$$

For values of R of interest, the van der Waals force between neutral atoms is about one order of magnitude smaller than the case of a neutral atom and an ion. A comparison of these two cases is shown in Fig. 7.2. As shown, in the range of interest, both are much smaller than the typical bonding energy of diatomic molecules — of the order of one electron volt.

7.2.3 Resonance energy as tunneling matrix element

As shown in Fig. 7.2, at a shorter proton–proton separation ($R<16$ a.u. or <8 Å), the electron in the $1s$ state in the vicinity of one proton has an appreciable probability of tunneling to the $1s$ state in the vicinity of another proton. The tunneling matrix element can be evaluated using the perturbation theory we presented in Chapter 2. A schematic of this problem is shown in Fig. 7.3. By defining a pair of one-center potentials, U_L and U_R, we define the right-hand-side states and the left-hand-side states. Because the potential U_L is different from the potential of a free proton, U_{L0}, the wavefunction ψ_L and the energy level E_0 are different from the $1s$ state of a free hydrogen atom. (The same is true for U_R and ψ_R.) We will come back to the effect of such a distortion later in this section.

In the following, we present a treatment of the hydrogen molecular ion problem using a time-dependent Schrödinger equation:

$$i\,\frac{\partial\Psi(\mathbf{r},\,t)}{\partial t} = \left(-\frac{1}{2}\,\nabla^2 + U\right)\Psi(\mathbf{r},\,t), \tag{7.6}$$

where U is the potential curve for the hydrogen molecular ion, as shown in Fig. 7.3. Similarly, for the left-hand-side and right-hand-side problems, we also look for solutions of corresponding time-dependent Schrödinger equations:

$$i\,\frac{\partial\Psi_L(\mathbf{r},\,t)}{\partial t} = \left(-\frac{1}{2}\,\nabla^2 + U_L\right)\Psi_L(\mathbf{r},\,t), \tag{7.7}$$

$$i\,\frac{\partial\Psi_R(\mathbf{r},\,t)}{\partial t} = \left(-\frac{1}{2}\,\nabla^2 + U_R\right)\Psi_R(\mathbf{r},\,t). \tag{7.8}$$

We denote the ground-state solutions of Equations (7.7) and (7.8) as:

$$\Psi_L(\mathbf{r},\,t) = \psi_L(\mathbf{r})\,e^{-iE_0 t}, \tag{7.9}$$

$$\Psi_R(\mathbf{r},\,t) = \psi_R(\mathbf{r})\,e^{-iE_0 t}. \tag{7.10}$$

According to the Wigner theorem (see Appendix A), because of time-reversal symmetry, the functions $\psi_L(\mathbf{r})$ and $\psi_R(\mathbf{r})$ can always be made real. Also, since the ground-state wavefunction does not have a node, we can always make the

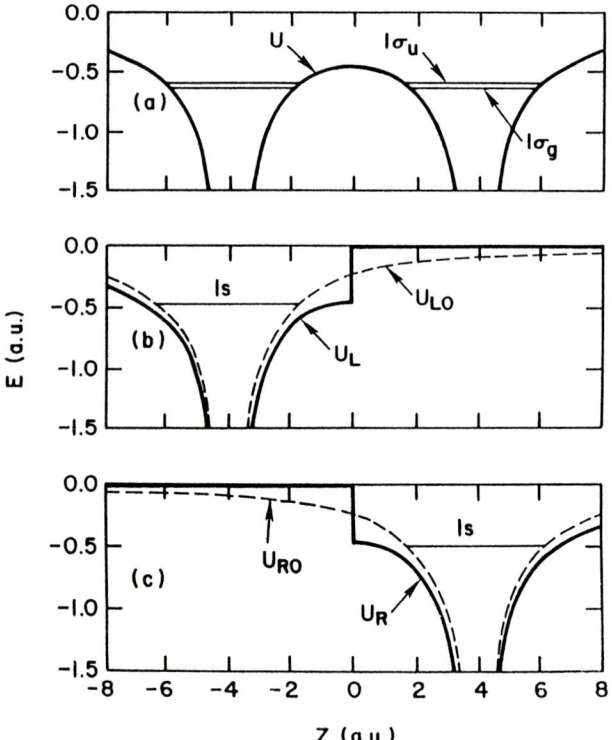

Fig. 7.3. Perturbation treatment of the hydrogen molecular ion. (a) The exact potential curve and the exact energy levels of the problem. (b) Solid curve, the left-hand-side potential for a perturbation treatment; dotted curve, the potential for a free hydrogen atom. (c) Solid curve, the right-hand-side potential for a perturbation treatment; dotted curve, the potential for a free hydrogen atom. (Reproduced from Chen, 1991c, with permission.)

values of wavefunctions everywhere positive. Now, we look for solutions of Eq. (7.6) that are linear combinations of the solutions of Equations (7.7) and (7.8). In other words, we make the following *Ansatz*:

$$\Psi(\mathbf{r}, t) = a_L(t)\psi_L(\mathbf{r}) \, e^{-iE_0 t} + a_R(t)\psi_R(\mathbf{r}) \, e^{-iE_0 t}. \tag{7.11}$$

Substituting Eq. (7.11) into Eq. (7.6), the problem is reduced to the same problem we encountered in Chapter 2. The potentials of the left-hand-side and right-hand-side problems satisfy Equations (2.21) and (2.22). Following the steps from Equations (2.23) through (1.20), we find that the time evolution of the coefficients satisfy the following equations:

$$\dot{a}_L(t) = iMa_R(t), \tag{7.12}$$

$$\dot{a}_R(t) = iMa_L(t), \tag{7.13}$$

where the transition matrix element M is the modified Bardeen integral, Eq. (1.20):

$$M = \frac{1}{2} \int (\psi_R \nabla \psi_L - \psi_L \nabla \psi_R) \cdot d\mathbf{S}, \tag{7.14}$$

which is evaluated on the separation surface, that is, the median plane (see Fig. 7.3).

A specific solution of Eq. (7.6) now depends on the initial condition. If at t=0 the electron is in the left-hand-side state, the solution is:

$$\Psi_1(\mathbf{r}, t) = [\cos Mt \, \psi_L(\mathbf{r}) + i \sin Mt \, \psi_R(\mathbf{r})] \, e^{-iE_0 t}. \tag{7.15}$$

This solution describes a back-and-forth migration of the electron between the two protons. At $t = 0$, the electron is revolving about the left-hand-side proton with a frequency $f = |E_0|/h$. Then the electron starts migrating to the right-hand side. At $t = \pi/|M|$, the electron has migrated entirely to the right-hand side; and at $t = 2\pi/|M|$, the electron comes back to the left-hand side, etc. In other words, the electron migrates back and forth between the two protons with a frequency $v = |M|/h$. Similarly, we have another solution:

$$\Psi_2(\mathbf{r}, t) = [\cos Mt \, \psi_R(\mathbf{r}) + i \sin Mt \, \psi_L(\mathbf{r})] \, e^{-iE_0 t}, \tag{7.16}$$

which starts with a right-hand-side state at $t=0$.

The linear combinations of the solutions, Equations (7.15) and (7.16), are also good solutions of the time-dependent Schrödinger equation, Eq. (7.6). For example, there is a state symmetric with respect to the median plane:

$$\Psi_g(\mathbf{r}, t) = (\Psi_1 + \Psi_2) = [\psi_L(\mathbf{r}) + \psi_R(\mathbf{r})] \, e^{-i(E_0 + M)t}, \tag{7.17}$$

as well as an antisymmetric state:

$$\Psi_u(\mathbf{r}, t) = (\Psi_1 - \Psi_2) = [\psi_L(\mathbf{r}) - \psi_R(\mathbf{r})] \, e^{-i(E_0 - M)t}. \tag{7.18}$$

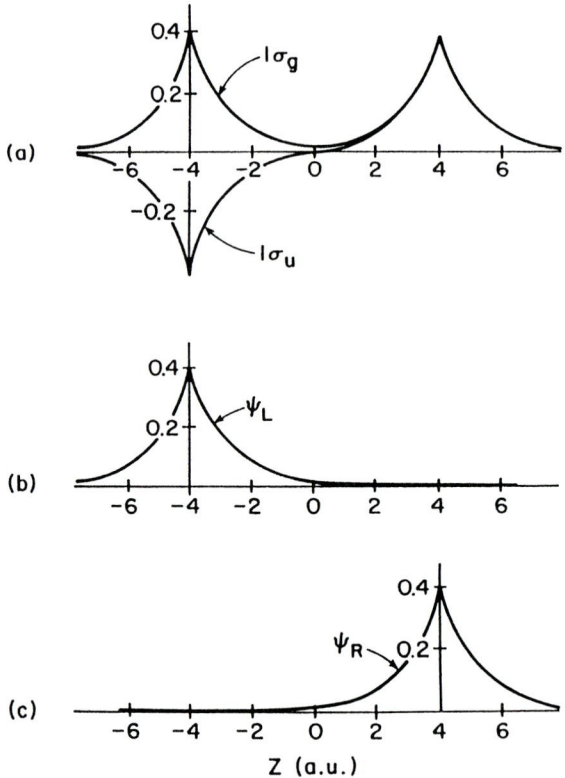

Fig. 7.4. Wavefunctions of the hydrogen molecular ion. (a) The exact wavefunctions of the hydrogen molecular ion. The two lowest states are shown. The two exact solutions can be considered as symmetric and antisymmetric linear combinations of the solutions of the left-hand-side and right-hand-side problems, (b) and (c), defined by potential curves in Fig. 7.3. For brevity, the normalization constant is omitted. (Reproduced from Chen, 1991c, with permission.)

Obviously, these solutions are *stationary states* of Eq. (7.6) with energy eigenvalues $(E_0 + M)$, and $(E_0 - M)$, respectively. Because both E_0 and M are negative, the symmetric state has a lower energy, which means an attractive force.

This discussion is a quantitative formulation of the concept of *resonance* introduced by Heisenberg (1926)   for treating many-body problems in quantum mechanics. Heisenberg illustrated this concept with a classical-mechanics model (see Pauling and Wilson 1935): two similar pendulums connected by a weak spring. Accordingly, the meaning of Eq. (7.15) is as follows: At $t < 0$, the right-hand-side pendulum is held still, and the left-hand-side pendulum is set to oscillate with a frequency $|E_0|/h$. At $t=0$, the right-hand-side pendulum is released. Because of the coupling through the weak spring, the left-hand-side pendulum gradually ceases to oscillate, transferring its momentum to the right-hand-side pendulum, which now begins its oscillation. At $t = \pi/|M|$, the right-hand-side pendulum reaches the maximum amplitude, and the left-hand-side pendulum stops. Then the process reverses. This mechanical system has two normal modes, with the two pendulums oscillating in the opposite directions or in the same direction, with frequencies $|E_0 + M|/h$ and $|E_0 - M|/h$, respectively. These two normal modes correspond to the symmetric and antisymmetric states of the hydrogen molecular ion, respectively, as shown in Fig. 7.4. The two curves in part (a) of Fig. 7.4 are the exact solutions of the two low-energy solutions of the H_2^+ problem. To a good approximation, these solutions can be represented by the symmetric and antisymmetric superpositions of the distorted hydrogen wavefunctions, as shown in parts (c) and (d) of Fig. 7.4. These wavefunctions are defined by the left-hand-side and right-hand-side potentials (Fig. 7.3).

7.2.4 Evaluation of the modified Bardeen integral

In this subsection, we show that by evaluating the modified Bardeen integral, Eq. (7.14), with the distortion of the hydrogen wavefunction from another proton considered, as shown by Holstein (1955), an accurate analytic expression for the exact potential of the hydrogen molecular ion is obtained.

Before we proceed to make an explicit evaluation of the Bardeen integral, we present a brief discussion of the effect of the distortion potentials, for example, $\Delta U = U_L - U_{L0}$. The value of ΔU at the center of the left-hand-side proton results in an increase of the total energy due to the repulsive force between the protons, which is exactly cancelled by the attractive force between the right-hand-side proton and the electron in its undistorted state. The gradient of ΔU in the z direction induces a shift of the center of the electron wavefunction, which is the origin of the van der Waals force, as we have discussed. For relatively large distances, to a good approximation, Eq. (7.4) should still be accurate. The distortion potential also increases the absolute value of the wavefunction on the medium plane, with respect to the wavefunctions of the free hydrogen atoms, which makes the tunneling matrix element, Eq. (7.15), larger than what would be expected from the wavefunction of free hydrogen atom.

The effect of distortion as well as the evaluation of Eq. (7.14) has been discussed by Holstein (1955) regarding the charge-exchange interaction between ions and parent atoms, as shown in Fig. 7.5. The exact time-independent Schrödinger equation for the electron, in atomic units, is

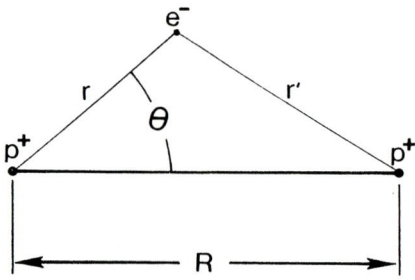

Fig. 7.5. Evaluation of the correction factor.

$$-\frac{1}{2} \nabla^2 \psi - \left(\frac{1}{r} + \frac{1}{r'} \right) \psi = E \, \psi. \tag{7.19}$$

where r' is the distance between the electron and second proton,

$$r' = \sqrt{R^2 + r^2 - 2Rr \cos \theta} \ . \tag{7.20}$$

In the absence of the second proton, Eq. (7.19) is the Schrödinger equation for the free hydrogen atom. The ground-state wavefunction is:

$$\psi_0 = \frac{1}{\sqrt{\pi}} \, e^{-r}, \tag{7.21}$$

and the energy eigenvalue is:

$$E = -\frac{1}{2} \ . \tag{7.22}$$

The presence of the second proton induces a perturbation to the wavefunction and the energy eigenvalue of the electron. For the perturbed wavefunction, we make the *Ansatz*

$$\psi = \psi_0 \, e^{-g}, \tag{7.23}$$

where the function $g(\mathbf{r})$ is to be determined. To a sufficiently accurate degree of approximation, the energy is

$$E = -\frac{1}{2} - \frac{1}{R},$$ (7.24)

which is equivalent to taking the energy of the system equal to that of the iso-lated atom plus a constant potential energy term, $-1/R$. The exact energy has an additional van der Waals term, $(9/4) r^{-4}$, the effect of which is much smaller than the first term. Inserting Equations (7.23) and (7.24) into Eq. (7.19), we obtain the equation for the correction function g:

$$\nabla r \bullet \nabla g - \frac{1}{R} + \frac{1}{r'} = (\nabla g)^2 - \nabla^2 g.$$ (7.25)

To make an approximate solution, we neglect the two terms on the right-hand side of the equation, because it is much smaller than the terms on the left hand side (see the following). Thus, we obtain the approximate equation for the correction function g:

$$\frac{\partial g}{\partial r} = \left(\frac{1}{R} - \frac{1}{r'} \right).$$ (7.26)

At $r = 0$, the correction function g should be independent of θ. Also, the cor-rections should have opposite signs for $+z$ and $-z$. To preserve normalization, it is accurate to set $g = 0$ at $r = 0$. Thus, Eq. (7.26) can be integrated imme-diately:

$$g = \int_0^r \left(\frac{1}{R} - \frac{1}{r'} \right) dr = \left(\frac{r}{R} - \ln \frac{r + r' - R \cos \theta}{R(1 - \cos \theta)} \right).$$ (7.27)

We are interested only in the values of the correction function near the medium plane, $r \approx R/2$. From Fig. 7.5, it is clear that

$$r = r' = \frac{R}{2} \sec \theta.$$ (7.28)

For small θ, we obtain

$$g = \frac{1}{2} - \ln(1 + \sec \theta) \approx \frac{1}{2} - \ln 2.$$ (7.29)

At $r \approx R/2$, the wavefunction gains a factor

$$e^{-g} = 2 e^{-1/2} \approx 1.213.$$ (7.30)

An alternative proof of this factor, which is more subtle, is given by Landau and Lifshitz (1977). On the far side of the molecule, the wavefunction gains a factor

$$e^{-g} = \frac{3}{2} e^{-1/2} \approx 0.9098.$$ (7.31)

Therefore, the net effect can be considered to be a p_z state, induced by proton B, superimposed on the $1s$ wavefunction.

By evaluating the integral in Eq. (7.14) explicitly, we find

$$M = -\frac{2}{e} R e^{-R}.$$ (7.32)

The total coupling energy is the sum of the van der Waals energy, Eq. (7.4), and the resonance energy, Eq. (7.32). For the $1\sigma_g$ state, it is:

$$\Delta E(1\sigma_g) = -\frac{9}{4R^4} - \frac{2}{e} R e^{-R},$$ (7.33)

and for the $1\sigma_u$ state,

$$\Delta E(1\sigma_u) = -\frac{9}{4R^4} + \frac{2}{e} R e^{-R}.$$ (7.34)

These results have been obtained with a time-independent perturbation method, described in Landau and Lifshitz (1977). They are actually the leading terms of the exact asymptotic expansion of the potential curves (Damburg and Propin, 1968; Cizek et al., 1986). Up to the third term, the exact result of the coupling energies for the $1\sigma_g$ and $1\sigma_u$ states is

$$\Delta E_{\pm} = -\frac{9}{4R^4} - \frac{15}{2R^6} - \frac{213}{4R^7}$$
$$\pm \frac{2R}{e} e^{-R} \left(1 + \frac{1}{2R} - \frac{25}{8R^2} \right).$$ (7.35)

For $R \geq 7$ bohr, the difference between the exact solution and the leading terms (Equations (7.33) and (7.34)) is less than 1 meV. Therefore, the interpretation of the resonance energy in terms of tunneling is verified quantitatively in the case of the hydrogen molecular ion. Furthermore, the comparison with the soluble case of the hydrogen molecular ion is also a verification of the accuracy of the perturbation theory presented in Chapter 2.

7.2.5 Repulsive force

As shown in Fig. 7.1, as the proton–proton separation becomes even smaller, the picture of the resonance becomes obscured, and the proton–proton repulsion is no longer screened by the electron. Slater (1963) showed that the Morse function can match the exact potential curve for the hydrogen molecular ion very precisely:

$$f = C\left(e^{-2\kappa(s-s_0)} - e^{-\kappa(s-s_0)}\right), \tag{7.36}$$

where s_0 is the equilibrium point, and C is a constant. As we have discussed in Section 1.5.2, the sum of Morse forces on a periodic surface has a finite analytic form.

7.3 Attractive atomic force and tunneling conductance

The goal of this section is to establish a correlation between the attractive atomic force and the tunneling conductance on metal surfaces, as a result of the equality between Bardeen's tunneling matrix element and Heisenberg's resonance energy. Regarding the total force, the equation derived in this section is valid only in the intermediate range of tip–sample distances where the van der Waals and the repulsive forces are insignificant. To obtain the total force for the entire range, the van der Waals force and the repulsive force must be added on.

To assess the relative importance of the van der Waals and the resonance contributions occurring within the normal operational distances in STM, it is instructive to compare the quantities occurring in real experiments versus those quantities in the exactly soluble H_2^+ problem.

First, the range of separation. The experimentally determined tip–sample separation is 1-4 Å before a mechanical contact (Dürig et al. 1988). For most metals, the normal nucleus–nucleus distance of a mechanical contact is 2.5–3 Å. Therefore, on the absolute tip–sample scale, it is 3.5–7 Å. By comparing with the case of H_2^+, we notice that this is the range where the resonance interaction dominates. In other words, under normal STM operation conditions, over a distance range of about 3 Å, resonance energy is almost solely responsible to the atomic force, and the distance dependence of the force should be approximately exponential.

Second, the sensitivity of AFM. In a typical AFM (Binnig et al., 1986), the force sensitivity is about 0.01 nN. In the range of 4–10 a.u., the resonance force in the hydrogen molecular ion is 4 nN to 0.01 nN. Therefore, the resonance force (attractive atomic force) of a single chemical bond, extended over a distance of 3 Å, can be detected. On the other hand, the van der Waals force of a pair of neutral atoms, when it is distinguishable from the total force, is always smaller than 0.01 nN, which is beyond the detection limit of the existing AFM. On the other hand, the *repulsive force*, when it is separable from the resonance force, can be as large as some tens of nN, which is always within the detection limits of AFM.

van der Waals force

The van der Waals force between two solid bodies is a well-studied problem (Lifshitz and Petaevskii, 1984). Because the van der Waals force is a valid description of the interatomic forces between two pieces of metals with a distance much larger than the diameter of a typical atom, a macroscopic approach is often used. The results show that the force between a pair of planes with separation l is proportional to l^{-3}, a result as if the force is additive with each pair of atoms exhibiting a van der Waals force, according to the r^{-6} law. This procedure gives the right order of magnitude. As we have shown for the case of H_2^+, although at larger distances ($R > 16$ a.u. or > 8 Å), the van der Waals force dominates; at distances relevant to STM and AFM experiments, namely, $R \approx 3.5-7$ Å, the van der Waals force is much smaller than the exchange force. However, the polarization does change the amplitude of the electron wavefunction at the separation surface. It results in an increase of the attractive force, similar to the correction factor $g = 2 \exp(-1/2)$ in the case of the hydrogen molecular ion (Holstein, 1955).

Resonance energy between metals

Although the van der Waals force works almost identically for metals and insulators, the exchange force, or resonance force, behaves differently. For metals, the electronic states near the Fermi level are half-filled. At the distances of normal STM operation, these unpaired states are responsible to the exchange force or resonance energy, which makes a net lowering of the total energy of the entire system. In this case, the one-electron picture of the resonance energy is applicable, and the tunneling matrix element provides an appropriate description of the resonance energy. On the other hand, for insulators, the exchange coupling results in bonding states as well as antibonding states across the boundary. The net effect of resonance on the total energy is zero. Therefore, only the van der Waals force and the repulsive force are effective.

To make a quantitative treatment, we define a system including a tip and a sample, as shown in Fig. 7.6. Independent electron approximation is applied. The Schrödinger equation is identical to Eq. (7.6), with the potential surface shown in Fig. 7.6. Similar to the treatment of hydrogen molecular ion, a separation surface is drawn between the tip and the sample. The exact position of the surface is not important. Define two subsystems, the sample S and the tip T, with potential surfaces U_S and U_T, respectively, as shown in Fig. 7.6 (c) and

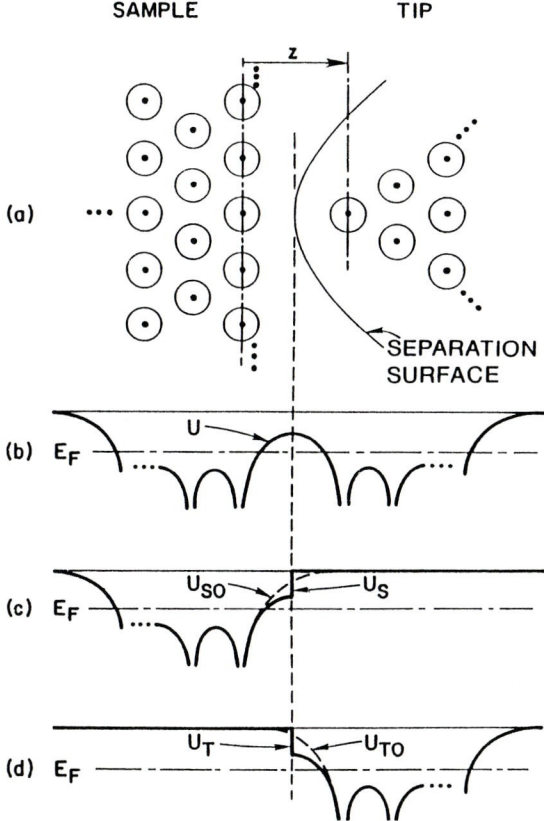

Fig. 7.6. Perturbation theory of the attractive atomic force in STM. (a) The geometry of the system. A separation surface is drawn between the tip and the sample. (b) The potential of the coupled system. (c) The potential surface of the unperturbed Hamiltonian of the sample, U_S, which may be different from the potential surface of the free sample, U_{S0}; (d) The potential surface of the unperturbed Hamiltonian of the tip, U_T, which may be different from the potential surface of the free tip, U_{T0}. The effect of the difference between the "free" tip (sample) potential and the "distorted" tip (sample) potential can be evaluated using the perturbation method; see Chapter 2. (Reproduced from Chen, 1991b, with permission.)

(d). The solution of the corresponding Schrödinger equation defines the stationary states of the sample and the tip, ψ and χ:

$$(-\frac{1}{2}\nabla^2 + U_S)\psi = E\psi, \qquad (7.37)$$

$$(-\frac{1}{2}\nabla^2 + U_T)\chi = E\chi. \tag{7.38}$$

Consider first a single state in the tip and a single state in the sample with the same energy eigenvalue E_0. In the absence of a magnetic field, the Hamiltonian exhibits time-reversal symmetry. Real wavefunctions can be chosen. The derivation of the tunneling matrix element and the resonance energy is almost identical to the case of H_2^+. If both electrodes are metals, near the Fermi level, every state in the sample side should have a state in the tip side that has the same energy eigenvalue. Therefore, resonance always exists even if the two sides are not identical. The resonance results in a pair of combined states, $2^{-1/2}(\psi + \chi)$, and $2^{-1/2}(\psi - \chi)$, as well as a splitting of the energy level to $(E + M)$ and $(E - M)$, where M is the Bardeen integral in terms of the modified wavefunctions defined in Chapter 2:

$$M = \frac{1}{2} \int (\chi\nabla\psi - \psi\nabla\chi) \cdot d\mathbf{S}. \tag{7.39}$$

For the case of the hydrogen molecular ion, the net effect of resonance is a lowering of the total energy. In the case of two pieces of solids, the net effect on the total energy of the coupled system depends on the position of the unperturbed energy level. If the energy level of the unperturbed states is much lower than the Fermi level, all the electronic states are occupied. After the resonance splitting, both bonding and antibonding states would be occupied. Therefore, the net energy change of the entire system is zero. For insulators, it is always the case. As a result, for insulators, the van der Waals force and the repulsive force are the major contributors to the observed atomic force. On the other hand, the electronic states on conductor surfaces consist of unpaired electrons. After resonance splitting, only the lower resonance states are occupied. Therefore, a net lowering of the total energy, that is, a net attractive force, occurs.

A measurable consequence

In the following, we will derive a relation between the measured tunneling conductance and the measured atomic force. The major uncertainty is the exact geometry and chemical nature of the end of the tip. Experimentally, the uncertainty is often observed. As we shall see, the geometrical arrangement of the atoms near the apex of the tip has a large influence on the magnitude of the force. Therefore, the relation we establish is of an order-of-magnitude and functional-dependence nature.

To account for the attractive force in the normal range of STM operation, we consider surface states on the tip near the Fermi level. Because the states above the Fermi level are normally unoccupied, therefore, they do not contribute to the force. For states much lower than the Fermi level, the decay length is much shorter, and the overlap is much weaker. For a single state, the attractive force is

$$F = - \frac{\partial |M|}{\partial z} .$$ (7.40)

For metals, the variation of the tunneling matrix element M and the density of states ρ over the valence band is small in comparison with their absolute values. A simple relation between the force and the tunneling current can be established. Assuming that the width of the valence band ϵ is the same for the tip and the sample, the density of states is $\rho = \epsilon^{-1}$ for both. The tunneling conductance is then

$$G = \frac{(2\pi)^2}{R_K \epsilon^2} |M|^2 ,$$ (7.41)

where $R_K = h/e^2 = 25812.8 \ \Omega$ is von Klitzing's constant. Experiments (Binnig et al., 1984) and theoretical studies (see Chapter 2) have shown that over the entire range of STM operation, the tunneling conductance varies exponentially with tip–sample distance, $G \propto \exp(-2\kappa z)$, where $\kappa = (2m_e\phi)^{1/2}/\hbar$ is the decay constant of the surface wavefunction, and ϕ is the work function of the material. Combining Equations (7.40) and (7.41), we obtain

$$F = - \frac{\kappa \epsilon}{2\pi} \sqrt{GR_K} .$$ (7.42)

To establish a quantitative relation between F and G for the entire tip and the entire sample, we have to consider all the states in the tip and the sample. A rigorous treatment is complicated. The following treatment is based on the approximate additivity of atomic force and tunneling conductance with respect to the atoms of the tip. In other words, the force between the entire tip and the sample can be approximated as the sum of the force between the individual atoms in the tip and the entire sample, so does tunneling conductivity. Because the tip is made of transition metals, for example, W, Pt, and Ir, the tight-binding approximation, and consequently, additivity, are reasonable assumptions. Under this approximation, the total force is

$$F = - \frac{\kappa \epsilon}{2\pi} \sum_n \sqrt{G_n R_K} ,$$ (7.43)

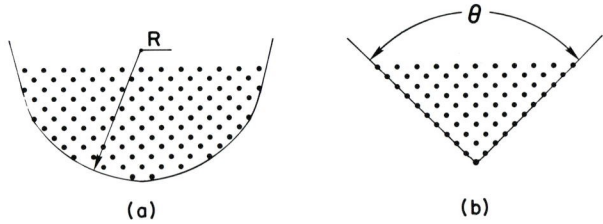

(a)   (b)

Fig. 7.7. The effect of tip geometry. (a) The end of a typical etched metal tip. (b) The end of a typical cleaved or cutout tip.

where the sum is over all the atoms in the tip. If the atomic resolution of the force is not a concern, the sum can be approximated as an integral over the volume of the tip. Denoting the distance from the plane of the topmost nuclei of the sample to the apex nucleus of the tip as z_0. For an etched tip, the shape of the end can be represented as a paraboloid (see Fig. 7.7). The cross section at z is $S(z) = 2\pi R(z - z_0)$, where R is the minimum radius of curvature. Denote the tunneling conductance per unit volume of the tip to the sample as $G_0 \exp(-2\kappa z)$, the total tunneling conductance is then

$$G = G_0 \int_{z_0}^{\infty} e^{-2\kappa z} S(z)\, dz$$

$$= \frac{\pi R G_0}{2\kappa^2} e^{-2\kappa z_0},$$

(7.44)

whereas the total force is

$$F = -\frac{\kappa\epsilon}{2\pi} \sqrt{G_0 R_K} \int_{z_0}^{\infty} e^{-\kappa z} S(z)\, dz$$

$$= -\frac{\kappa\epsilon}{2\pi} \sqrt{G_0 R_K} \frac{2\pi R}{\kappa^2} e^{-\kappa z_0}$$

(7.45)

$$= -\frac{2\kappa\epsilon}{\pi} \sqrt{G R_K} .$$

For tips cut mechanically (for example, with a surgical blade) or cleaved from a brittle material that exhibits a conical end (see Fig. 7.7), $S(z) \propto (z - z_0)^2$. A similar calculation gives

$$F = -\frac{4\kappa\epsilon}{\pi} \sqrt{G R_K} .$$

(7.46)

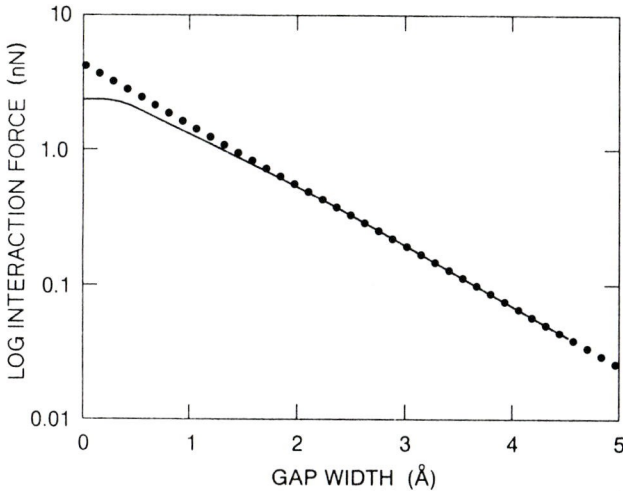

Fig. 7.8. Comparison of the theoretical and measured forces in STM. The solid curve is the measured dependence of the attractive force by Dürig et al. (1988). The dotted curve represents Eq. (7.47). Parameters used for curve fitting: work function ϕ= 4 eV, width of valance band ϵ=5 eV, and $f\approx1$. The origin of the abscissa corresponds to $G = 10^{-5}\ \Omega^{-1}$. At very short tip–sample distances, the repulsive force occurs, which reduces the net attractive force.

A *shape factor f* , which is defined as $f = 2/\pi \approx 0.637$ for tips with a paraboloidal end and $f = 4/\pi \approx 1.27$ for tips with a conical end, Eq. (7.45) and Eq. (7.46) can be combined to give:

$$F = -f\kappa\epsilon\sqrt{GR_K}$$
$$\approx -0.257f\kappa\epsilon\sqrt{G}\ \text{nN,} \qquad (7.47)$$

in commonly used units: length in Å, energy ϵ in eV, conductance G in $10^{-6}\ \Omega^{-1}$, and force in nN. The actual shape of the tip end might be in between. Thus, in general, the shape factor should be a dimensionless number of order 1. Taking typical numbers, $\kappa^{-1}\approx1$ Å^{-1}, $\epsilon\approx$ 4 eV, at $G\approx10^{-6}\ \Omega^{-1}$, the attractive force should be about 1 nN. As the tunneling conductance varies about one order of magnitude per Å, the force should vary about one order of magnitude per 2 Å.

A comparison with experiments is shown in Fig. 7.8. The solid curve is the measured attractive force, by Dürig et al. (1988). The dotted curve represents Eq. (7.47), with parameters as follows: work function ϕ=4 eV, width of conduction band ϵ=5 eV.

Repulsive force

When the tip is brought even closer to the sample, the repulsive force appears as a result of core–core interaction. A reduction of the net attractive is observed, as shown by the experimental data in Fig. 7.8. To describe the attractive force as well as repulsive force, the Morse function is a convenient choice (Slater 1963). In terms of interactions energy, and choose the equilibrium point as the zero point for z, it has the form

$$E = U_0 \left(e^{-2\kappa z} - 2e^{-\kappa z}\right), \tag{7.48}$$

where U_0 is the binding energy and κ is a constant. As an example of its application, Soler et al. (1986) used the Morse function to explain the large corrugations observed on graphite surfaces. One of the advantages of using the Morse function to represent interatomic forces is that for periodic surfaces, the sum of Morse force has a simple analytic form. We will explain this point in Chapter 8. Also, we will show in Chapter 8 that by assuming a Morse force and a deformable tip, the measured data of the variation of the apparent barrier height versus tip–sample distance can be reproduced with reasonable accuracy (Chen and Hamers, 1991a).

The analytic form for the force curves and potential curves is not unique. In describing the potential curves of diatomic molecules, many analytic formulas have been proposed. For a review, see Vashni (1957). Although the Morse function is one of the most widely used, there are many other possible approximate analytic forms. Dürig et al. (1990) has successfully fitted the observed interaction energy as a function of the z piezo displacement in the entire range (including the repulsive-force regime) with a Rydberg function (Vashni, 1957; Rose, Ferrante, and Smith, 1981). In terms of interaction energy E, the Rydberg function has the form

$$E = -U_0(1 + \kappa' z)e^{-\kappa' z}, \tag{7.49}$$

where κ' is another constant. Actually, by letting $\kappa' = 1.35\kappa$, in the entire range of interest ($\kappa' z = -0.3 \sim +\infty$), the difference between the Morse function and the Rydberg function is less than 1% of U_0 (notice that the values of κ and κ' are different). Therefore, whenever a set of experimental data can be fitted with a Rydberg function, it can be fitted equally well with a Morse function, by using a different constant in the exponent. In addition, in the region of strong attractive force and repulsive force, the deformation of the gap plays an important role in the interpretation of experimental data (Teague, 1987;

Coombs and Pethica, 1985; Chen and Hamers, 1991a). The error caused by force and deformation is much larger than the difference between the Morse function and the Rydberg function.

CHAPTER 8

TIP–SAMPLE INTERACTIONS

8.1 Local modification of sample wavefunctions

First-principles calculations of an STM, including a real tip and a real sample, clearly show that within the normal tip–sample distances (3.–6 Å from nucleus to nucleus), in the gap region, the local electronic density resembles neither that of the tip nor that of the sample. Substantial local modifications are induced by the strong interaction. An example is the system of an Al sample with an Al tip, calculated by Ciraci, Baratoff, and Batra (1990a), as shown in Fig. 8.1. As the tip–sample distance is reduced to 8 bohr, the electron density begins to show a substantial concentration in the middle of the gap. This phenomenon becomes much more pronounced when the tip–sample distance is reduced to about 7 bohr. These distances are exactly the normal distances where atom-resolved images are obtained.

The local electronic modification can be understood from the elementary example of the hydrogen molecular ion problem. As we have discussed in Chapter 7, when a hydrogen atom in its ground state and a hydrogen nucleus approach each other, the electronic state of the neutral hydrogen is modified by the other hydrogen nucleus. There are two ways to look at it: The first view is that a p_z state is induced along the nucleus–nucleus line, which is superimposed on the s state. The existence of the p_z state increases the tunneling probability. Another view is that the magnitude of the wavefunctions in the middle of the gap is increased by a factor $g > 1$, as defined in Eq. (7.23). With this point of view, the enhancement of the tunneling matrix element can be evaluated. It gains a factor $4/e \approx 1.472$, independent of the internucleus distance.

The modification of an s-wave sample state due to the existence of the tip is similar to the case of the hydrogen molecule ion. For nearly free-electron metals, the surface electron density can be considered as the superposition of the s-wave electron densities of individual atoms. In the presence of an exotic atom, the tip, the electron density of each atom is multiplied by a numerical constant, $4/e \approx 1.472$. Therefore, the total density of the valence electron of the metal surface in the gap is multiplied by the same constant, 1.472. Consequently, the corrugation amplitude remains unchanged.

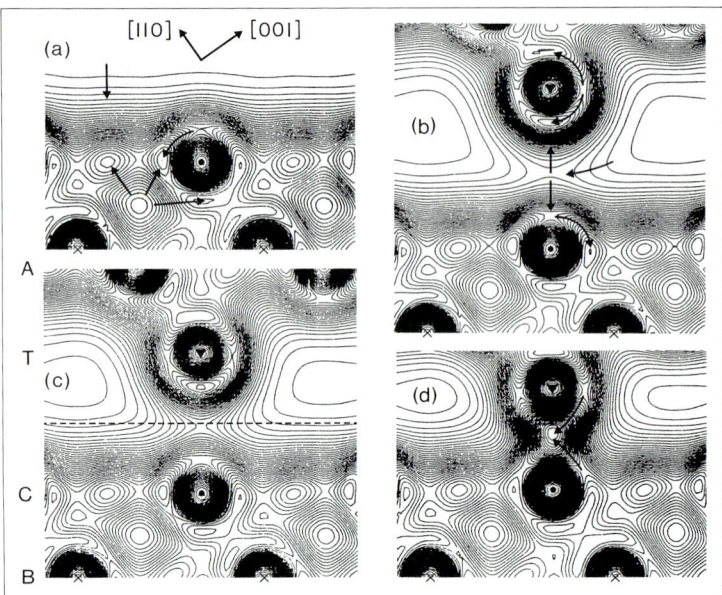

Fig. 8.1. Local modification of the electronic structure in the gap. Charge-density contours of a system with an Al tip and an Al sample. (a) Free Al(111) surface. (b) At a tip–sample distance 8 bohr (4.2 Å); (c) 7 bohr (3.6 Å); and (d) 5 bohr (2.6 Å). (Reproduced from Ciraci et al., 1990a, with permission.)

8.1.1 The point of view of momentum space

The effect of the tip-induced electronic modifications on the corrugation amplitude of STM images can be understood from a k-space point of view. The effective range of a local modification of the sample states caused by an s-wave tip state has a radius proportional to the tip–sample distance z, and has the same order of magnitude. Therefore, it is equivalent to induce a localized state of radius z superimposed on the original state of the solid surface. The magnitude of that state is about $(4/e - 1) \approx 0.472$ times the local magnitude of the electron density roughly in the middle of the gap. This superimposed local state cannot be formed by Bloch states with the same energy eigenvalue. Because of dispersion (that is, the finite value of $\partial E/\partial k_x$ and $\partial E/\partial k_y$) that tip-induced local state (TILS) is a superposition of Bloch states with a range of energy values. The following is an estimation based on the uncertainty principle:

For metals that can be described approximately as free-electron metals, the dispersion relation is

$$E = E_0 + \frac{\hbar^2 k^2}{2m^*} , \tag{8.1}$$

where m^* is the effective mass and E_0 is the bottom of the valance band. The Fermi wave vector is

$$k_F = \frac{\sqrt{2m^*(E_F - E_0)}}{\hbar} . \tag{8.2}$$

It follows from the uncertainty relation

$$\Delta x \, \Delta k \geq \frac{1}{2} \tag{8.3}$$

that

$$\Delta E \geq \frac{\hbar^2 k_F}{2m^* z} = \frac{E_F - E_0}{z k_F} . \tag{8.4}$$

Again, for nearly free-electron metals, the inclusion of off-Fermi-level Bloch states does not change the STM image appreciably, because the Fermi-level LDOS is almost proportional to the total charge density. The corrugation remains the same. However, for surfaces whose band structure has a singularity at the Fermi level, the tip-induced local states would have a dramatic effect. A typical example, graphite, is treated using the first-principles method by Batra and Ciraci (1988).

8.1.2 Uncertainty principle and scanning tunneling spectroscopy

The local modification of sample wavefunctions due to the proximity of the tip, and consequently the involvement of the Bloch functions outside the "energy window" $E_F \pm eV$ in the tunneling process, has an effect on the limit of the energy resolution of scanning tunneling spectroscopy. This effect is discussed in detail by Ivanchenko and Riseborough (1991). First, if the tunneling current is determined by the bare wavefunctions of the sample and the tip, the process is linear, and there is no effect of quantum uncertainty. The effect of quantum uncertainty is due to the modification or distortion of the sample wavefunction due to the existence of the tip. Here, we present a simple treatment of this problem in terms of the MBA.

From Eq. (8.4), the energy uncertainty due to the proximity of tip and sample is

$$\Delta E \geq \frac{E_F - E_0}{z k_F}, \tag{8.5}$$

with z the tip–sample distance. For s states, z is the apparent radius of the image of an individual atom. For other $m = 0$ states, the lateral range of tip–sample interactions and the apparent radius of the image are narrowed down in proportion. Therefore, it is more convenient to relate the energy resolution to the apparent radius of the image of an individual atomiclike state, r,

$$\Delta E \geq \frac{E_F - E_0}{r k_F}. \tag{8.6}$$

For example, for $(E_F - E_0) = 4$ eV and $\kappa_F = 1\,\mathring{A}^{-1}$, when atomic resolution is achieved, that is, $r \leq 2\,\mathring{A}$, the energy resolution is 0.5 eV. Even for the reconstructed surfaces, such as Si(111) 7×7, when the adatoms are barely resolved, that is, $r \approx 4\,\mathring{A}$, the energy resolution is 0.25 eV, a magnitude much larger than the broadening due to the thermal effect at room temperature, $4k_B T \approx 0.1$ eV. Therefore, when the fine details in the tunneling spectra are interpreted, care must be taken not to make any claims which would violate the uncertainty principle.

8.2 Deformation of tip and sample surface

As we have shown in Chapter 1, under normal STM operating conditions, there is a sizeable force between the tip and the sample. Because of the extremely small dimension of the tip, the force will generate a measurable deformation. The deformation, in turn, produces measurable consequences in the tunneling current, sometimes very dramatic. The first case of the dramatic effect of atomic forces was identified by Teague (1978), a few years before the invention of STM (see Section 1.8.2). After the invention of STM, the first identification of the effect of force in tunneling was by Coombs and Pethica (1986), who attributed the observed low barrier height in STM to the repulsive force between the tip and the sample surface during contact. Soler et al. (1986) interpreted the giant corrugation observed on graphite as due to the amplification of the corrugation by the deformation of the sample surface. The theory of Soler et al. (1986) was confirmed and extended by Mamin et al. (1986) and Mate et al. (1989) to the cases with a contamination layer between the tip and the sample. Chen and Hamers (1991) show that the measured distance dependence of apparent barrier height is affected by force and deformation under well-defined experimental conditions.

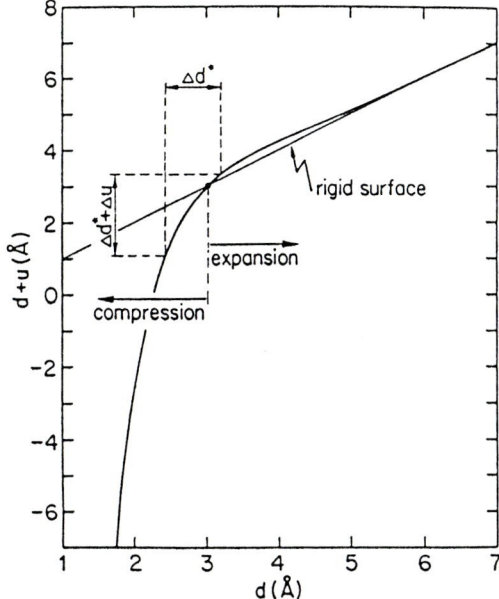

Fig. 8.2. Amplification of corrugation amplitude by deformation. Due to the softness of the graphite surface, the observed mechanical displacement of the z piezo, as the sum of the displacement of the gap width and the displacement of the sample surface, can be orders of magnitude greater than the displacement of the gap width. (Reproduced from Soler et al., 1986, with permission.)

8.2.1 Mechanical amplification of corrugation amplitudes

Binnig et al. (1986) observed that the corrugation amplitude on graphite with STM can be as large as a few Å, whereas the lateral dimensions of the corrugations are as expected from the crystallographic data. The observed large corrugation was interpreted by Soler et al. (1986) as originating from the deformation of the graphite surface, which amplifies the electronically based corrugations we have discussed in the previous section.

The interpretation of Soler et al. (1986) was based on the experimental fact that the corrugation amplitudes depends dramatically on tunneling conductance (Binnig et al., 1986; Soler et al., 1986; Mamin et al., 1986), as shown in Fig. 8.2. Qualitatively, their interpretation is as follows: During the experiment, the tip actually in mechanical contact with graphite surface. In other words, there is a repulsive force between the tip and the sample, which generates a compression of the graphite surface. By moving the z piezo, the

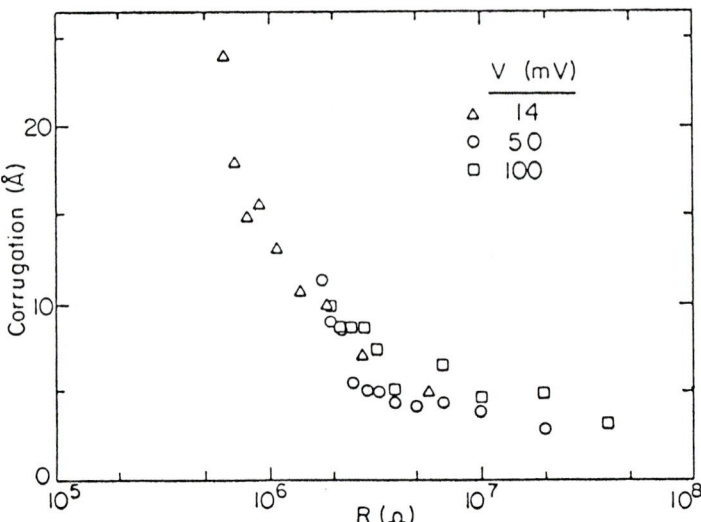

Fig. 8.3. Dependence of corrugation on tunneling conductance. A systematic study shows that the corrugation amplitude is determined by tunneling *conductance* but has no direct connection with the *bias*. It indicates that the tunneling characteristics of graphite are metallike. (Reproduced from Mamin et al., 1986, with permission.)

tip moves relative to the *back side* of the sample. Because graphite is very soft, the graphite surface in the vicinity of the tip deforms. The change of the true gap width, d, is much smaller than the observed z displacement of the piezo, Δz, which is the sum of the gap displacement, Δd, and the deformation of the graphite surface, Δu, as shown in Fig. 8.2. Because of the softness of graphite, the corrugation *observed from the displacement of the z piezo* is much larger than the true corrugation, that is, the displacement of the true gap width. In the attractive-force regime, the situation could be the opposite. As shown in Fig. 8.2, the attractive force generates an expansion of the graphite surface. The retraction of the tip results in a reduction of the attractive force. The expansion of the graphite surface is reduced. The displacement of the true gap is then greater than the observed z-piezo displacement. In other words, in the attractive-force regime, the observed corrugation amplitude is reduced by the elastic deformation. Quantitatively, Soler et al. (1986) expressed the force between the tip (a carbon cluster) and the graphite surface with a Morse curve,

$$F(z) = 2\kappa U_0 \, (e^{-2\kappa(z-z_0)} - e^{-\kappa(z-z_0)}), \tag{8.7}$$

and the constants are obtained from first-principle calculations of the forces between graphite planes.

The theory of Soler et al. (1986) was confirmed and extended by Mamin et al. (1986). While explaining the large corrugations in terms of an amplification of the tip motion arising from surface deformation, they proposed that while graphite is imaged in air, there is a contamination layer between the tip and the sample. The force, which is the origin of the corrugation amplification, is mediated by that contamination layer. Tunneling proceeds through a miniature tip protruding through the contamination. Also, a systematic study of the dependence of apparent corrugation amplitude on tunneling resistance is conducted. A sharp dependence of the corrugation amplitude on tunneling conductance is found, as shown in Fig. 8.3. The existence and value of the force between the tip and the graphite surface during tunneling experiments were directly measured by Mate et al. (1989). The force can be as large as a few hundreds of nanonewtons.

The same phenomenon is observed on other layered materials, for which the deformation perpendicular to the cleavage surface is relatively easy. When the large corrugation amplitudes are observed, the apparent barrier height becomes very low, indicating a nearly synchronous motion of the sample surface with the tip.

8.2.2 The Pethica mechanism

Pethica (1986) proposed a mechanism by which a periodic corrugation in the STM image can be obtained, without a true atomic resolution. This mechanism is most clearly illustrated by imaging graphite with the STM. Because graphite is composed of thin sheets, when the tip touches its surface, a flake of graphite may be transferred to the tip. The graphite flake on the tip has the same atomic periodicity as the sample surface. During imaging, the tip and the graphite surface are in mechanical contact, as evidenced by the very low apparent barrier height (Coombs and Pethica, 1986). By moving the tip (with the graphite flake) laterally on the sample surface, a periodic variance of conductivity occurs. This periodicity equals exactly the periodicity of the sample surface. An image with atomic-scale corrugation is observed, without any true atomic resolution. The signature of such a process is the lack of point defects altogether on very large area. Such a phenomenon is observed by Colton et al. (1988). Even by using a pencil lead as the STM tip, STM images with atom-scale periodic corrugation are observed.

The Pethica mechanism is important in imaging diagnosis. If an abnormally low apparent barrier height and a completely defect-free periodic STM image are observed, the possibility of a Pethica mechanism must be considered seriously.

8.2.3 Corrugation of attractive force and its effect

The effect of the repulsive atomic force in the STM imaging process is well-established experimentally and theoretically. It is natural to inquire about the effect of the corrugation of attractive forces in the STM imaging process (Wintterlin et al., 1989). The problem was studied using a first-principles numerical method by Ciraci, Baratoff, and Batra (1990a). The major conclusions are as follows:

1. At the operating tip–sample distances of STM, the attractive force prevails.

2. The magnitude of the attractive force at the hollow site (H) is larger than that at the top site (T). In other words, there is a reversed corrugation for the attractive force.

3. The deformation of the tip would reduce the observed corrugation of the constant-current topographic image.

4. For most metals, the effect of deformation on the corrugation in the topographic image is much smaller than the observed values.

The conclusions of Ciraci et al. (1990a) can be understood with a simple independent-atom model, that is, by considering the tip as a single atom, and the total force is the sum of the forces on every atom on the sample surface. Assuming that the force between individual pairs of atoms can be represented as a Morse curve, the z component of the force to the nth atom at a point in space, $\mathbf{r}$, is:

$$f_z = 2\kappa U_0\!\left(e^{-\kappa(r-r_0)} - e^{-2\kappa(r-r_0)}\right)\cos\theta, \tag{8.8}$$

where $r = |\mathbf{r} - \mathbf{r}_n|$ is the distance from the point in space to the center of the nth atom on the sample surface, and θ is the angle between the norm of the surface and the line $\overline{\mathbf{r}\,\mathbf{r}_0}$.

For a crystalline surface which has one atom at each lattice point with primitive vectors $\mathbf{a}_1$ and $\mathbf{a}_2$, the total force at point $\mathbf{r}$, $F(\mathbf{r})$, is

$$
\begin{aligned}
F(\mathbf{r}) &= \sum_{n,m=-\infty}^{\infty} f_z(\mathbf{r} + n\,\mathbf{a}_1 + m\,\mathbf{a}_2) \\
&= \sum_{j,k=-\infty}^{\infty} \tilde{F}_{jk}(z)\, e^{i(j\,\mathbf{b}_1 + k\,\mathbf{b}_2)\bullet\mathbf{x}}.
\end{aligned}
\tag{8.9}
$$

The Fourier coefficients are

$$\tilde{F}_{jk}(z) = \frac{1}{|\mathbf{a}_1 \times \mathbf{a}_2|} \int d^2\mathbf{x}\, F(\mathbf{r})\, e^{-i(j\,\mathbf{b}_1 + k\,\mathbf{b}_2)\bullet\mathbf{x}}, \tag{8.10}$$

where $\mathbf{b}_1 = 2\pi(\mathbf{a}_2 \times \mathbf{k})/[(\mathbf{a}_1 \times \mathbf{a}_2)\bullet\mathbf{k}]$ and $\mathbf{b}_2 = 2\pi(\mathbf{k} \times \mathbf{a}_1)/[(\mathbf{a}_1 \times \mathbf{a}_2)\bullet\mathbf{k}]$ are primitive vectors of the reciprocal lattice, and the integration extends over the entire surface.

Consider now the lowest non-trivial Fourier components only. For elementary crystalline surfaces with hexagonal symmetry, similar to the arguments leading to Eq. (6.51), the force distribution is

$$F(\mathbf{r}) = \tilde{F}_0 + 9\tilde{G}_1\, \Phi^{(6)}(b\mathbf{x}), \tag{8.11}$$

where the function $\Phi(\mathbf{X})$ has maximum value of 2/3 at the T sites and value of 1/3 at the H sites. The Fourier coefficients can be evaluated using an identity in Appendix D. The results are:

$$\tilde{F}_0(z) = -\frac{4\pi z U_0}{\sqrt{3}\, a^2}\left(2e^{-\kappa(z-r_0)} - e^{-2\kappa(z-r_0)}\right), \tag{8.12}$$

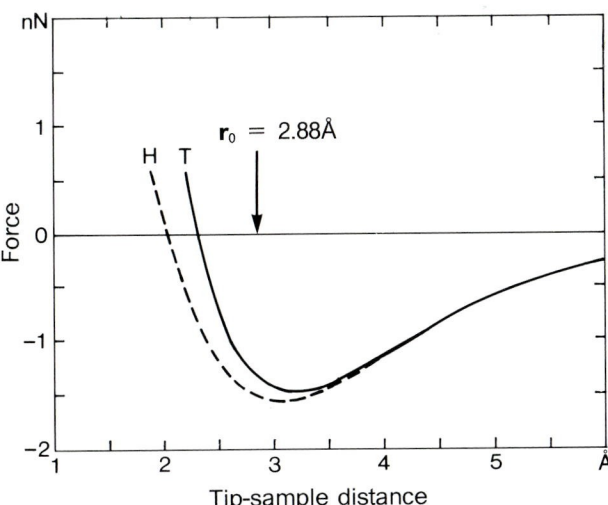

Tip-sample distance

Fig. 8.4. Atomic force between an Al tip and an Al sample. The magnitude of the attractive force on top of an Al atom (T site) is smaller than that on top of a hole (H site). (1 eV/1 Å=1.60 nN.)

$$\tilde{F}_1(z) = -\frac{4\pi z \kappa U_0}{\sqrt{3}\, a^2} \left(\frac{2}{\beta_1}\, e^{-\beta_1 z + \kappa r_0} - \frac{1}{\beta_2}\, e^{-\beta_2 z + 2\kappa r_0} \right), \tag{8.13}$$

where

$$\beta_1 = \sqrt{\kappa^2 + b^2}\,, \tag{8.14}$$

$$\beta_2 = \sqrt{(2\kappa)^2 + b^2}\,. \tag{8.15}$$

At a T site, the force is

$$F_T(z) = \tilde{F}_0(z) + 6\tilde{F}_1(z), \tag{8.16}$$

and at an H site, the force is

$$F_H(z) = \tilde{F}_0(z) - 3\tilde{F}_1(z). \tag{8.17}$$

An example of an Al(111) surface is shown in Fig. 8.4. From the measured work function, $\phi=3.5$ eV, we find $\kappa = 0.96\,\text{Å}^{-1}$, The atomic distance is $a = 2.88\,\text{Å}$, which also equals the parameter r_0 in the Morse formula. An estimation of the parameter U_0 in the Morse formula, or the binding energy per pair of Al atoms, can be made as follows: The evaporation heat of aluminum is 293 kJ/mol, which is 3.0 eV per atom. Aluminum is an fcc crystal, where each atom has 12 nearest neighbors. Therefore, the binding energy per pair of Al atoms is about 0.5 eV. Substitute these numbers into Eq. (8.16) and Eq. (8.17), we find the forces at the T site and the H sites which reproduce the result of first-principle calculations by Ciraci et al. (1990a).

8.2.4 Stability of STM at short distances

Because of the existence of force, under certain circumstances, the STM gap becomes unstable. This problem has been studied extensively by Pethica and Oliver (1987) and Pethica and Sutton (1988), using the classical theory of continuum elasticity. In spite of its simplicity, the theory reproduces the basic features of a large number of the observed phenomena.

As we have discussed in Chapter 7, the force on the entire range of tip–sample distances under normal STM operating conditions can be described by the Morse curve (Slater, 1963):

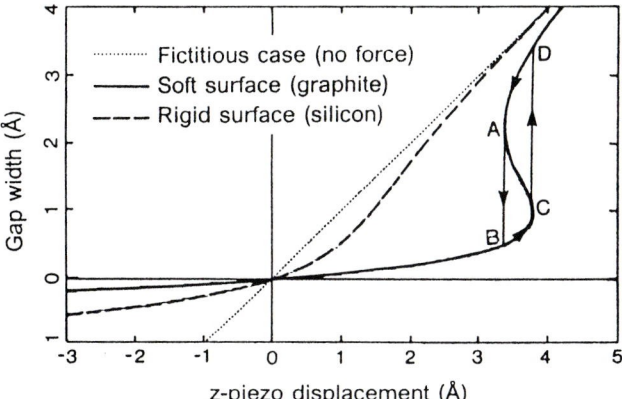

Fig. 8.5. Stability of STM and the rigidity of surfaces. If the tip and the sample surface are relatively rigid, the system is stable. If the tip is not plunging into the sample surface too deep, the approaching of the tip into the repulsive-force region is reversible. If the sample surface is soft, hysteresis occurs. However, if the lateral strength of the sample surface is high, there could be no permanent destruction.

$$F = -2\kappa U_0 \left(e^{-\kappa(s - s_0)} - e^{-2\kappa(s - s_0)} \right), \tag{8.18}$$

where U_0 is the binding energy of chemisorption, and s_0 is the equilibrium distance, i. e., the distance at which the net force is zero. In Chapter 7, we have shown that the constant κ in the expression of the Morse curve is equal to that of the sample wavefunction based both on experimental evidence and theoretical arguments.

Following Pethica and Sutton (1988), the mechanical loop of STM responses to the force and exhibits an elastic deformation. By formally introduce an elastic constant α, the deformation is

$$\delta z = \alpha F \tag{8.19}$$

For a well-designed, rigid STM, the deformation takes place predominately near the end of the tip. In this case, the elastic constant α is

$$\alpha = \frac{1}{a_0} \left(\frac{1}{E_T^*} + \frac{1}{E_S^*} \right), \tag{8.20}$$

where E_S^* is Young's modulus of the sample surface, which may have to be corrected by a factor very close to 1 (see Pethica and Oliver, 1987; Pethica and Sutton, 1988); and E_T^* is that for the tip. For many materials, the value

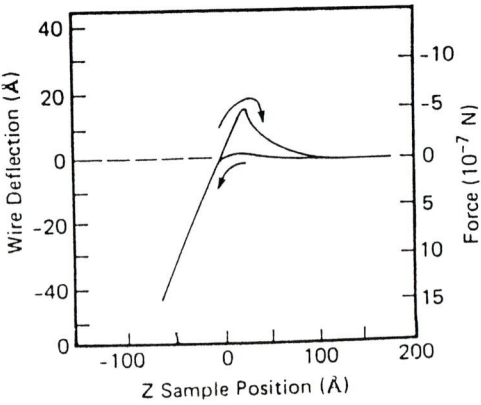

Fig. 8.6. Hysteresis observed in the STM imaging of graphite. (Reproduced from Mate et al., 1989, with permission.)

of E^* is of the order of 10^{11} Pa or 10^{12} dyn/cm^2. The length scale a_0 is a characteristic radius of the tip end.

The observed z-piezo displacement is then the sum of $\delta z = \alpha F$ and the true tip–sample displacement δs. Using Equations (8.18) and (8.19), we find

$$z = s + 2\alpha\kappa U_0 \left(e^{-\kappa(s - s_0)} - e^{-2\kappa(s - s_0)}\right). \tag{8.21}$$

Taking derivatives with respect to s on both sides, it becomes

$$\frac{dz}{ds} = 1 - \gamma\left(e^{-\kappa(s - s_0)} - 2e^{-2\kappa(s - s_0)}\right). \tag{8.22}$$

The dimensionless quantity $\gamma = 2\alpha\kappa^2 U_0$ is a measure of the relative stiffness of the STM system with respect to the force in the gap.

In order to have a stable system, the quantity dz/ds should be a positive number. In other words,

$$\gamma\left(e^{-\kappa(s - s_0)} - 2e^{-2\kappa(s - s_0)}\right) < 1. \tag{8.23}$$

The above function has the value $-\gamma$ at $s = s_0$, and 0 at $s = \infty$. The maximum is at a value of s satisfying

$$\gamma\kappa\left(e^{-\kappa(s - s_0)} - 4e^{-2\kappa(s - s_0)}\right) = 0. \tag{8.24}$$

It implies

$$e^{\kappa(s - s_0)} = 4. \tag{8.25}$$

Therefore, the condition of stability is

$$\gamma \leq 8. \tag{8.26}$$

For most metals and semiconductors, Eq. (8.26) is satisfied. For soft materials such as graphite, γ could be much larger than 8. Figure 8.5 shows both cases. The dashed curve is the relation of z and s with $\gamma=2$. The gap is stable. The solid curve is the relation of z and s with $\gamma=12$. At about $z = 3.5\,\text{Å}$, dz/ds becomes negative. Hysteresis takes place. As the tip approaches the sample surface, at point A, ds/dz becomes infinite. The sample surface (and the tip) expands to make a jump of about 2 Å. The tip touches the sample at point B. By retracting the tip from mechanical contact with the sample surface, at point C, ds/dz again becomes infinity. The sample surface (and the tip) restores to make an increase of separation of about 2 Å, to reach point D. On graphite, the hysteresis has been observed by Mamin et al. (1986) and Mate et al. (1989). The occurrence of hysteresis depends not only on the stiffness of the materials, but also on surface conditions, as shown in Fig. 8.6.

8.2.5 Force in tunneling barrier measurements

For a large number of materials, the stability condition, Eq. (8.26), is satisfied. In addition, in most cases, the STM images are taken under attractive-force conditions. The dramatic amplification effect does not occur. However, the effect of force and deformation is still observable. By measuring the apparent barrier height using ac method, based on Eq. (1.13),

$$\phi_{\text{app}} \equiv 0.95 \left(\frac{d \ln I}{dz} \right)^2, \tag{8.27}$$

even a slight variation of gap distances can be detected. In fact, if the apparent barrier height in terms of the actual displacement of gap width δs is a constant,

$$\phi_0 \equiv 0.95 \left(\frac{d \ln I}{ds} \right)^2 = \text{const.}, \tag{8.28}$$

but due to the force and deformation,

$$\frac{dz}{ds} \neq 1, \tag{8.29}$$

an apparent variation of the measured barrier height, Eq. (8.27), should be observed. In the strong repulsive-force regime, $(dz/ds) >> 1$, a very small apparent barrier height should be observed. This phenomenon was first described and analyzed by Coombs and Pethica (1986), and can be easily observed on graphite and other layered materials, such as the case of VSe_2 as shown in Fig. 8.7. In the attractive-force regime, $(dz/ds) < 1$, an increase of the the measured value of apparent barrier height should be observed. Actually, this was first reported by Teague (1987) in an MIM tunneling experiment, and then reported by Binnig et al. (1984) using STM: at a very short tip–sample distance, where the actual barrier collapses, the current becomes even higher than what was expected from an exponential dependence on the distance.

A systematic study was conducted on clean Si(111) surface with a clean W tip (Chen and Hamers, 1991). The entire curve of the dependence of the measured apparent barrier height, Eq. (8.27), with z-piezo displacement, was recorded. The experiment was performed under the condition that a clear 7×7 pattern was observed, which indicated that both the tip (near the apex atom) and the sample were clean. By carefully moving the tip back and forth, so as not to press into the sample surface too deeply, the entire process is completely reversible. The experimental barrier height measurements were performed using an ac modulation method, by applying a small 0.05 Å modu-

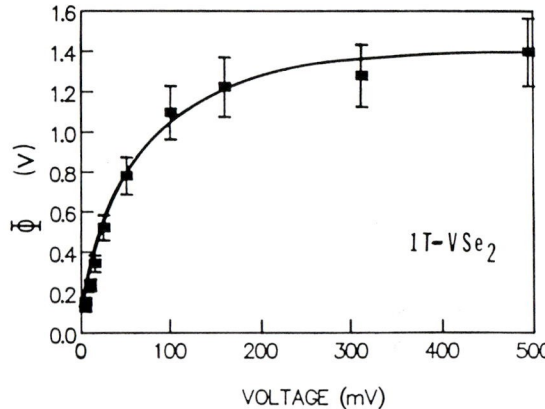

Fig. 8.7. Apparent barrier height measured on 1T-VSe$_2$. At low bias, the tip is in contact with the sample surface. The deformation of the sample surface with the displacement of the tip makes the apparent barrier height much smaller than the definition in Eq. (8.28). (Reproduced from Giambattista et al., 1990. with permission.)

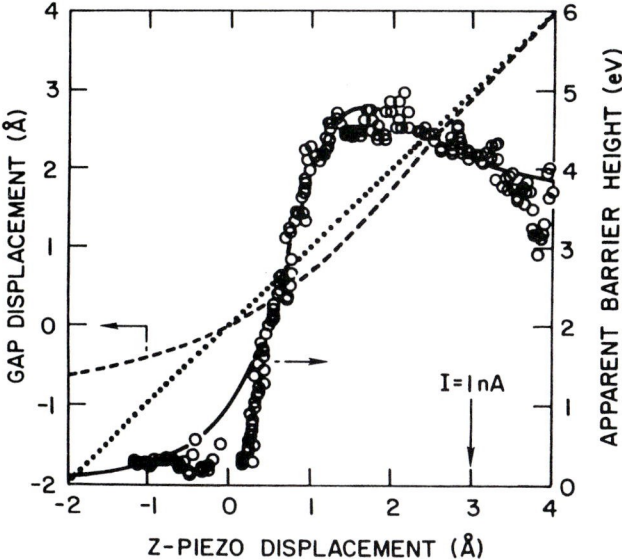

Fig. 8.8. Variation of the measured apparent barrier height with distance. Circles are data points. The solid curve is derived from Eq. (8.26). The dashed curve is the actual gap displacement as a function of the measured z-piezo displacement. The dotted curve, the fictitious gap displacement in the absence of force, is included for comparison. (Reproduced from Chen and Hamers, 1991a, with permission.)

lation to the z piezo at a frequency $\omega_{mod} \approx 2$ kHz. The ac method provides better accuracy than the dc method. The data points in Fig. 8.8 are the experimentally measured apparent barrier height as a function of tip–sample separation obtained in this manner on a clean Si(111)-7 × 7 sample with a sample bias of -1 V with respect to the tip. As a function of tip–sample separation, the barrier height versus distance can be separated into four distinct regimes: (1) At large separations the barrier height is approximately 3.5 eV, which is roughly equal to the average work functions of tungsten and silicon. (2) As the tip–sample separation is decreased, the barrier height first exhibits a small *increase* to about 4.8 eV. (3) Further decreasing the tip–sample separation causes the barrier height to plummet by more than a factor of ten with only a 1 Å change in tip–sample separation. (4) Pushing the tip toward the sample even further produces only a small modulation of the current dI/dz, leading to an apparent barrier height of near zero, a phenomenon reported first by Coombs and Pethica (1986). In both cases, the observed behavior of the apparent barrier height is *continuous and reversible* if the tip is not pushed too deeply.

The observed variation of apparent barrier height can be understood quantitatively by assuming that the force follows a Morse curve. Thus, the relation between the z piezo reading and the true gap displacement s follows Eq. (8.22),

$$\frac{dz}{ds} = 1 - \gamma\left(e^{-\kappa(s-s_0)} - 2e^{-2\kappa(s-s_0)}\right). \tag{8.30}$$

The solid curve in Fig. 8.8 is drawn with $\gamma=0.95$ and assuming the actual apparent barrier height is 3.5 eV throughout the entire region. The accurate fit indicates that the model is reasonable.

Using Eq. (8.19), the characteristic radius of the end of the tip can be estimated. Assuming $U_e = 5$ eV, with the elastic constants of tungsten, $E = 34 \times 10^{11}$ dyn/cm^2 and $\nu = 0.26$, we find $a_0 \approx 5$ Å. This is a reasonable value for tips that exhibit atomic resolution.

The normal tip–sample distance in STM experiments can be obtained accurately from this experiment. In Fig. 8.8, the equilibrium distance, where the net force is zero, is taken as the origin of z. As shown for the case of aluminum, because the attractive force has a longer range than the repulsive force, the absolute equilibrium distance between the apex atom and the counterpart on the sample surface is slightly less than the sum of the atomic radii of both atoms, which is about ≈ 2 Å. The normal topographic images on Si(111) are usually taken at $I = 1$ nA, corresponding to a distance of ≈ 3 Å from the equilibrium point, or ≈ 5 Å from nucleus to nucleus.

PART II

INSTRUMENTATION

I suppose that when the bees crowd round the flowers it is for the sake of honey that they do so, never thinking that it is the dust which they are carrying from flower to flower which is to render possible a more splendid array of flowers, and a busier crowd of bees, in the years to come. We cannot, therefore, do better than improving the shining hour in helping forward the cross-fertilization of the sciences.

Lecture given at Cambridge, May 14, 1878
while directing a demonstration of the telephone

James Clerk Maxwell
Distinguished Professor of
Experimental Physics
Cambridge University

In view of the extreme simplicity of the STM as an instrument, one may wonder why it was not invented many decades ago. Probably the reason is that the STM is a hybrid of several different branches of science and technology that generally have very little communication with each other. These parent areas are classical tunneling experiments (a branch of low-temperature physics), surface science (a branch of vacuum physics), and microscopy. The technological implementation of STM required skills and knowledge in different disciplines such as mechanical, electronic, and control engineering. The invention of the STM was a result of the cross-fertilization of different branches of science (Binnig and Rohrer, 1987), much as advocated by Maxwell more than a century ago.

In Part II, we discuss the essential elements of STM instrumentation. Except for a few cross-references, all the chapters can be read independently. The chapter on piezodrives starts with an introduction to piezoelectricity and piezoelectric ceramics at the general physics level. Three major types of piezodrives, the tripod, the bimorph, and the tube, are analyzed in detail. The chapter on vibration isolation starts with general concepts and vibration measurements. Various vibration isolation devices used in STM are described and analyzed. The chapter on electronics and control is presented at the college electronics course level. The analysis of the transient responses of feedback circuits requires some knowledge of the Laplace transform, but the presentation is aimed at general scientists rather than specialists in this field. The actual mechanical design of the STM largely depends on the coarse-positioning mechanism, which is discussed in the same chapter. Tip preparation and characterization is a crucial, but still not well understood, technique in STM. This topic also has a dedicated chapter. It is followed by a chapter on scanning tunneling spectroscopy, where tip treatment and tip-electronic-structure characterization are essential. The chapter on atomic-force microscopy only touches the repulsive-force mode, which provides atomic resolution, as STM's next of kin. The applications of STM are so broad that a number of books are required to describe them. In the final chapter of this book, a number of illustrative applications are presented.

The theme of this book is atomic-scale imaging through tunneling. For it to be realized, all the instruments must be well orchestrated. The limited size of this book prevents the author from including all the variations. After proper **instrumentation,** may you play and develop the theme in your sphere: be it physics, chemistry, electrochemistry, biology, materials science, or any engineering science.

CHAPTER 9
PIEZOELECTRIC SCANNER

The heart of STM is a piezoelectric scanner, sometimes called *piezodrive* or simply *piezo*. In this chapter, we provide a brief summary of the basic physics of piezoelectricity and piezoelectric ceramics relevant to the applications in STM. Three major types of piezodrives, the tripod, the bimorph, and the tube, are analyzed in detail.

9.1 Piezoelectric effect

The piezoelectric effect was discovered by Pierre Curie and Jacques Curie (1880), about 100 years before the invention of the STM.[4] A sketch of their pioneering experiment is shown in Fig. 9.1. A long, thin quartz plate, cut from a single crystal, was sandwiched between two tin foils. While one tin foil is grounded, another tin foil was connected to an electrometer. By applying a weight to generate vertical tension, an electrical charge was detected by the electrometer.

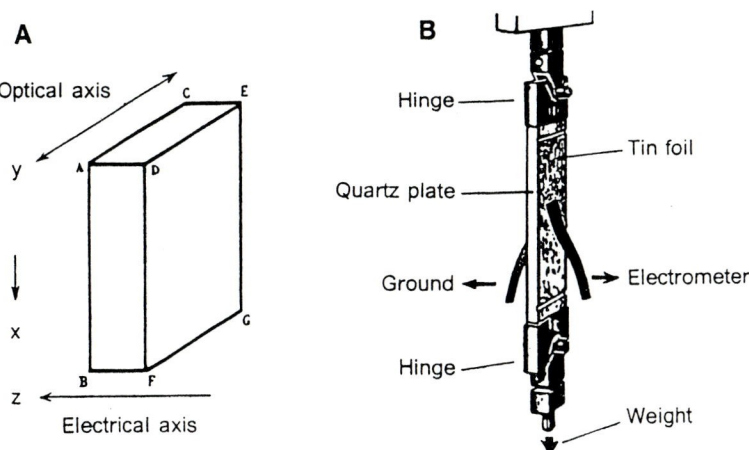

Fig. 9.1. Piezoelectric effect. (A) A quartz plate, cut from a single crystal. (B) By stressing the quartz plate, an electrical charge is generated. (After Curie, 1889.)

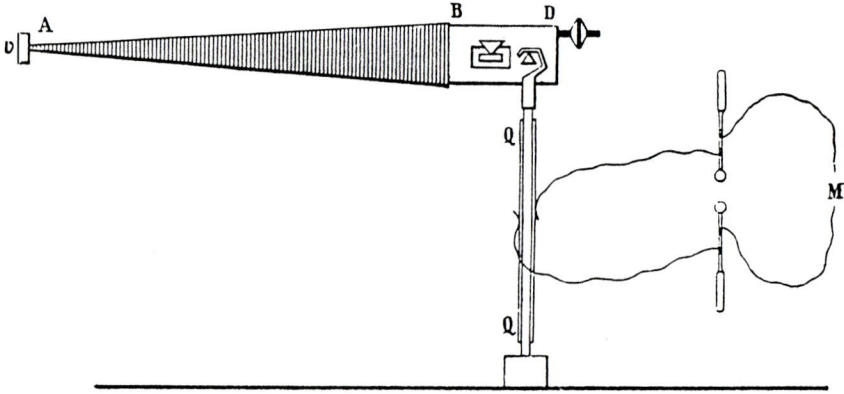

Fig. 9.2. The inverse piezoelectric effect. A thin and long quartz plate, QQ, is sandwiched between two tin foils. By applying a voltage to the tin foils, the quartz plate elongates or contracts according to the polarity of the applied voltage. To measure the very small displacement, Curie (1889a) used a lever ABD with a small piece of glass v attached at its end, the displacement of which is then measured with an optical microscope. (After Curie, 1889a.)

A few months later, Lippman (1881) predicted the existence of the *inverse piezoelectric effect:* By applying a voltage on the quartz plate, a deformation should be observed. This effect was soon confirmed by the Curie brothers (Curie and Curie, 1882), who designed a clever experiment to measure the tiny displacement, as shown in Fig. 9.2. Here, a light-weight lever with an arm of about 1:100 amplifies the displacement by two orders of magnitude. An optical microscope further amplifies it by two orders of magnitude. The displacement is then measured by an eyepiece with a scale.

The definitions of the parameters for describing piezoelectric effect are shown in Fig. 9.3. A voltage V is applied on a rectangular piece of piezoelectric material. Inside it, the electrical field intensity is

$$E_3 = \frac{V}{z} . \tag{9.1}$$

---

4 The word *piezoelectric* is a combination of two Greek words, πιεζειν, which means "to press," and ηλεκτρον, which means "amber." Literally, *piezoelectric* means "using pressure to generate the phenomenon previously generated by rubbing a piece of amber." In Webster's Dictionary, the pronunciation of the prefix is marked as [pai'i:zou-]. In the STM community and the transducer industry, everyone says [pi'ezou-], or even simpler, ['pi:zou-]. Nevertheless, the pronunciation of a word is merely a convention within a community, which has no scientific significance. Let's follow the crowd!

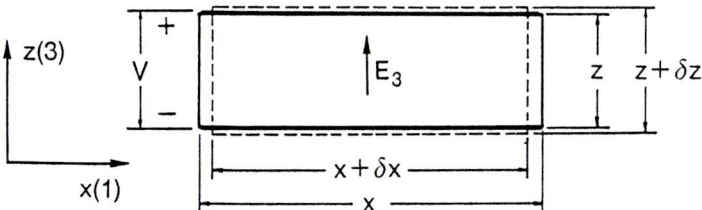

Fig. 9.3. Definition of piezoelectric coefficients. A rectangular piece of piezoelectric material, with a voltage V applied across its thickness, causes a strain in the x as well as the z directions. A piezoelectric coefficient is defined as the ratio of a component of the strain with respect to a component of the electrical field intensity.

In the standard convention, the directions x, y, and z are labeled as 1, 2, and 3, respectively. As a result, a strain is generated. The xx component of the strain tensor is (see Fig. 9.3)

$$S_1 \equiv \frac{\delta x}{x}, \tag{9.2}$$

and the zz component is

$$S_3 \equiv \frac{\delta z}{z}. \tag{9.3}$$

The *piezoelectric coefficients* are defined as the ratios of the strain components over a component of the applied electrical field intensity, for example,

$$d_{31} \equiv \frac{S_1}{E_3}, \tag{9.4}$$

and

$$d_{33} \equiv \frac{S_3}{E_3}. \tag{9.5}$$

Because strain is a dimensionless quantity, the piezoelectric coefficients have dimensions of meters/volt in SI units. Their values are extremely small. In the literature, the unit 10^{-12} m/V is commonly used. For applications in STM, a natural unit is Å/V, or 10^{-10} m/V. Using primitive means as shown in Fig. 9.2, the Curie brothers (Curie and Curie, 1882) obtained a value of 0.021

Å/V for the parallel piezoelectric coefficient for quartz, which matches accurately the results of modern measurements (Cady 1946).

From the very beginning of their experiments, the Curie brothers realized that the linear piezoelectric effect only exists in anisotropic crystals, such as quartz, tourmaline, and Rochelle salt (Curie and Curie, 1880). In fact, as seen from Fig. 9.3, if the $+z$ and $-z$ directions of the plate are equivalent, by reversing the direction of the electric field E_3, the strain should be the same. In such cases, the strain should be proportional to $|E_3|^2$ instead of E_3. In other words, there should be no linear piezoelectric effect.

The reasoning of Lippman (1881) for the inverse piezoelectric effect is based on thermodynamics. Consider the experiment shown in Fig. 9.1. The increment of the total energy due to an applied voltage V is

$$dE = T\,dS + F\,d(\delta x) + Q\,dV, \tag{9.6}$$

where F is the force in the vertical direction, x, and Q is the electrical charge on the surface. The reversibility of the process implies that the Gibbs free energy G is also a function of the state, that is,

$$dG = -S\,dT - (\delta x)\,dF + Q\,dV, \tag{9.7}$$

from which we obtain immediately the Maxwell relation

$$\left(\frac{\partial Q}{\partial F}\right)_{T,\,V} = \left(\frac{\partial(\delta x)}{\partial V}\right)_{T,\,F}. \tag{9.8}$$

Using Eq. (9.1) and Eq. (9.2), note that the polarization in the z direction is $P_3 = Q/xy$ and the stress in the x direction is $\sigma_1 = F/yz$, where y is the width of the quartz plate, Eq. (9.8) can be rewritten as

$$\left(\frac{\partial P_3}{\partial \sigma_1}\right)_{T,\,E_3} = \left(\frac{\partial S_1}{\partial E_3}\right)_{T,\,\sigma_1}. \tag{9.9}$$

The left-hand side is the forward piezoelectric coefficient, in units of coulombs/newton. The right-hand side is the reverse piezoelectric coefficient, in units of meters/volt. They are equal. The coexistence of forward and reverse piezoelectric effects provides a simple method to test the piezodrive used in STM, which is discussed in Section 9.6.

The piezoelectric effect was one of the crucial elements for Pierre Curie and Marie Skladowska Curie in their discovery of radioactivity. However, there was no technological application for over 30 years, until it was used in radio transmitters and ultrasonics on the 1910s. Currently, the largest application of piezoelectric devices is in wrist watches and clocks. The original piezoelectric material discovered and studied by the Curie brothers, quartz, is still the material of choice. Each year, many millions of "quartz watches" and "quartz clocks" are made and sold.

9.2 Lead zirconate titanate ceramics

The piezoelectric materials used in STM are various kinds of lead zirconate titanate ceramics (PZT).[5] The mechanism of piezoelectricity in PZT is somewhat different from single-crystal piezoelectric materials, such as quartz. Pure $PbZrO_3$ and $PbTiO_3$ and their solid solutions are *ferroelectric* or *antiferroelectric,* and exhibit a permanent electric dipole *even in the absence of external electric field.* PZTs have never been used in single-crystal form. Those are useful only in ceramic form. The as-made ceramics do not exhibit a piezoelectric effect. Macroscopically, they are isotropic due to the random arrangement of the electrical dipoles. A *poling process* is then applied to produce a permanent electric polarization, similar to the process of making a permanent magnet from a piece of hard ferromagnetic material. After poling, the dipoles are aligned with the poling field. Macroscopically, the piece of material becomes anisotropic. A strong piezoelectric effect is generated. One of the advantages of the piezoelectric ceramics over single-crystal materials is that it can be shaped easily and poled at a desired direction. However, inherently, its piezoelectric parameters are not as stable and reproducible as single crystals, such as quartz. The first practical material in this category is barium titanate, $BaTiO_3$, which had been extensively used in ultrasonic transducers in the 1940s and 1950s (Cherry and Adler, 1947). One of the disadvantages of $BaTiO_3$ is that its Curie temperature is too low, 120°C. Once being heated above this temperature, its piezoelectricity is permanently lost. In the 1950s, Jaffe et al. (1954) discovered that the ceramics based on a mixture of $PbZrO_3$ and $PbTiO_3$, after a similar poling process, exhibit excellent piezoelectric properties. The Curie temperatures of such ceramics range from 200 to 400°C, and are therefore much more stable than barium titanate ceramics.

The PZT ceramics are made by firing a mixture of $PbZrO_3$ and $PbTiO_3$ together with a small amount of additives at about 1350°C under strictly controlled conditions. The result is a solid solution. Macroscopically, it is isotropic. Microscopically, it consists of small ferroelectric crystals in random orientations. After machining and metallization, a high electric field is applied for a sufficiently long period of time, for example, 60 kV/cm for 1 h. A strong

---

5 PZT is the trade name of the lead zirconate titanate piezoelectric ceramics of one of its largest producer, Vernitron. It is also commonly used in the scientific literature as a standard acronym.

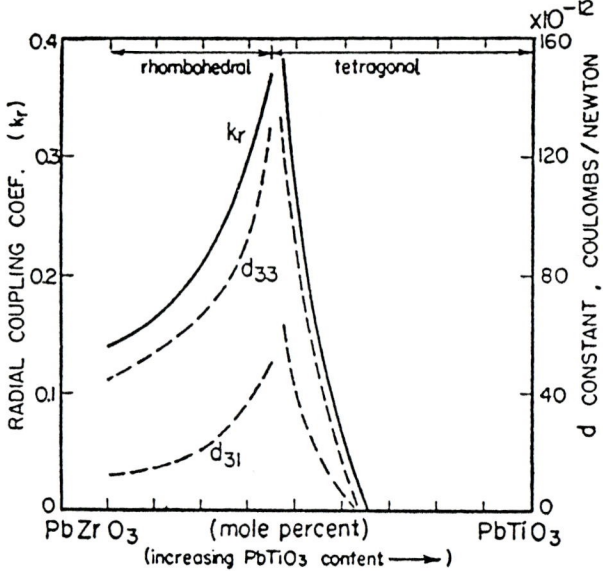

Fig. 9.4. Dependence of piezoelectric properties of PbZrO$_3$–PbTiO$_3$ on composition. The zirconate-rich phase is rhombohedral, whereas the titanate-rich phase is tetrahedral. The piezoelectric coefficients reach a maximum near the morphotropic phase boundary, approximately 45% PbZrO$_3$ and 55% PbTiO$_3$. (After Jaffe et al., 1954.)

piezoelectricity is generated. As a convention, the direction of the poling field is labeled as the 3 direction, or the positive z direction.

The crystallographic and piezoelectric properties of the ceramics depend dramatically on composition. As shown in Fig. 9.4, the zirconate-rich phase is rhombohedral, and the titanate-rich phase is tetragonal. Near the morphotrophic phase boundary, the piezoelectric coefficient reaches its maximum. Various commercial PZT ceramics are made from a solid solution with a zirconate–titanate ratio near this point, plus a few percent of various additives to fine tune the properties for different applications.

In addition to the parameters discussed in the previous section, that is, the piezoelectric coefficients d_{31}, d_{33}, and the velocity of sound, c, there are several other parameters that are important for applications in STM.

Curie point

As a ferroelectric material, each piezoelectric ceramic is characterized by a Curie point or Curie temperature, T_c (Jaffe et al., 1971). Above this temperature, the ferroelectricity is lost. An irreversible degradation of the

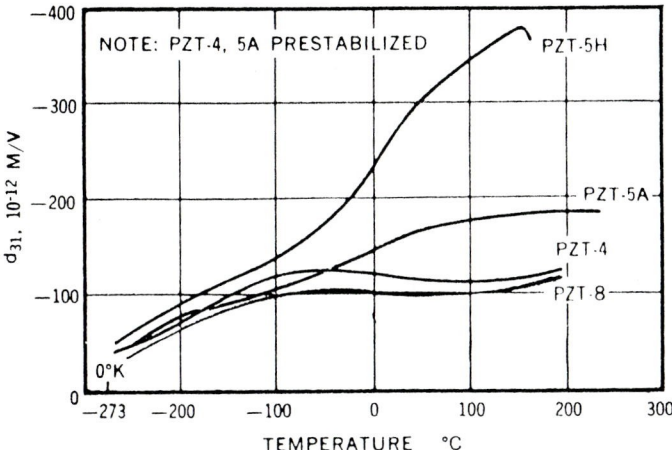

Fig. 9.5. Variation of piezoelectric coefficient with temperature. (By courtesy of Morgan Matroc, Inc., Vernitron Division, Bedford, Ohio.)

piezoelectric property occurs if heated above such a temperature. At temperatures close to but still much lower than the Curie temperature, serious degradation may occur. Therefore, each piezoelectric material has a well-defined maximum operating temperature, which is much lower than its Curie point.

Temperature dependence of piezoelectric coefficients

While designing or using STM at low or high temperatures, the variation of piezoelectric coefficients with temperature has to be considered seriously. The variation differs for different PZT materials. Fig. 9.5 shows measured variations of d_{31} for several commonly used PZT materials with temperature.

Depoling field

As we have mentioned, the piezoelectric properties of ceramics are generated by a poling process. Apparently, if a strong electric field in a direction other than the poling direction is applied, the piezoelectric property is altered or lost. The safe value of an ac field to avoid causing depoling, E_d, can be found in the product specifications.

Mechanical quality number

This quantity, usually denoted Q_M, is a measure of the internal mechanical energy loss of the material. Roughly speaking, it is the number of vibrations it can sustain without substantial amplitude reduction. The larger the number Q_M is, the smaller the internal loss.

Coupling constants

Probably the best measure of the effectiveness of a piezoelectric material is its electromechanical coupling constant, k, defined as

$$k = \frac{\text{electrical energy converted to mechanical energy}}{\text{input electrical energy}}. \qquad (9.10)$$

This is the efficiency of energy conversion between mechanical and electrical forms. For PZTs, it ranges from 0.5 to 0.7, which are the most efficient of all known piezoelectric materials (see Table 9.1). For quartz, the coupling constant is about 0.1.

Aging

In contrast to piezoelectric single crystals, such as quartz, the piezoelectricity of PZT ceramics decays with time due to relaxation. Experimentally it is found that on a large time scale (for example, months and years), the aging process can be accurately described by a logarithmic law. For example, the coupling constant k varies with time t as

Table 9.1. Important properties of PZT ceramics commonly used in STM. By courtesy of Morgan Matroc Inc., Vernitron Division, Bedford, Ohio.

Item	Unit	PZT-4D	PZT-5H	PZT-7D	PZT-8
d_{31}	Å/V	-1.35	-2.74	-1.00	-0.97
d_{33}	Å/V	3.15	5.93	2.25	2.25
Y	10^{10} N/m^2	7.5	6.1	9.2	8.7
ρ	g/cm^3	7.6	7.5	7.6	7.6
c	km/sec	3.3	2.8	2.9	3.4
T_c	°C	320	195	325	300
E_d	kV/cm (rms)	>10	4	>10	>15
Q_M	-		600	65	500	1000
k_p	-		-.60	-.65	-.48	-.51
Aging	k_p/time decade	-1.7%	-0.2%	-0.006%	-2.3%

$$k(t) = k(0)(1 + \Delta \log_{10} t), \qquad (9.11)$$

where Δ is the relative variation of the coupling constant per time decade. The zero point of time is the completion of poling.

Table 9.1 shows selected parameters of several PZT ceramics commonly used in STM. PZT-5H has by far the highest sensitivity (that is, the largest piezoelectric coefficient d_{31}). However, its Curie temperature is low and its internal friction is high, which means it has a serious hysteresis problem. Also, its properties sharply depend on temperature. On the other hand, PZT-7D, with a lower sensitivity, exhibits very low hysteresis and a very small aging effect. The properties of PZT-4D have very low temperature variation near room temperature. At cryogenic temperatures, PZT-8 has the lowest temperature variations. Also, PZT-8 has a high depoling field and a high Curie temperature, which is suitable for high-temperature applications.

9.3 Tripod scanner

In the early years of STM instrumentation, tripod piezoelectric scanners were the predominant choice, as shown in Fig. 9.6. The displacements along the x, y, and z directions are actuated by three independent PZT transducers. Each of them is made of a rectangular piece of PZT, metallized on two sides. Those three PZT transducers are often called x piezo, y piezo, and z piezo, respectively. By applying a voltage on the two metallized surfaces of a piezo, for example, the x piezo, the displacement is

$$\Delta x = d_{31} V \frac{L}{h}, \qquad (9.12)$$

where V is the applied voltage, L the length, and h the thickness of the piezo. The quantity

$$K \equiv \frac{dx}{dV} = d_{31} \frac{L}{h} \qquad (9.13)$$

Fig. 9.6. Tripod scanner. Three PZT bars to control the x, y, and z displacements, respectively. The tip is mounted at the vertex of the tripod. (Reproduced from Binnig and Rohrer, 1987, with permission.)

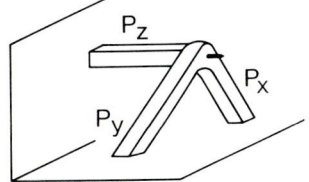

is called a *piezo constant.* For example, with $L = 20$ mm, $h = 2$ mm, and using PZT-4D, $d_{31} = -1.35$ Å/V, Eq. (9.13) gives $K = 14$ Å/V.

The accurate calculation of the resonance frequencies of a tripod scanner is a complicated problem. The flexing modes are effectively coupled with the stretching modes. An evaluation of the lowest resonance frequency of the flexing mode provides an order-of-magnitude estimation of the lowest resonance frequency of the tripod scanner. For a piezo made of PZT-5A, 20 mm long and 2 mm thick, the radius of gyration is 2 mm/$\sqrt{12}$ = 0.577 mm . The speed of sound is about 2.8 km/sec. Using Eq. (9.44), the resonance frequency is found to be 3.3 kHz, which is close to the values often observed experimentally.

Because the piezo constant is proportional to L/h, whereas the resonance frequency is proportional to h/L^2, it is clear that by reducing the length and thickness in proportion, the piezo constant remains the same, and the resonance frequency will increase. This is in general true. The natural limit of such a reduction is the depoling field of the material.

9.4 Bimorph

If larger displacements are required, an arrangement as shown in Fig. 9.7, the *bimorph*,[6] can be applied. The principle is similar to the bimetal thermometer. Two thin plates of piezoelectric material are glued together. By applying a voltage, one plate expands and the other one contracts. The composite flexes.

A common arrangement of a PZT bimorph is to pole both plates in the same direction (normal to the plane). When it is used, both outer electrodes are grounded. The driving voltage is applied to the center electrode. Such an arrangement reduces the stray field of the applied voltage to a minimum, and simplifies mounting.

The following is the treatment of bimorphs presented by Curie (1889a). As shown in Fig. 9.8, immediately after the application of a voltage V, a strain $d_{31}V(2/h)$ is generated. It in turn generates a stress σ_1 in the x direction:

$$\sigma_1 = \pm Yd_{31}V\frac{2}{h},\qquad(9.14)$$

where Y is Young's modulus. A torque occurs that flexes the element. As shown in Fig. 9.8(B), the bimorph bends to produce a distribution of stress to

---

6 Bimorph is a trade name of Vernitron for such flexing-type piezoelectric elements. Historically, it was developed by the Curie brothers in the 1880s (see Curie, 1889a). No specific term was suggested by the Curie brothers. Currently, the term bimorph is used frequently in the literature for the flexing-type piezoelectric element.

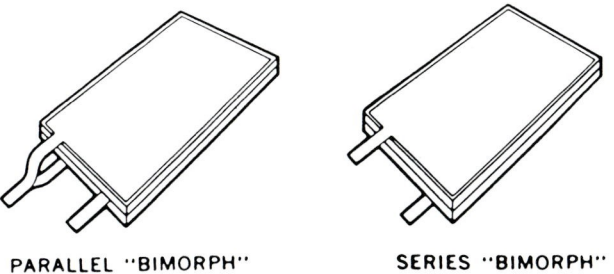

PARALLEL "BIMORPH"   SERIES "BIMORPH"

Fig. 9.7. Bimorph. Although the word bimorph is a trade name of Vernitron, Inc., it was described and analyzed by Curie (1889a). The most common mode of connection is to ground the two outer electrodes and to apply a voltage to the center electrode. The operation is similar to the bimetal thermometer. (By courtesy of Morgan Matroc, Inc., Vernitron Division, Bedford, Ohio.)

compensate the torque. It then reaches equilibrium. The additional stress is assumed to be linear in z,

$$\sigma(z) = \sigma_1 - \alpha z. \tag{9.15}$$

The constant α is determined by the condition of equilibrium, that is, with zero torque M:

$$M = \int z\sigma(z)\, dz = 0. \tag{9.16}$$

Substituting Eq. (9.15) for Eq. (9.16), we obtain

$$\alpha = \frac{3\sigma_1}{h}. \tag{9.17}$$

Substituting Eq. (9.16) for Eq. (9.15), the position of the neutral plane [where $\sigma(z) = 0$,] is found to be at $z = h/3$. Using elementary geometry and Eq. (9.14), we obtain the radius of curvature of the bimorph R

$$R = \frac{h^2}{6d_{31}V}. \tag{9.18}$$

Another application of elementary geometry gives the deflection Δz at the end of the bimorph,

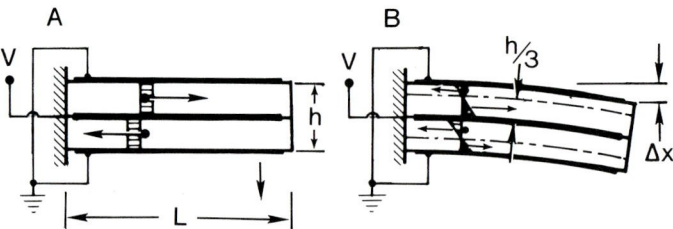

Fig. 9.8. Deflection of a bimorph. Two long, thin plates of piezoelectric material are glued together, with a metal film sandwiched in between. Two more metal films cover the outer surfaces. Both piezoelectric plates are poled along the same direction, perpendicular to the large surface, labeled z. (A) By applying a voltage, stress of opposite sign is developed in both plates, which generates a torque. (B) The bimorph flexes to generate a stress to compensate the torque. The neutral plane, where the stress is zero, lies at $h/3$ from the central plane.

$$\Delta z = 3 d_{31} V \frac{L^2}{h^2}. \tag{9.19}$$

As an example, a bimorph with the same material and dimensions as used in the last section, PZT-4D, 20 mm long and 2 mm thick, has a sensitivity of 2.74 Å/V. This is substantially higher than the stretching mode, 17 Å/V.

However, it is very difficult to construct a three-dimensional scanner based on bimorphs. An example of such a design is reported by Muralt et al. (1986). It is much more complicated than the tripod scanner.

9.5 Tube scanner

In this section, we describe the tube scanner, which has high piezo constants as well as high resonance frequencies. Moreover, it is much simpler than both the tripod scanners and bimorph-based scanners. The tube scanner soon became the predominant STM scanner after its invention by Binnig and Smith (1986).

The original design of the tube scanner (Binnig and Smith, 1986) is shown in Fig. 9.9. A tube made of PZT, metallized on the outer and inner surfaces, is poled in the radial direction. The outside metal coating is sectioned into four quadrants. In their original arrangement, the inner metal coating is connected to the z voltage, and two neighboring quadrants are connected to the varying x, y voltages, respectively. The remaining two quadrants are connected to a certain dc voltage to improve linearity. The tip is attached to the center of one of the dc quadrants. The first tube scanner was made by the EBL Division of Stanley Sensors, using PZT-5H as the piezoelectric mate-

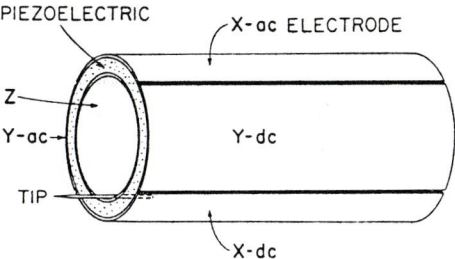

Fig. 9.9. The tube scanner. (Reproduced from Binnig and Smith, 1986, with permission.)

rial. The tube, 12.7 mm long, 6.35 mm in diameter, and 0.51 mm thick, has a measured piezo constant 50 Å/V in the x and y directions. The resonance frequency in the x and y directions is 8 kHz, and in the z direction, 40 kHz. Thus, the overall performance is much better than that of the tripod scanners.[7]

One of the problems with this design is that the motions driven by the x, y, and z voltages are nonlinear and not precisely orthogonal. This can be taken care of by proper programming in the control system, as we will discuss it in Chapter 11. Because the simplicity of the mechanical and electrical design, this arrangement is still widely used. Substantial improvement can be achieved by using bipolar, symmetric x and y voltages, and by placing the tip at the center of the tube. In this case, a $+ V_x$ and $- V_x$ voltages are applied on the opposite x quadrants, whereas a $+ V_y$ and $- V_y$ voltages are applied on the opposite y quadrants. This linearity is much improved.

First, we present a treatment of the deflection of a piezoelectric tube with quartered electrodes in the symmetric-voltage mode, as shown in Fig. 9.10. The wall thickness of the piezoelectrics is usually much smaller than the diameter. Therefore, the variation of strain and stress over the wall thickness can be neglected. As shown in Fig. 9.10, two voltages, equal in magnitude and opposite in sign, are applied to the two y quadrants. The inner metal coating and the two x quadrants are grounded. Immediately after the onset of y voltages, a strain in the z direction, $S_3 = d_{31}V/h$, is generated. It in turn creates a stress $\sigma_3 = YS_3$ in the z direction, where Y is Young's modulus. The torque of this pair of forces causes the tube to bend. The bending of the tube generates a torque in the opposite direction. At equilibrium, the total torque in any cross

---

7   A similar tube piezoelectric phonograph pickup device was described by Germano (1959), which has four narrow metal strips on the periphery of the tube to collect the voltage signal, instead of the complete metallization separated by four narrow gaps for an efficient actuation in the tube scanners. Germano (1959) also made an analysis of the tube piezoelectric phonograph pickup device. However, this analysis is not useful to the understanding of the functions of the tube scanners invented by Binnig and Smith (1986).

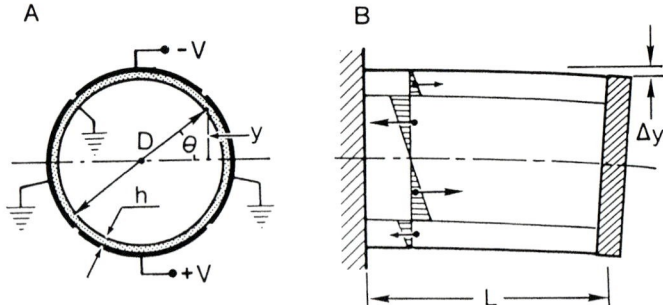

Fig. 9.10. Deflection of a tube scanner. (A) Opposite and equal voltages are applied to the y electrodes of a tube scanner. The x, z electrodes are grounded. A positive stress (pressure) is generated in the upper quadrant, and a negative stress (tension) is generated in the lower quadrant. (B) At equilibrium, a distribution of stress and strain is established such that the total torque at each cross section is zero. This condition determines the deflection of the tube scanner in the y direction. (Reproduced from Chen, 1992, with permission.)

section should be zero. By virtue of the symmetry of the problem, it is sufficient to consider one quarter of the circle. Assuming that the stress generated by the bending is linear with respect to y, the total stress $\sigma(\theta)$ in the piezoelectrics, as a function of angle θ, is

$$
\begin{aligned}
0 < \theta < \frac{\pi}{4}, \quad & \sigma(\theta) = -\alpha \sin \theta; \\
\frac{\pi}{4} < \theta < \frac{\pi}{2}, \quad & \sigma(\theta) = \sigma_3 - \alpha \sin \theta.
\end{aligned}
\tag{9.20}
$$

The constant α is to be determined by the condition of zero torque. The torque can be evaluated easily by integrating over the angle θ; see Fig. 9.10. The condition of zero torque is

$$
\int_0^{\frac{\pi}{4}} (-\alpha \sin \theta) \sin \theta \, d\theta + \int_{\frac{\pi}{4}}^{\frac{\pi}{2}} (\sigma_3 - \alpha \sin \theta) \sin \theta \, d\theta = 0,
\tag{9.21}
$$

which gives

$$
\alpha = \frac{2\sqrt{2}\,\sigma_3}{\pi}.
\tag{9.22}
$$

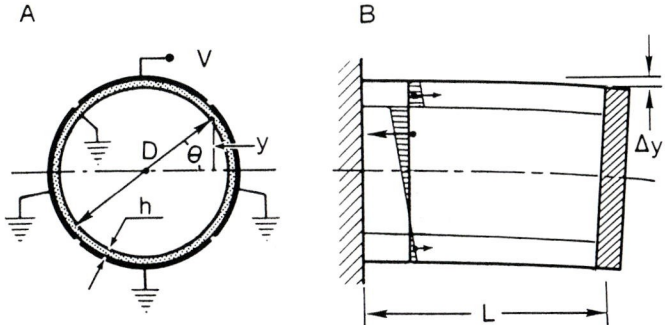

Fig. 9.11. Deflection of a tube scanner in the unipolar arrangement. (A) A voltage is applied to only one quadrant of the outer metal coating. All other quadrants and the inner metal coating are grounded. (B) At equilibrium, a distribution of stress and strain is established such that the total force and torque at each cross section is zero. This condition determines the y displacement. (Reproduced from Chen, 1992, with permission.)

The neutral plane, where the stress is zero, lies *beyond the periphery of the tube:*

$$y_0 = \frac{\pi D}{4\sqrt{2}} \approx 0.555D, \tag{9.23}$$

which is 11% larger than the radius r. Using elementary geometry, we find the curvature of bending:

$$R = \frac{\pi D h}{4\sqrt{2}\, d_{31} V}. \tag{9.24}$$

Again, using elementary geometry, the deflection is found to be

$$\Delta y = \frac{L^2}{2R} = \frac{2\sqrt{2}\, d_{31} V L^2}{\pi D h}. \tag{9.25}$$

It is convenient to define a *piezo constant* as

$$K_y \equiv \frac{dy}{dV} = \frac{2\sqrt{2}\, d_{31} L^2}{\pi D h}. \tag{9.26}$$

The formula for the x deflection is identical.

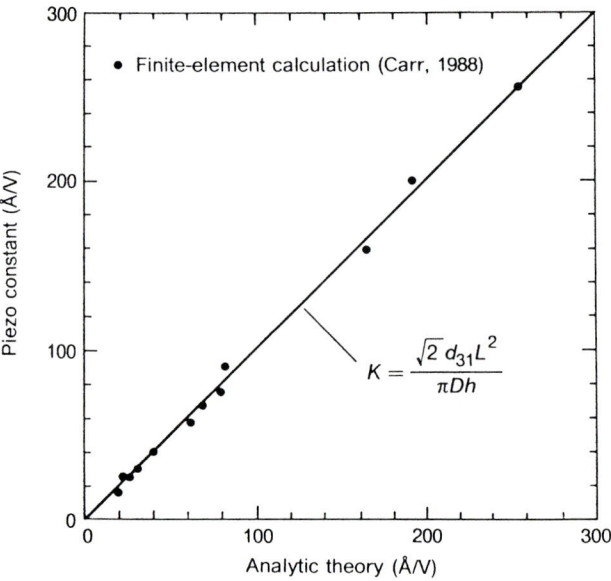

Fig. 9.12. Accuracy of the analytic expression. A comparison of Eq. (9.30) with the finite-element calculation by Carr (1988). (Reproduced from Chen, 1992, with permission.)

In the original design of Binnig and Smith (1986), the deflection voltage is applied to only one of the four quadrants. The stress is no longer antisymmetric with respect to the $y = 0$ plane, as shown in Fig. 9.11. The general form of the stress should have an additional term:

$$0 < \theta < \frac{\pi}{4}, \quad \sigma(\theta) = -\beta - \alpha \sin \theta;$$
$$\frac{\pi}{4} < \theta < \frac{\pi}{2}, \quad \sigma(\theta) = \sigma_3 - \beta - \alpha \sin \theta. \tag{9.27}$$

The two constants are determined by two independent conditions: the total force on any plane to be zero, and the total torque on any plane to be zero. A calculation similar to Eq. (9.21) gives

$$\beta = \frac{\sigma_3}{4}, \tag{9.28}$$

and

$$\alpha = \frac{\sqrt{2}\,\sigma_3}{\pi}\,.$$ (9.29)

Using similar arguments based on elementary geometry, we find

$$K_x = K_y = \frac{\sqrt{2}\,d_{31}L^2}{\pi D h}\,.$$ (9.30)

It is not surprising that the deflection by applying one voltage to a single quadrant, Eq. (9.30), exactly equals one-half of the deflection by applying two equal and opposite voltages to two opposite quadrants, Eq. (9.26). It is simply a consequence of symmetry.

The accuracy of Eq. (9.30) is verified by comparing it with the results of finite-element calculations by Carr (1988), as shown in Fig. 9.12.

9.6 In situ testing and calibration

The geometry of the tube scanner also provides a purely electrical method to self-test and self-calibrate, especially for measuring the piezoelectric constants in a cryogenic environment. The piezoelectric constant varies with temperature in a complicated manner, and also with the particular batch of materials by the manufacturer and time (the aging effect).

Consider a simple case: a piezoelectric tube with one end fixed to a base plate, and another end fixed to an end block. The two ends, however, are allowed to experience relative displacements freely. Thus, the strain patterns at different cross sections are identical.

First, we consider the case that an ac voltage $V = V_0 \sin 2\pi f t$ is applied on both y quadrants, as shown in Fig. 9.13. In this case, only uniform elongation or contraction is present. The frequency f is assumed to be much lower than the resonance frequency of the tube. At each instant, the tube is in equilibrium. In response to the instantaneous voltage V, a strain in the two y quadrants is generated:

$$S_3 = \frac{d_{31}V}{h}\,,$$ (9.31)

where d_{31} is the piezoelectric coefficient, and h is the wall thickness of the tube. Because every cross section must remain planar, the x quadrants must deform in the same direction by one-half of the value, and the strain in the y quadrant must be reduced by one-half. This requires a stress in the y quadrants,

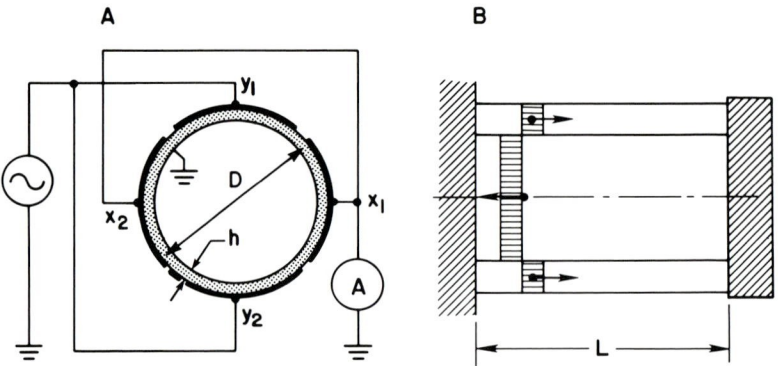

Fig. 9.13. Double piezoelectric response of a tube scanner with symmetric connections. (A) The two y quadrants are connected to an ac voltage source. The two x quadrants are connected to the ground through an ac ammeter. (B) The stress in the x quadrants of the piezoelectric ceramics is equal in magnitude and opposite in sign to the y quadrants. (Reproduced from Chen, 1992a, with permission.)

$$\sigma_3 = -\frac{d_{31}VY}{2h},\tag{9.32}$$

where Y is Young's modulus. (To be exact, it is Young's modulus of the piezoelectric ceramics in the lateral direction under constant electrical field, Y^E_{11}. The superscript and subscript are omitted for convenience.) In the x quadrants, the stress is the negative value of Eq. (9.32),

$$\sigma_3 = \frac{d_{31}VY}{2h}.\tag{9.33}$$

The stress in the x quadrants generates a polarization $P = d_{31}\sigma_3$. The total surface charge on one of the two x quadrants is then

$$Q = \frac{\pi DL(d_{31})^2VY}{8h}.\tag{9.34}$$

By connecting that x quadrant to the ground (the inner metallization), an ac current is generated:

$$I = \frac{dQ}{dt} = \frac{\pi^2 DL(d_{31})^2 Y}{4h}V_0\cos 2\pi ft.\tag{9.35}$$

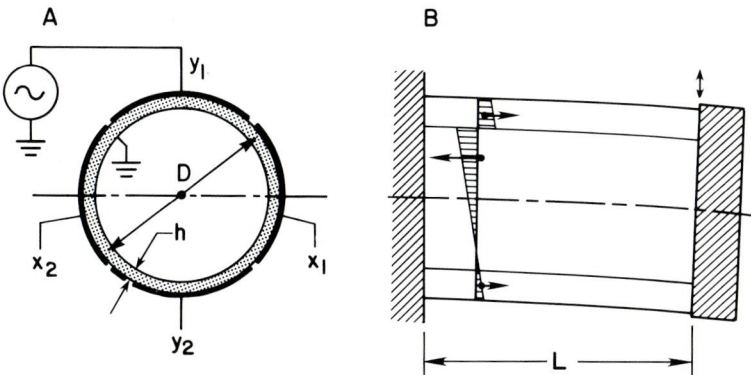

Fig. 9.14. Double piezoelectric response of a tube scanner with asymmetric connections. (A) The y_1 quadrant is connected to an ac voltage source. The current from other quadrants to the ground is measured. (B) Stress in the cross section has a linear distribution with respect to y. (Reproduced from Chen, 1992a, with permission.)

In terms of rms values, $\overline{V}$ and $\overline{I}$, the current is

$$\overline{I} = \frac{\pi^2 f D L (d_{31})^2 Y}{4h} \, \overline{V}. \tag{9.36}$$

If both x quadrants are connected together, the current is doubled:

$$\overline{I} = \frac{\pi^2 f D L (d_{31})^2 Y}{2h} \, \overline{V}. \tag{9.37}$$

The ac current is generated by a combination of piezoelectric effect and inverse piezoelectric effect. In other words, it is a *double piezoelectric response*.

By applying an ac voltage to only one of the quadrants, the distribution of stress on a cross section can be obtained using the conditions of zero force and zero torque, as described in Section 9.5. As shown in Fig. 9.14, if only the y_1 quadrant is activated, the stress distribution is

$$\sigma_3(\theta) = \left(-\frac{1}{4} - \frac{\sqrt{2}}{\pi} \sin \theta \right) \frac{d_{31} V}{h}. \tag{9.38}$$

The surface charge density is $d_{31}\sigma(\theta)$. By integrating over θ, we find the double piezoelectric response on one of the x quadrants to be

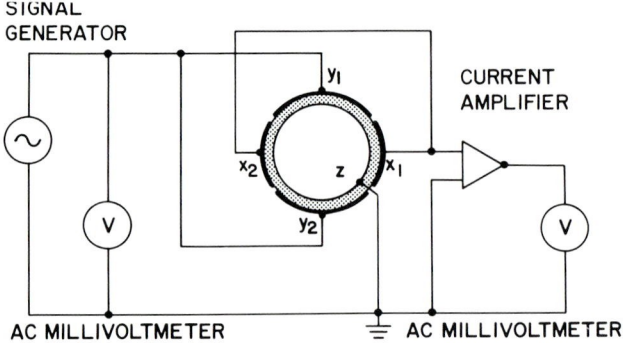

Fig. 9.15. Measuring circuit for the double piezoelectric response of a tube scanner. A sinusoidal signal from a sinusoidal signal generator is connected to the two *y* quadrants. The rms voltage of the sinusoidal signal is measured by an ac voltmeter. The *x* quadrants are connected to the ground through the input of a current amplifier. The output of the current amplifier is measured by another ac voltmeter. (Reproduced from Chen, 1992a, with permission.)

$$\bar{I} = \frac{\pi^2 f DL (d_{31})^2 Y}{8h} \bar{V},$$

(9.39)

and the double piezoelectric response of the y_2 quadrant,

$$\bar{I} = \left(\frac{2}{\pi} - \frac{\pi^2 f DL (d_{31})^2 Y}{8h} \right) \bar{V}.$$

(9.40)

A typical measuring circuit is shown in Fig. 9.15. A signal generator supplies a sinusoidal signal with 600 Ω output impedance. The current is amplified at sensitivity of 10^{-6} A/V. The ac voltages are measured by ac digital voltmeters. The experiment is performed with a PZT-4 tube, provided by EBL, Inc., with $L = 25.4$ mm, $D = 12.7$ mm, $h = 0.50$ mm, and $Y = 7.5 \times 10^{10}$ N/m². The lowest resonance frequency is 5 kHz. The results of measurements are shown in Fig. 9.16. The current, about 1 μA, can easily be measured with 1% accuracy. The current from the two *x* quadrants agrees well with that from the two *y* quadrants. In terms of the units mentioned, the piezoelectric coefficient d_{31} can be obtained from directly measurable quantities as:

$$d_{31} = 45.0 \sqrt{\frac{\bar{I} h}{f \bar{V} DLY}} \ .$$

(9.41)

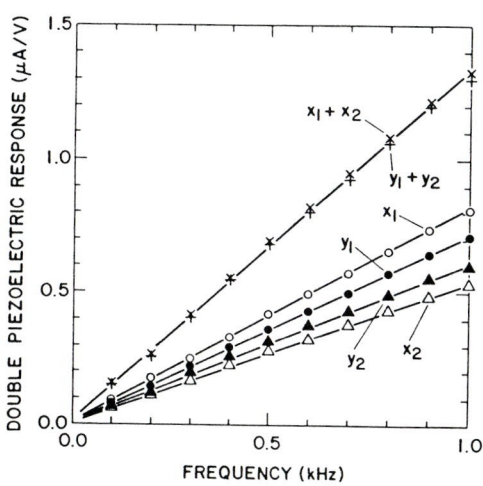

Fig. 9.16. Measured double piezoelectric response of a tube scanner. A linear frequency response is observed. The responses of the two x quadrants are off by about 40%. The responses of the two y quadrants are off by about 16%. (Reproduced from Chen, 1992a, with permission.)

From Fig. 9.16, we obtain $d_{33} \approx 1.05$ Å/V, a value consistent with the value listed in the catalog (1.27 Å/V). The value might be somewhat lower than the true value because the bonding of the tube ends is not perfectly rigid. If one end of the tube is free, or both ends are free, the deformation pattern varies significantly at the end(s). The net end effect is to reduce the value of the double piezoelectric response. Even if the end-bonding condition is unknown, an accurate measurement of the temperature or time variation of the piezoelectric constant can still be achieved. In other words, if the piezoelectric scanner is calibrated by a direct mechanical measurement or by the scale of images at one temperature, then its variation can be precisely determined by the electrical measurements based on double piezoelectric responses.

Actual measurements of the double piezoelectric response also indicated that the double piezoelectric responses from individual quadrants vary significantly. As shown by the example in Fig. 9.16, the currents from the two x quadrants differ by about 40%. The currents from two y quadrants differ by about 16%. Therefore, the double piezoelectric response provides a sensitive method for testing the tube scanner.

9.7 Resonance frequencies

For small and slow signals, in STM, the piezoelectric coefficients are the only relevant parameters. At relatively high frequencies, the dynamic response of the piezoelectric materials becomes important. The lowest resonance frequencies of the piezodrive are the limiting factor for the scanning speed.

A piece of elastic solid has a large number of vibrational modes and vibrational frequencies, which can be determined either by direct experiments or by numerical calculations. Approximate formulas for estimating the lowest resonance frequencies of piezodrives occurring in STM are given in the following subsections.

Stretching mode

The lowest resonance frequency usually corresponds to the standing sound wave with longest wavelength. The velocity of sound c is

$$c = \sqrt{\frac{Y}{\rho}} \, , \tag{9.42}$$

where Y is Young's modulus, and ρ is the density of the material. The lowest vibrational mode of a rod with one end clamped and the other end free is equivalent to a sound wave with wavelength four times the length of the rod L, that is,

$$f = \frac{c}{4L} \, . \tag{9.43}$$

As long as the materials and the cross section of the rod are uniform, the resonance frequency of the stretching mode is independent of the cross section of the rod.

Bending mode

The lowest resonance frequency in the bending mode (that is, flexing mode) of a long and uniform rod, clamped at one end and free at another end, was derived by Lord Rayleigh (1895). The derivation can be found in Appendix F:

$$f = \frac{0.56\kappa c}{L^2} \, , \tag{9.44}$$

again, c is the velocity of sound given by Eq. (9.42), L is the length, and κ is the radius of gyration about the neutral axis of the cross section, which is defined as

$$\kappa^2 = \frac{1}{A} \int z^2 \, dA, \qquad (9.45)$$

where A is the area of the cross section. For a rectangular bar, $\kappa = h/\sqrt{12}$, where h is the thickness in the plane of vibration. For a round bar, $\kappa = D/8$, where D is its diameter. For a tube, $\kappa = (D^2 + d^2)^{1/2}/8$, where D and d are the outer and inner diameters, respectively.

9.8 Repoling a depoled piezo

Under the influence of a high temperature or a high applied voltage, or even due to the aging effect, a piezoelectric ceramic can become depoled. Such depoled piezoelectric ceramics can be repoled in an ordinary laboratory at room temperature using the following procedure:[8]

1. Attach a dc power supply into the convention chosen. For piezo tubes, the common convention is that the outer electrode is positive.

2. Apply a voltage according to the thickness of the piezoelectric element. A rule of thumb is 10 volts per 25 micrometers.

3. Hold the voltage (at room temperature) for more than 8 hours.

4. Detach the power supply and check the piezoelectric constant.

---

[8] Private information, by courtesy of Mr. Russell Petrucci of Stavely Sensors, Inc., Hartford, CT 06108.

CHAPTER 10

VIBRATION ISOLATION

Effective vibration isolation is one of the critical elements in achieving atomic resolution by STM (Binnig et al., 1983). The typical corrugation amplitude for STM images is about 0.1 Å. Therefore, the disturbance from external vibration must be reduced to less than 0.01 Å, or one picometer (1 pm = 10^{-12} m). Analyses of vibration isolation in STM have been conducted by Pohl (1986), Okano et al. (1987), Park and Quate (1987), Kuk and Silverman (1989), and Tiedje and Brown (1990). Vibration and the vibration isolation problem is ubiquitous in mechanical engineering, and there are excellent textbooks about it (Timoshenko, Young, and Weaver, 1974; Frolov and Furman, 1990). We start this chapter with a description of the basic concepts in vibration isolation through the analysis of a one-dimensional system, followed by a discussion of environmental vibration and various examples of vibration isolation systems for STM and AFM.

10.1 Basic concepts

Much of the physics of vibration isolation in STM can be illustrated by a vibrating system with one degree of freedom, as shown on Fig. 10.1 (Frolov and Furman, 1990; Park and Quate, 1987). Also, the formalism developed in this section will be useful for the understanding of the feedback system we will discuss later.

The frame for the instrument always has vibrations transmitted from the ground and the air. The displacement of the frame is described by a function of time, $X(t)$. The STM is represented by a mass M, mounted on the frame. The problem of vibration isolation is to devise a proper mounting to minimize the vibration transferred to the mass, that is, to minimize its displacement of the mass M, $x(t)$. The basic method for vibration isolation is to mount the mass to the frame through a soft spring, as shown in Fig. 10.1. The restoring force of the spring acting on the mass is

$$f = -k(x - X), \tag{10.1}$$

where k is the stiffness of the spring. In addition, a viscous (damping) force is acting between the frame and the mass,

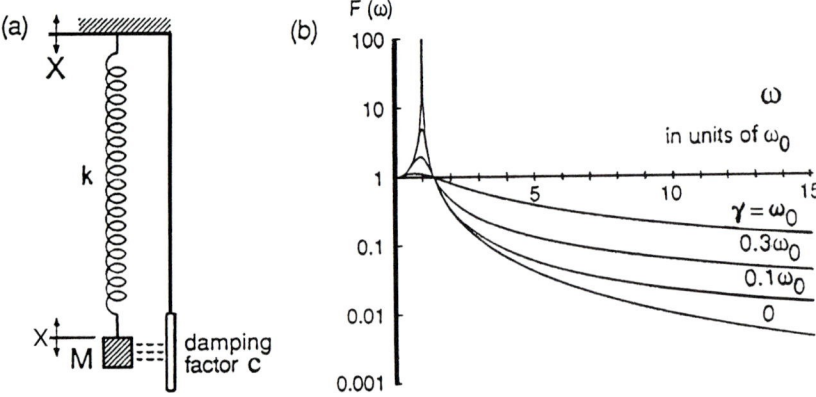

Fig. 10.1. A vibrating system with one degree of freedom and its transfer function. (a) The vibrating system. A mass M is connected to the frame through a spring and a viscous damper. Regarding STM, there are two realizations of this model. First, the frame represents the floor, and the mass represents the STM. Second, the frame represents the base plate (with the sample) in STM, and the mass represents the tip assembly. (b) The transfer function, which is the ratio of the vibration amplitude of the mass to that of the frame at different frequencies. (After Park and Quate, 1987.)

$$f = -c(\dot{x} - \dot{X}). \tag{10.2}$$

By introducing standard parameters, the natural frequency f_0 (or the natural circular frequency $\omega_0 = 2\pi f_0$) and the damping constant γ,

$$\omega_0 = 2\pi f_0 = \sqrt{\frac{k}{M}}, \tag{10.3}$$

$$\gamma = \frac{c}{2M}, \tag{10.4}$$

Newton's equation for the mass is reduced to

$$\ddot{x} + 2\gamma\dot{x} + \omega_0^2 x = f(t) = 2\gamma\dot{X} + \omega_0^2 X. \tag{10.5}$$

The right-hand side, $f(t)$, represents the effect of force transmitted from the frame to the mass.

First, we examine the solution of Eq. (10.5) in the absence of external force, that is, when $f(t) = 0$. The standard method is to make the *Ansatz*

$$x(t) = Ae^{st}. \tag{10.6}$$

Eq. (10.5) reduces to an algebraic equation for s,

$$s^2 + 2\gamma s + \omega_0^2 s = 0. \tag{10.7}$$

If $\omega > \gamma$, the roots of Eq. (10.7) are

$$s = -\gamma \pm i\sqrt{\omega_0^2 - \gamma^2} . \tag{10.8}$$

Introducing a circular frequency with damping,

$$\omega_d = \sqrt{\omega_0^2 - \gamma^2} , \tag{10.9}$$

the general solution of Eq. (10.5) with $f(t) = 0$ is

$$x(t) = Ae^{-\gamma t} \cos(\omega_d t + \alpha). \tag{10.10}$$

The two constants, the amplitude A and the phase angle α, are determined by the initial conditions. It represents a damped oscillation. The formula for the frequency with damping, Eq. (10.9), indicates that the existence of damping slows down the oscillation.

If $\gamma = \omega_0$, the characteristic equation has coincident roots, $s = -\omega_0$. The general solution of Eq. (10.5) is

$$x(t) = (A + Bt)e^{-\omega_0 t}. \tag{10.11}$$

There is no more oscillation; the system has the *critical damping*.

If $\gamma > \omega_0$, the characteristic equation Eq. (10.7) has two negative real roots. The viscous resistance is large that the mass creeps to its equilibrium position from any initial condition.

It is convenient to define a dimensionless number to characterize the "quality" of the oscillator, a number used extensively in mechanical engineering and electrical engineering: the Q factor or Q number,

$$Q \equiv \frac{\omega_0}{2\gamma} . \tag{10.12}$$

Roughly speaking, the Q factor is the number of oscillations the oscillator can sustain after an initial push. The stronger the damping, the smaller the quality factor.

The general solution of Eq. (10.5) with an external force is a superposition of the general solution, Eq. (10.10), plus a term representing the response of the mass to the external force,

$$x(t) = \frac{1}{\omega_d} \int_0^t f(\tau) e^{-\gamma(t-\tau)} \sin \omega_d(t-\tau) \, d\tau. \tag{10.13}$$

A proof of this result is given in Appendix G. For a sinusoidal vibration,

$$X(t) = X_0 e^{i\omega t}, \tag{10.14}$$

at the steady state, the motion of the mass should also be sinusoidal,

$$x(t) = x_0 e^{i\omega t}. \tag{10.15}$$

Substituting Eq. (10.14) and Eq. (10.15) into Eq. (10.5), we obtain

$$\frac{x_0}{X_0} = \frac{\omega_0^2 + 2i\gamma\omega}{\omega_0^2 - \omega^2 + 2i\gamma\omega}. \tag{10.16}$$

The ratio of the amplitudes is the *transfer function* or the *response function* of the vibration isolation system:

$$K(\omega) \equiv \left| \frac{x_0}{X_0} \right| = \sqrt{\frac{\omega_0^2 + 4\gamma^2\omega^2}{(\omega_0^2 - \omega^2)^2 + 4\gamma^2\omega^2}}. \tag{10.17}$$

In the engineering literature, the *decibel* (db) unit is frequently used. The transfer function in terms of decibels is

$$Z = 20 \log_{10} K(\omega) \ (\text{db}). \tag{10.18}$$

Figure 10.1 shows the frequency dependence of the transfer function for the one-stage system. An efficient vibration isolation means a small $K(\omega)$. The qualitative features of such a vibration isolation system can be visualized by considering the following special cases:

1. At high frequencies, if the damping is negligible, then the transfer function is inversely proportional to the excitation frequency:

$$K(\omega) \approx \left(\frac{\omega_0}{\omega} \right)^2 = \left(\frac{f_0}{f} \right)^2. \tag{10.19}$$

2. If the excitation frequency is close to the natural frequency of the system, then the transfer function can be greater than unity, that is, the vibration is worsened. Actually, at $\omega = \omega_0$,

$$K(\omega_0) = \sqrt{1 + \frac{\omega_0^2}{4\gamma^2}}$$

$$\approx \frac{\omega_0}{2\gamma} = Q$$

$$(10.20)$$

If the Q factor is too large, the external vibration at ω_0 would be amplified tremendously. To avoid such *resonance excitation*, appropriate damping must be applied.

3. Damping worsens the vibration isolation at higher frequencies. When $\omega > Q\omega_0$, Eq. (10.17) becomes

$$K(\omega) \approx \frac{1}{Q}\frac{\omega_0}{\omega} = \frac{1}{Q}\frac{f_0}{f}.$$

$$(10.21)$$

The dependence of the transfer function on the ratio (f_0/f) becomes linear instead of quadratic. Therefore, a compromise between the suppression of resonance and the suppression of high-frequency vibrations has to be made. A Q value of 3–10 is usually chosen.

It is clear that the lower the natural frequency ω_0 is, the better the vibration isolation. The natural frequency is the prime parameter of a vibration isolation system. In a realistic STM system, the reduction of natural frequency is not unlimited. If a suspension spring is used, the elongation of the suspension spring with stiffness k due to the weight of the mass M is

$$\Delta L = \frac{Mg}{k},$$

$$(10.22)$$

where $g \approx 9.8$ m/ sec^2 is the gravitational acceleration. The natural frequency of the system is $f_0 = 2\pi\sqrt{k/M}$, or

$$f_0 = 2\pi\sqrt{\frac{g}{\Delta L}} \approx \frac{5.0}{\sqrt{\Delta L(\text{cm})}}.$$

$$(10.23)$$

To make a suspension-spring system with a natural frequency of 1 Hz, the weight of the mass should stretched the spring by 25 cm. Notice that Eq. (10.23) is exactly the formula for the natural frequency of a simple pendulum with length ΔL. To isolate the horizontal vibration, a pendulum is the natural choice. The suspension spring then acts as the isolation device for both vertical and horizontal environmental vibrations.

There is another incarnation for the model in Fig. 10.1. by interpreting the frame as the base plate (with the sample) of the STM, and the mass as the tip assembly, the model describes the influence of external vibration on the relative displacement of the tip versus the sample, which is the quantity we want to reduce. A good STM design means a high resonance frequency. When the excitation frequency is much lower than the natural frequency of the STM, then the tip assembly moves closely with the frame. In fact, when $f \ll f_0$, Eq. (10.16) is reduced to

$$\frac{x_0 - X_0}{X_0} \approx \frac{f^2}{f_0^2}. \tag{10.24}$$

By choosing a rigid STM design, the low-frequency vibration does not affect the relative motion inside the STM.

From Eq. (10.19) and Eq. (10.24), we can reach a general concept for a good STM vibration isolation: By denoting the natural frequency of the STM as f_S and the natural frequency of the vibration isolation system as f_I, then for vibrations with intermediate frequencies f,

$$f_S > f > f_I, \tag{10.25}$$

the overall transfer function is

$$K(f) = \left(\frac{f}{f_S}\right)^2 \left(\frac{f_I}{f}\right)^2 = \left(\frac{f_I}{f_S}\right)^2. \tag{10.26}$$

For example, if the natural frequency of the STM is 2kHz, and a suspension-spring with natural frequency 2Hz is used for vibration isolation, the overall transfer function for intermediate frequencies is a constant, 10^{-6}, or 120db. Therefore, *the rigidity of the STM unit is the most important factor in vibration isolation.*

10.2 Environmental vibration

Before a vibration isolation device is designed, the vibration characteristics of the laboratory has to be measured. A typical instrument for measuring vibration is shown in Fig. 10.2.

The simple mechanical system analyzed in the previous section that can be used as an instrument for this measurement is a seismometer. Figure 10.2

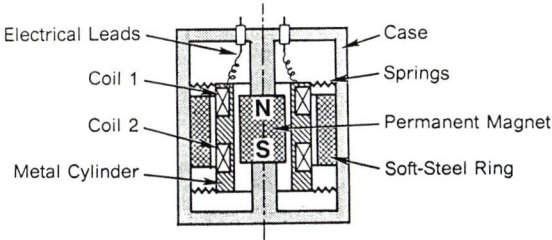

Fig. 10.2. Schematic and working mechanism of the Hall–Sears seismometer, HS-10-1. A metal cylinder, typically 1 kg in weight, is hung from the case with a set of springs. The springs are highly anisotropic; that is, it only allows the cylinder to be mobile in the vertical direction. Two coils are arranged on the mobile cylinder. A permanent magnet in the center generates strong radial magnetic fields. The instantaneous vertical velocity of the metal cylinder generates an instantaneous electromotive force, which can be measured by an oscilloscope or a frequency analyzer. The metal cylinder is also the vibration damper through the eddy current.

is the schematic of a typical seismometer, the Hall–Sears geophone.[9] The function of the seismometer can be understood in terms of the simple model of the previous section.

As shown in Fig. 10.2, a metal cylinder with two coils is suspended by a set of springs. The springs allow the cylinder to vibrate in the vertical direction with a typical natural frequency of 1 Hz. Therefore, for environmental vibrations with frequency higher than 1 Hz, the vibration amplitude of the metal cylinder is negligibly small. The relative vertical motion of the cylinder versus the case is virtually equal to the environmental vibration. A permanent magnet and the pole piece (a soft steel ring) create a strong radial magnetic field. The motion of the coil then generates an electromotive force (emf) that can be detected by an oscilloscope or a frequency analyzer. The quantity measured by the seismometer is the instantaneous velocity. The sensitivity of the seismometer depends on the model. It ranges from 1.2 to 25 V/(cm/sec). The typical frequency of environmental vibration ranges from 10 to 200 Hz (which are also the most harmful frequencies for STM operation). The actual amplitude and frequency distribution of floor vibration depends on the structure of the building, the location of the room inside the building, and the source of vibration. Pohl (1986) observed that the peak frequency of floor vibration in the IBM Zurich laboratory is about 17 Hz, which is probably the resonance frequency of the building. Secondary peaks at 50, 75, and 100 Hz are also observed. Okano et al. (1987) observed that in the Electrotechnical

9   The Hall-Sears Geophone is currently available from OYO Geospace Corporation, 7334 North Gessner Road, Houston, TX 77040.

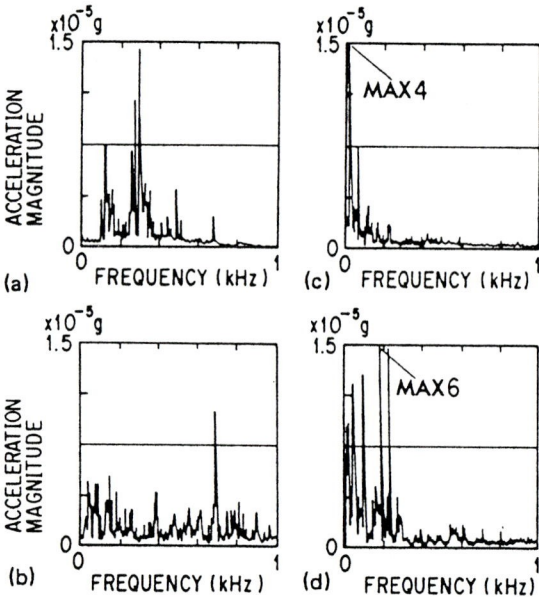

Fig. 10.3. Vibration spectra of laboratory floors, measured in four locations in the Electrotechnical Laboratory, Ibaraki, Japan. (a) Basement, (b) first floor, (c) first floor, another location, and (d) third floor. (From Okano et al., 1987.)

Laboratory of Japan, the peak frequency of the floor vibration is at 180 Hz; see Fig. 10.3. To achieve atomic resolution, an overall transfer function of 10^{-6} or better is needed.

10.3 Suspension springs

In this section, we will discuss vibration isolation systems based on suspension springs with eddy-current damping. To date, it is probably the most efficient vibration isolation system. The design and choice of springs are also discussed. An elementary theory of helical springs, sufficient for all the applications in STM and AFM, is presented in Appendix F.

If the STM is rigid enough, a single-stage suspension spring stage, as described in Section 10.1, is sufficient. Hansma et al. (1988) described a vibration isolation system that consists of a concrete block hung from the ceiling with rubber tubes. The rubber tubes are at the same time springs and dampers. The STM, a very rigid unit (the nanoscope, see Section 12.2), is placed directly on the concrete block.

Fig. 10.4. A two-stage suspension-spring vibration isolation system. Two masses are hung from the frame via two springs and two damping mechanisms. The ratio between the vibration amplitudes of the frame and of the second mass (the transfer function) is calculated. The efficiency of its vibration isolation is much better than the single-stage system. Analysis shows that one damping mechanism alone is sufficient.

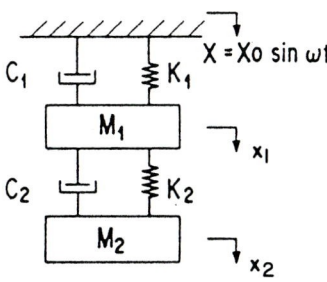

10.3.1 Analysis of two-stage systems

If the STM is not rigid enough, or in UHV and cryogenic environments, more sophisticated vibration isolation systems are needed. To date, vibration isolation systems using a two-stage suspension spring with eddy-current damping is probably the most efficient one. The following analysis is based on the study of Okano et al. (1987).

For the two-stage system shown in Fig. 10.4, Newton's equations for the two masses are:

$$M_1\ddot{x}_1 + c_1\dot{x}_1 + k_1x_1 + c_2(\dot{x}_1 - \dot{x}_2) + k_2(x_1 - x_2) = c_1\dot{X} + k_1X,$$
$$M_2\ddot{x}_2 + c_2(\dot{x}_2 - \dot{x}_1) + k_2(x_2 - x_1) = 0. \tag{10.27}$$

For a sinusoidal external excitation,

$$X = X_0e^{i\omega t}, \tag{10.28}$$

the equations can be brought into matrix form,

$$[A]\,\mathbf{x} = \mathbf{X}, \tag{10.29}$$

where

$$[A] = \begin{bmatrix} k_1 + k_2 - M_1\omega^2 & -k_2 \\ -k_2 & k_2 - M_2\omega^2 \end{bmatrix} + i\omega\begin{bmatrix} c_1 + c_2 & -c_2 \\ -c_2 & c_2 \end{bmatrix}, \tag{10.30}$$

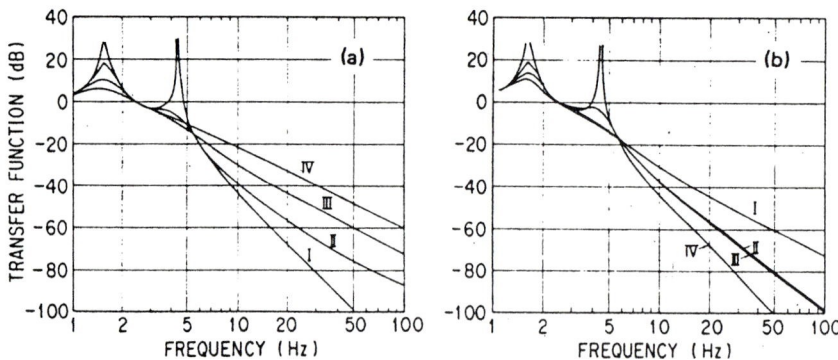

Fig. 10.5. Transfer functions for two-stage vibration isolation systems. Parameters for both (a) and (b): $M_1=2.4$ Kg, $M_2=2.9$ Kg, $k_1=800$ N/m, $k_2=700$ N/m. In (a), the damping stages are equally arranged, $c_1 = c_2 = c$.   I, $c=0$.   II, $c=10$ Ns/m.   III, $c=20$ Ns/m.   IV. $c=50$ Ns/m.   In (b), the effect of different arrangements of damping is illustrated.   I, $c_1 = c_2 = 20$ Ns/m.   II, $c_1=20$ Ns/m, $c_2=0$.   III, $c_1= 0$, $c_2=20$ Ns/m.   IV, $c_1 = c_2=0$.   (After Okano et al., 1987.)

$$\mathbf{x} = \begin{pmatrix} x_1 \\ x_2 \end{pmatrix}, \qquad (10.31)$$

$$\mathbf{X} = \begin{pmatrix} X \\ 0 \end{pmatrix}, \qquad (10.32)$$

The transfer function, in units of decibels, is

$$Z \equiv 20 \log_{10} \left| \frac{x_2}{X} \right|. \qquad (10.33)$$

It is straightforward to obtain an analytic formula for the transfer functions from the matrix equation, Eq. (10.29). Using a desktop computer, the curves can be easily displayed. Typical results are shown in Fig. 10.5. In general, the efficiency of vibration isolation using a two-stage system is much better than a single-stage system. As in the case of a single-stage system, in the absence of viscous damping, the vibration isolation for higher excitation frequencies is optimized. However, the resonances at the natural frequencies of the isolation system are huge. By introducing various amount of viscous damping, the resonance is suppressed. The efficiency of vibration isolation at higher frequencies is affected. An important fact observed from Fig. 10.5(b) is that a single viscous damping mechanism is enough to suppress the resonance. This is much simpler than two damping mechanisms from design point of view.

10.3.2 Choice of springs

A proper choice of the parameters for the extension springs can provide the desired stiffness, endurance, and to minimize the dimension of the instrument. A simple theory of the deflection and maximum stress of springs is presented in Appendix F. The relation between the axial load P and the deflection F of a spring with coil diameter D, number of coils n, and wire diameter d is:

$$P = \frac{Gd^4}{8nD^3} F, \qquad (10.34)$$

where G is the shear modulus of elasticity of the wire. The quantity $(GD^4/8nD^3)$ is often called the *stiffness* or *rate* of a spring.

The maximum shear stress in the wire is

$$\tau_{max} = \frac{8D}{\pi d^3} P. \qquad (10.35)$$

In choosing a spring, the maximum shear stress must not exceed the endurance limit of the material. Table 10.1 lists these two parameters for commonly used spring materials.

Table 10.1. Typical properties of common spring materials

Material	Modulus of rigidity G	Yield strength τ_{max}
Music wire	8.0×10^{10} N/m²	3.9×10^4 N/m²
Stainless steel	6.7×10^{10} N/m²	5.8×10^4 N/m²

From Eq. (10.34) we see that to make a soft spring, it is not necessary to use thin wires. The stress in the thin wire might exceed its yield strength, and permanent deformation might occur. A larger coil diameter and a larger number of coils also make a softer spring. According to Eq. (10.35), the maximum shear stress in a relatively thick wire is much smaller.

We have shown in Eq. (10.23) that the natural frequency of a spring–mass system is only linked with the stretch ΔL under gravity. In any

STM design, it is better to reduce both the total length of the spring and the resonance frequency. This can be achieved by choosing extension springs that are preloaded, or with an initial tension. The extension spring will not be stretched unless a weight greater than the initial tension is applied.

In the actual design work, the springs are chosen from catalogs of manufacturers.[10]

10.3.3 Eddy-current damper

When a conductor moves in a magnetic field, damping forces are generated by eddy currents induced in the conductor. The magnetic damper, with its reliability and thermal stability, has been utilized in various branches of engineering. A recent analysis of it is given by Nagaya and Kojima (1984).

A schematic of an eddy-current damper is shown in Fig. 10.6. A copper block with resistivity ρ is under the influence of the magnetic field B of a permanent magnet. The relative motion of the magnet and the copper block in the x direction causes a force in the $-x$ direction. By solving Maxwell's equation numerically, Nagaya (1984) calculated the force to be

$$F_x = -cV_x = -C_0 \left(\frac{B^2 \pi a^2 t}{\rho} \right) V_x. \tag{10.36}$$

This equation shows the general characteristics of the damping: The force is proportional to the area πa^2, thickness t, velocity V_x, conductivity of the copper block $1/\rho$, and *square* of the magnetic field intensity. Therefore, by using strong permanent magnets, for example, Co_5Sm magnets, the damping

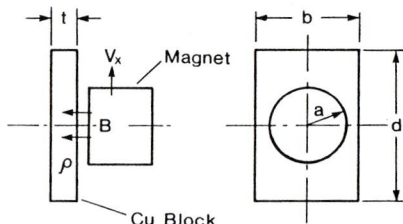

Fig. 10.6. Eddy-current damper. A magnet is placed against a metal plate of resistivity ρ. When there is a relative velocity between them, a viscous force occurs. The force constant depends on the strength of the magnetic field and the dimensions of the plate and the magnet. (After Okano et al., 1987.)

---

10 For example, Lee Spring Company, 1462 62nd Street, Brooklyn, New York 11219.

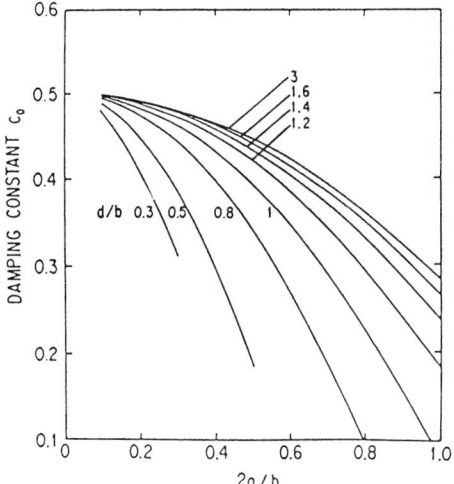

Fig. 10.7. Dimensionless constant in the calculation of the damping constant. While using Eq. (10.36) to calculate the damping constant, the dimensionless parameter C_0 can be chosen from this chart. (After Nagaya, 1984.)

force can be made very strong. If the size of the copper block is much larger than the diameter of the magnet, $C_0 = 0.5$. If the sizes are comparable, the dimensionless constant C_0 depends on the geometry. Nagaya (1984) gave the values of C_0 as a function if the ratios $2a/b$ and d/b, as shown in Fig. 10.7. In a design by Okano et al. (1987), $B= 0.2$ T, $a=2$ cm, $b=6$ cm, $d=6$ cm, $t=2$ cm, and for copper, $\rho = 1.56 \times 10^{-8}$ Ω/m. From Fig. 10.7 one finds $C_0 =0.31$. Eq. (10.36) gives $c \approx 20$ Ns/m. It fits reasonably well with their measurements.

10.4 Stacked plate–elastomer system

The suspension-spring vibration isolation system has a disadvantage: It is large. Gerber et al. (1985) developed a vibration isolation system using a stack of metal plates separated by rubber pieces. It was first used in the pocket-size STM, which will be described in Section 13.1. By increasing the number of metal plates, reasonably good vibration isolation can be achieved in a modest space (see Fig. 10.8).

An analysis of the transfer function of this system can be made using the matrix method described by Okano et al. (1987). However, the stiffness of the rubber pieces is highly nonlinear. Okano et al. (1987) found that the measured transfer function does not fit theoretical predictions based on a constant stiffness. A nonlinear elastic behavior must be taken into account. Another problem with the metal-stack system is that the resonance frequency is around

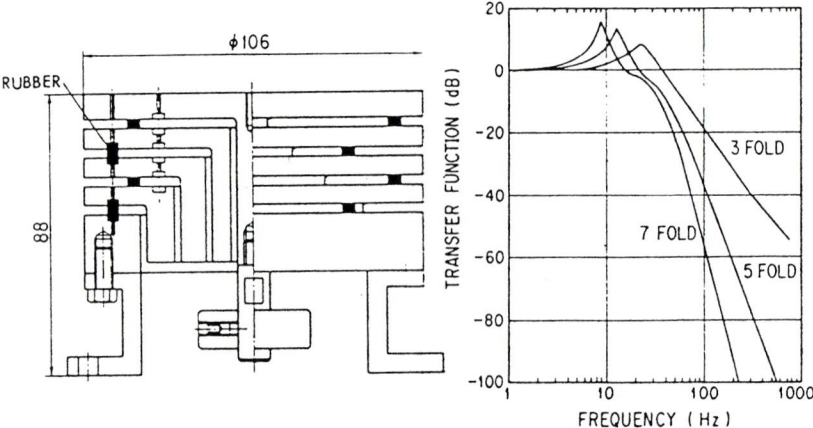

Fig. 10.8. Stacked plate system and its transfer function. (a) A fivefold stacked-plate vibration isolation system, with four sets of viton pieces between metal plates. (b) Solid curve, measured transfer function. Dashed curve, transfer function calculated with a constant stiffness. Dash-dotted curve, transfer function calculated with the measured nonlinear transfer function. (After Okano et al., 1987.)

10 Hz *because of its small size*. The isolation is effective only for vibrations with relatively high frequencies (>50 Hz). To suppress the low-frequency vibrations, an additional suspension-spring stage is needed.

10.5 Pneumatic systems

Numerous pneumatic vibration isolation systems are commercially available. The prime market of these systems are for optical benches. The typical natural frequency is 1–2 Hz. For vibrations with frequencies larger than 10 Hz, a transfer function of 0.1 can be achieved. Some systems provide effective vibration isolation only in the vertical direction, whereas others are effective for horizontal directions as well. All those systems are fairly bulky. If the STM cannot be isolated from the chamber in which it resides, the entire chamber has to be vibration isolated. In this case, the commercial pneumatic system is the choice.

CHAPTER 11

ELECTRONICS AND CONTROL

11.1 Current amplifier

The tunneling current occurring in STM is very small, typically from 0.01 to 50 nA. The current amplifier is thus an essential element of an STM, which amplifies the tiny tunneling current and converts it into a voltage.[11] The performance of the current amplifier, to a great extent, influences the performance of the STM. There are natural limits for the overall performance of current amplifiers, as determined by the thermal noise, stray capacitance, and the characteristics of the electronic components. In this section, we will present these issues by analyzing several typical current amplifier circuits, which can be easily made and used in actual STMs.

Figure 11.1 shows two basic types of current amplifiers. The first one is called a feedback picoammeter (Keithley et al., 1984). The simplest circuit consists of two components, an operational amplifier A, and a feedback resistor R_{FB}. An operational amplifier (usually abbreviated as op-amp) has a very high input impedance, a very high voltage gain, and a very low output impedance (see Appendix H for details). To a very good approximation, the output voltage should provide a feedback current through the feedback resistance R_{FB} to compensate the input current such that the net current entering the inverting input of the op-amp is zero. The noninverting input is grounded, and the voltage at the inverting input should be equal to ground. This implies

$$V_{OUT} = - I_{IN} R_{FB}.$$ (11.1)

The minus sign indicates that the phase is reversed. For $R_{FB} = 100\,M\Omega$, one nanoampere of input current results in an output voltage of 100 mV. In some cases, the sample must be grounded, then the noninverting input of the op-amp may be connected to a fixed dc voltage as the bias. The output voltage is the

---

11 Strictly speaking, it is a transimpedence amplifier, because the output is voltage instead of current. However, we will use the term *current amplifier* when no ambiguity occurs.

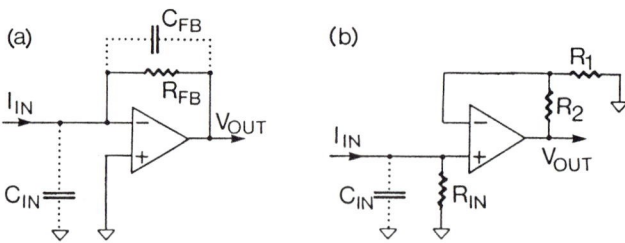

Fig. 11.1. Two basic types of current amplifiers. (a) Feedback picoammeter. It consists of two components, an operational amplifier (op-amp) A, and a feedback resistor R_{FB}. A typical value of the feedback resistor used in STM is $10^8\ \Omega$. The stray capacitance C_{FB} is an inevitable parasitic element in the circuit. In a careful design, $C_{FB} \approx 0.5$ pF. The input capacitance C_{IN} is also an inevitable parasitic element in the circuit. Those parasitic capacitors, the thermal noise of the feedback resistor, and the characteristics of the op-amp are the limiting factors to the performance of the picoammeter. (b) An electrometer used as a current amplifier (the shunt current amplifier). The voltage at the input resistance R_{IN} is amplified by the circuit, which consists of an op-amp and a pair of resistors R_1 and R_2. The parasitic input capacitance C_{IN} limits the frequency response, and the Johnson noise on R_{IN} is the major source of noise. Also, the input resistance for this arrangement is large.

sum of the fixed dc voltage and the voltage corresponding to the current. The fixed dc voltage is then subtracted from the output.

There are several factors which impose natural limits for the performance of that current amplifier, such as the stray capacitance parallel to the feedback resistance, the stray capacitance parallel to the input terminals, which will be discussed in the following subsections.

Another possible circuit is a voltage amplifier, with a shunt resistor to convert the input current to input voltage (see Fig. 11.1. This type of current amplifier has more disadvantages than the picoammeter: The input capacitance is always much larger than the stray capacitance across the feedback resistance, which seriously affects the frequency response; the input impedance of that current amplifier is vary large, which makes the actual bias different from the applied bias. Therefore, we will concentrate on the picoammeter.

11.1.1 Johnson noise and shot noise

The thermal motion of electrons in a resistor results in noise that is independent of the nature of the resistance. The power of such thermal noise, within a frequency interval Δf, is

$$P = 4k_B T\, \Delta f. \tag{11.2}$$

It is often called the *Johnson noise,* after its discoverer (see, for example, Ott, 1976). From Eq. (11.2), we find that the rms Johnson voltage noise in a resistor R is

$$E = \sqrt{4k_B T R \, \Delta f} \, , \tag{11.3}$$

and the rms Johnson current noise through a resistor R is

$$I = \sqrt{\frac{4k_B T \, \Delta f}{R}} \, . \tag{11.4}$$

The peak-to-peak noise value is approximately 8 times the rms value. For example, at room temperature, over a frequency interval of 3 kHz, on a 100 MΩ resistor, the Johnson current noise is

$$I \approx 8 \times \sqrt{\frac{4 \times 1.38 \times 10^{-23} \times 3 \times 10^3}{10^8}} \approx 0.3 \text{ pA}. \tag{11.5}$$

The larger the feedback resistor, the smaller the current noise. By using a 1 MΩ feedback resistor, the theoretical noise becomes 2 pA, a tangible value when very small tunneling current is measured.

The noise of an actual resistance is always higher than the theoretical limit. While for metal resistors the noise level is close to the theoretical limit, the noise level in carbon resistors is much higher. The resistance of the tunneling junction, which is parallel to the feedback resistor, should be taken into account when its value is comparable to that of the feedback resistor.

In addition to the thermal noise, the discrete nature of the current results in *shot noise,* whose rms value over a frequency interval Δf is of the form (Ott, 1976):

$$I_{\text{shot}} = \sqrt{2e\bar{I} \, \Delta f} \approx 5.66 \times 10^{-10} \sqrt{\bar{I}} \, , \tag{11.6}$$

where e is the electron charge and $\bar{I}$ is the average dc current, in units of amperes. Except for an extremely low current ($\bar{I} < 1$ pA), the shot noise is negligible.

In reducing Johnson noise, to incorporate all the amplification in one single stage is the best choice. However, a stray capacitance C_{FB} is always present. In this case, the relation between the input current and the output voltage is determined by the differential equation

$$I_{\text{IN}} = \frac{V_{\text{OUT}}}{R_{\text{FB}}} + C_{\text{FB}} \frac{dV_{\text{OUT}}}{dt} \, . \tag{11.7}$$

The solution is

$$V_{\text{OUT}} = R_{\text{FB}} I_{\text{IN}} \left[1 - \exp\left(-\frac{t}{R_{\text{FB}}C_{\text{FB}}} \right) \right]. \tag{11.8}$$

The existence of the stray capacitance results in a time delay $R_{\text{FB}}C_{\text{FB}}$. The frequency response of the amplifier is reduced. In a practical design, a compromise between bandwidth and noise is often to be made, as explained in the following subsection.

11.1.2 Frequency response

First, we analyze the effect of the stray capacitance in parallel with the feedback resistance. Consider a sinusoidal input current with frequency f. From Fig. 11.1, the magnitude of the output voltage is

$$|V_{\text{OUT}}| = \frac{I_{\text{IN}}R_{\text{FB}}}{\sqrt{1 + (2\pi f R_{\text{FB}}C_{\text{FB}})^2}}. \tag{11.9}$$

The output voltage drops by a factor of $1/\sqrt{2}$, or -3 dB, at a frequency

$$f = \frac{1}{2\pi R_{\text{FB}}C_{\text{FB}}}. \tag{11.10}$$

Even if a very careful layout design is made, a stray capacitance of $C_{\text{FB}} = 0.5$ pF is common. With $R_{\text{FB}} = 100$ MΩ, using Eq. (11.10), the -3 dB cutoff frequency is estimated to be $f \approx 3$ kHz. For most applications in STM, a gain of 1V/1nA is desirable. A one-stage current amplifier requires a feedback resistance of 1GΩ. The -3 dB cutoff frequency would be about 0.3 kHz, which is too low.

To design current amplifiers with a broader frequency range, and at the same time to have a sufficient gain, a standard method is to have multiple stages (that is, cascade amplifier). The same goal can be achieved by using a resistance network and a single op-amp, as shown in Fig. 11.2. The frequency is inversely proportional to the immediate feedback resistor R_1. The Johnson noise, according to Eq. (11.4), is proportional to the square root of the feedback resistance R_1. Therefore, either by using cascade amplifiers or using the resistance network shown in Fig. 11.2, the improvement of the frequency response is accompanied by a sacrifice of noise level.

A method to improve frequency response without sacrificing noise level is to introduce an RC feedback network, also shown in Fig. 11.2. A simple calculation (we leave it as an exercise for the reader) shows that for a

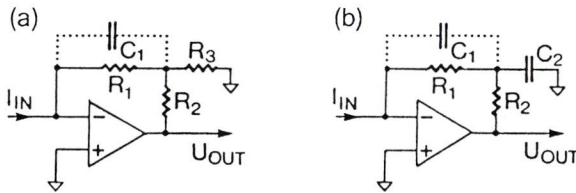

Fig. 11.2. Broad-band current amplifiers. (a) By replacing the feedback resistor in Fig. 11.1 with a resistor network, the cutoff frequency of the amplifier can be greatly increased, but the Johnson noise is increased. (b) Broad-band current amplifier with a compensation capacitor. By introducing a condensation capacitor C_2, the effect of C_1 can be reduced. Under the condition $C_1 R_1 = C_2 R_2$, the frequency range is substantially expended. The Johnson noise is not affected.

sinusoidal signal of circular frequency ω, the relation between the input current and the output voltage is

$$V_{OUT} = I_{IN}(R_1 + R_2) \frac{1 + i\omega R_2 C_2}{1 + i\omega R_1 C_1}. \tag{11.11}$$

If $R_1 C_1 = R_2 C_2$, then there is no phase difference between I_{IN} and V_{OUT}. The optimum value of C_2 can be determined by trial and error. Taking $R_1 = 100\ \mathrm{M\Omega}$ and $R_2 = 1\ \mathrm{M\Omega}$, the typical value of C_2 is 10–20 pF. The compensation is never perfect, because there are other stray capacitances in the circuit, which are not compensated by this simple RC network. Nevertheless, a bandwidth of 100 kHz is attainable.

Another limiting factor is the bandwidth of the op-amp. On the factory specifications, the commonly used indicator is the gain–bandwidth product. The nominal dc gain is valid up to a cutoff frequency f_c which is typically 10 Hz. Above that frequency, the gain g is inversely proportional to the frequency. The product of gain and frequency, the gain–bandwidth product f_T, is typically 1 MHz. The input impedance of the amplifier increases with frequency:

$$Z_{IN} = \frac{R_{FB}}{g} = R_{FB} \frac{f}{f_T}. \tag{11.12}$$

To generate a voltage on the noninverting input to compensate the input current, For example, if $R_{FB} = 100\ \mathrm{M\Omega}$, at $f = 100\ \mathrm{kHz}$, $g = 10$, then $Z_{IN} = 10\ \mathrm{M\Omega}$. At I=1nA, it creates a 10 mV offset voltage.

Fig. 11.3. The influence of the input capacitance on output noise. To make a simple estimation, the input noise of the op-amp is represented by an ac source at the noninverting input end. The smaller the input impedance, the larger the noise at the output end. Therefore, the input capacitance generates a large high-frequency noise.

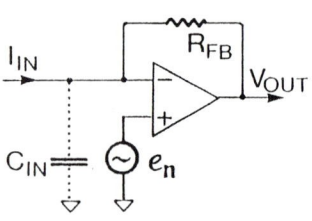

11.1.3 Microphone effect

If a coaxial cable is used to connect the tunneling tip to the input of the amplifier, a major source of noise is the capacitance of the coaxial cable itself. The typical capacitance between the center conductor and the shielding is 100 pF per meter. In response to the acoustic noise in the room, the coaxial cable deforms. The capacitance changes. The current is

$$I = \frac{dQ}{dt} = C\frac{dV}{dt} + V\frac{dC}{dt}. \tag{11.13}$$

For example, if the voltage on the coaxial cable is 10 mV, a noise of 1 kHz makes a periodic change of capacitance with an amplitude of 1 pF, and the noise current is 60 pA, a tangible value. The phenomenon is the same as the principle of the capacitance microphone used in almost every portable tape recorder. To avoid such a microphone effect, the best way is to connect the current amplifier as close to the current source as possible and eliminate the coaxial cable. Almost every commercial STM uses such an arrangement.

The coaxial cable is the major source of noise for yet another reason. At the input of the op-amp, there is always a small voltage noise, which is amplified by the op-amp and appears at the output end. To make a simplified analysis, let the input noise be represented by an ac source at the noninverting input end. The output voltage is (see Fig. 11.3)

$$V_{OUT} = e_n\left(1 + \frac{Z_{OUT}}{Z_{IN}}\right)$$
$$= e_n(1 + 2\pi f C_S R_{FB}). \tag{11.14}$$

For example, if $R = 10^8$ Ω and $C_s = 100$ pF (a 1-m coaxial cable), at 16 Hz the noise is doubled from its low-frequency limit, and at 1.6 kHz the noise is increased by a factor of 100. The noise amplification continues until the frequency approaches the gain–bandwidth product. It is very easy to demonstrate such a noise. By attaching a one-meter coaxial cable to the input of an Ithaco 1311 or a Keithley 427 current amplifier set at 10^9 V/A, the noise will increase suddenly by more than an order of magnitude.

It should be noted that all the elements in the circuits of Fig. 11.2, including the circuit board, in ultra-high-vacuum (UHV) compatible form, are available commercially. Therefore, the entire broad-band current amplifier can be enclosed in a UHV chamber and located as close as possible to the tip.

11.1.4 Logarithmic amplifier

As we have described in Section 1.2, the tunneling characteristics, I_T versus gap width s, are highly nonlinear:

$$I_T \propto e^{-2\kappa s}. \tag{11.15}$$

To make the entire electronic response linear with respect to tunneling gap s, a logarithmic amplifier is attached at the output of the current amplifier. A logarithmic amplifier can be made from a feedback amplifier, by replacing the feedback resistor with a diode, as shown in Fig. 11.4. The current–voltage characteristics of a good-quality, forward-biased silicon diode follow an exponential law over more than five orders of magnitude:

$$I = I_0 e^{eV/2k_B T}. \tag{11.16}$$

The condition that, at the inverting input of the op-amp, the current and the voltage are zero, implies

$$U_{OUT} = \text{const.} + \frac{2k_B T}{e} \ln I_{IN}. \tag{11.17}$$

For every decade of input, the output changes about 120 mV. In commercial logarithmic amplifiers, a subsequent linear amplification stage further amplifies the output voltage to a preset value. A typical input–output characteristic of a commercial logarithmic amplifier is shown in Fig. 11.4.

By using an input resistor, the logarithmic amplifier accepts voltage input (see Fig. 11.4). The logarithmic amplifier only accepts one polarity of input current. Because the tunneling current can be both positive and negative, an electronic switch is needed to reverse the polarity if the input voltage is of the wrong sign.

It is possible to use a logarithmic amplifier as the first-stage current amplifier, which may become the only stage before being compared with the set point of the tunneling current. The disadvantages are: The leakage current

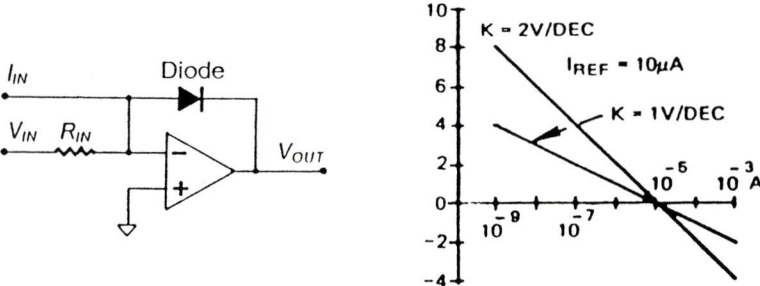

Fig. 11.4. Logarithmic amplifier. (a) Schematic of a logarithmic amplifier. A diode is used as the feedback element in a current amplifier. The current–voltage characteristics are exponential. The output voltage is then proportional to the logarithm of the input current. (b) The transfer curve of a typical logarithmic amplifier, AD757N from Analog Devices. The reference current is internally set to be 10 μA. It is accurate up to six decades.

of a typical diode is comparable to or larger than the typical tunneling current, which reduces the accuracy; the parasitic capacitance of a diode is of the order of tens of picofarads, which significantly reduces the bandwidth.

11.2 Feedback circuit

The output of the logarithmic amplifier is compared with a reference voltage, which represents the set point of the tunneling current. The error signal is then sent to the feedback circuit, a judicially designed amplifier, which sends a voltage to the z piezo. The phase of the collection of all the amplifiers is chosen to constitute a negative feedback: If the tunneling current is larger than the preset current, then the voltage applied to the z piezo tends to withdraw the tip from the sample surface, and vice versa. Therefore, an equilibrium z position is established through the feedback loop. As the tip scans in the x direction, the contour height z also changes with time. The function of the feedback circuit is to make the tip accurately follow the constant tunneling current contour at the highest possible speed. We first analyze the steady-state response of the feedback mechanism, then study its transient response.

11.2.1 Steady-state response

A schematic of the feedback system of an STM is shown in Fig. 11.5. The tunneling current, after the current amplifier and the logarithmic amplifier, is compared with a reference voltage, which represents the tunneling current set point. During a scan, the instantaneous height of the constant tunneling current surface contour changes with time,

$$z(x) = z(x_0 + vt), \tag{11.18}$$

where v is the velocity of scanning. The tunneling gap is $s = z_T - z(x)$, where z_T is the height of the tip, controlled by the z piezo. Thus, the tunneling current is

$$I_T = I_0 e^{-2\kappa[z_T - z(x)]}. \tag{11.19}$$

The minus sign indicates that the further away the tip is from the surface, the smaller the tunneling current. After the current is amplified by the current amplifier and the logarithmic amplifier, the signal becomes

$$V_L = K_V \ln I_T = \text{const.} - 2\kappa K_V[z_T - z(x)]. \tag{11.20}$$

The constant K_V is determined by the logarithmic amplifier. For a 2 V/decade logarithmic amplifier, $K_V = 2$ V/ $\ln 10 \approx = 0.87$ V. If $\kappa = 1$ Å^{-1}, then $2\kappa K_V \approx 1.74$ V/Å. Using a 1 V/decade logarithmic amplifier, $2\kappa K_V \approx 0.87$ V/Å. This voltage is then compared with the current setpoint, V_P, which is also represented by a voltage. It is always convenient to define the zero of the z piezo that the setpoint is equal to the "const." term in Eq. (11.20). The error signal is then

$$\epsilon = -2\kappa K_V[z_T - z(x)]. \tag{11.21}$$

Two points are worth noting here. First, the gain of the current amplifier does not appear in Eq. (11.21). Actually, it only affects the definition of the zero point of the z piezo, which has nothing to do with the overall gain of the circuit. However, an appropriate gain is essential to ensure that the logarithmic amplifier is working in its best range. Second, the decay constant κ is an essential element in the overall gain of the circuit. This is expected and understandable — and should be kept in mind.

The error signal is further amplified by the feedback electronics,

$$V_Z = K_F \epsilon, \tag{11.22}$$

where K_F is a dimensionless number, typically $K_F = 10-200$. The output voltage drives the z piezo:

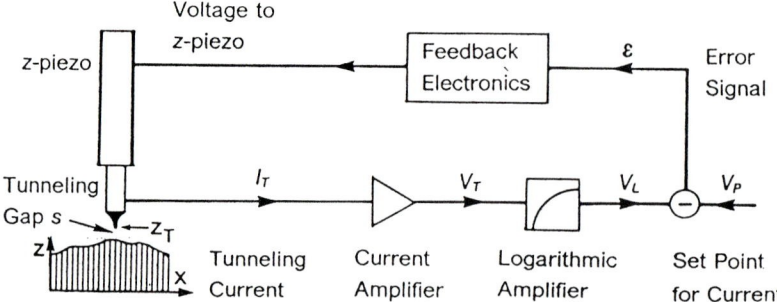

Fig. 11.5. A schematic of the feedback loop in an STM. The tunneling current, after the current amplifier and the logarithmic amplifier, is compared with a predetermined voltage, which represents the current setpoint. The error signal is processed by the feedback electronics, which typically contains an amplifier and an integration circuit. The output of the feedback electronics is applied to the z piezo, to keep the error between the actual tunneling current and the reference current very small. The voltage applied to the z piezo is recorded as the topographic image.

$$z_T = K_Z V_Z. \tag{11.23}$$

where K_Z is the piezo constant, typically of the order of 10–100 Å/V. Combining Equations (11.21), (11.22), and (11.23), we find

$$z_T = -G[z_T - z(x)], \tag{11.24}$$

where the dimensionless positive number G is the *open-loop gain,*

$$G = 2\kappa K_V K_F K_Z. \tag{11.25}$$

Using the typical values given, we estimated that the open-loop gain G can be adjusted from 10^2 to 10^4. From Eq. (11.23) we find

$$z_T = \frac{G}{1+G} z(x) \approx \left(1 - \frac{1}{G}\right) z(x). \tag{11.26}$$

Because usually $1/G < 10^{-2}$, the tip closely follows the contour of the constant tunneling current surface of the sample. The higher the open-loop gain G, or the stronger the feedback, the more accurately the tip follows the constant-I_T contour.

It is instructive to see, qualitatively, the effect of positive feedback. For positive feedback, Eq. (11.26) becomes

$$z_T = \frac{1}{1 - \dfrac{1}{|G|}} z(x). \tag{11.27}$$

Qualitatively, the gain G can be considered as a frequency-dependent quantity. At a certain frequency, it may happen to have $G = 1$. The amplitude of the motion of the tip becomes infinitely large regardless of the value of $z(x)$. This indicates an oscillation. Further, if at low frequencies there is a negative feedback, it does not guarantee that the system does not oscillate at higher frequencies. At least, the reaction of the piezo is not instantaneous. The time delay causes the feedback to become positive at certain frequencies. Furthermore, the time delay of the components in the feedback loop also sets stringent limits on the transient response of the STM. The following two sections are devoted to the understanding of those problems, that is, the stability and the transient response of STM.

11.2.2 Transient response

During a scan, the instantaneous height of the surface contour changes. The voltage to the z piezo responds. In any application, we wish the response to be as fast as possible. However, there are delays in the loop. The piezodrive does not respond instantaneously because of its mechanical inertia. The current amplifier has a response time τ. On the other hand, we wish the response to be as accurate as possible, which requires a large gain G. Nevertheless, the gain cannot be raised indefinitely. Because of the delays in the circuit, at one point, the circuit will start to oscillate — it becomes unstable. The analysis of such a problem leads to high-order differential equations. In the case of STM, as in many problems in feedback control engineering, it is appropriate to simplify it to differential equations with *constant coefficients*. The standard method to solve these kinds of differential equations is the Laplace transform, which converts the differential equations to algebraic equations. For a function of time $f(t)$, defined for $t \geq 0$, its Laplace transform $F(s)$, a function of the complex variable s, is defined as

$$F(s) = \int_0^\infty e^{-st} f(t) dt. \tag{11.28}$$

A brief summary of Laplace transforms necessary for understanding the problems in STM is compiled in Appendix G.

To formulate the problem in a convenient way, we consider that during a scan, the contour of the sample surface has a unit jump at $t = 0$. In other words, the height is assumed to be a unit step function,

$$z(t) = \begin{cases} 0, & t < 0, \\ 1, & t \geq 0. \end{cases} \tag{11.29}$$

We do not lose generality by considering such a unit step function. Because the differential equation is linear, by making superposition of the step function, the response from any surface contour can be treated. The Laplace transform of a step function is

$$Z(s) = \frac{1}{s}. \tag{11.30}$$

The problem is to find the response of the tip, $z_T(t)$ by first finding its Laplace transform, $Z_T(s)$.

The error voltage $\epsilon(t)$ relates to the tunneling gap by

$$\left(1 + \tau \frac{d}{dt}\right)\epsilon(t) = -2\kappa K_V[z_T(t) - z(t)]. \tag{11.31}$$

The Laplace transform of Eq. (11.31) is

$$(1 + \tau s)E(s) = 2\kappa K_V\left(Z_T(s) - \frac{1}{s}\right). \tag{11.32}$$

From the definition of the Laplace transform, Eq. (11.28), it is straightforward to show that it replaces a differential operator d/dt by the Laplace variable s (see Appendix G for details). The feedback circuit is typically an amplifier with an RC network, as shown in Fig. 11.6. The RC network is used for compensation, which will be explained here. By denoting the Laplace transform of the voltage on the z piezo, $V_z(t)$, by $U(s)$, the Laplace transform of the feedback circuit is

$$(1 + RCs)U(s) = K_F E(s). \tag{11.33}$$

The Laplace transform of the piezodrive is

$$\left(\frac{s^2}{\omega_0^2} + \frac{s}{Q\omega_0} + 1\right)Z(s) = K_Z U(s). \tag{11.34}$$

Using elementary algebra, the response of the tip, in the form of Laplace transform, is obtained:

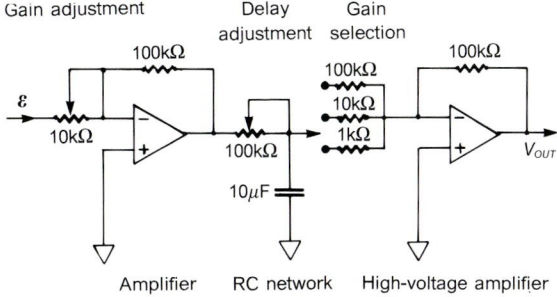

Fig. 11.6. Simple feedback electronics with integration compensation. The first op-amp amplifies the error signal with a variable gain. An *RC* network provides an integration compensation. A high-voltage op-amp provides an output of ±100 V or more, to drive the z piezo.

$$Z_T(s) = F(s)\,\frac{1}{s}$$

$$= \frac{G}{G + (1 + \tau s)(1 + RCs)\left(1 + \dfrac{s}{Q\omega_0} + \dfrac{s^2}{\omega_0^2}\right)}\,\frac{1}{s}, \qquad (11.35)$$

where the gain G is defined in Eq. (11.25) The response function $F(s)$, defined by Eq. (11.35), is a rational (algebraic) function, which can be expanded into partial fractions. In principle, by taking the inverse Laplace transform, the solution in the time domain can be obtained immediately. The form of the denominator in Eq. (11.35) is a fifth-order polynomial. Using any standard procedure in elementary algebra, Eq. (11.35) can be written in terms of the following partial fraction expansion,

$$Z_T(s) = \sum_{n=1}^{5} \frac{C_n}{s - s_n}, \qquad (11.36)$$

where s_n are roots of the denominator, or the *poles* of $F(s)$ on the complex s plane. For simplicity, we assume that there are no duplicate poles. The response of the tip with time is then

$$z_T(t) = \sum_{n=1}^{5} C_n e^{s_n t}. \qquad (11.37)$$

This general solution, known for more than a century, can be found in any standard textbook on control engineering, for example, DiStefano et al. (1967). The entire problem can be solved by a desktop computer, with the response curves displayed. The parameters can be chosen by correlating with the response curves.

Before the widespread use of computers, the direct numerical solution was too time consuming. A number of semiqualitative methods were developed to determine the condition of stability. These methods are very effective in providing guidance to improve the design and to choose the parameters. Readers who are familiar with these methods can apply them to the problem of STM with very little effort. In the following, we describe a method for approximating the problem as a second-order differential equation, to provide physical insight of the transient response in STM. No previous knowledge of control theory is required.

First, we assume that there is no delay in the current amplifier and the feedback circuit; that is, $\tau = 0$ and $RC = 0$. From the partial fraction expansion of Eq. (11.35), we find the poles are at

$$
\begin{aligned}
s_1 &= -\beta + i\omega_1, \\
s_2 &= -\beta - i\omega_1, \\
s_3 &= 0.
\end{aligned}
\tag{11.38}
$$

The tip response can be obtained by using the short Laplace transform table in Appendix G:

$$
z_T(t) = \frac{G}{G+1} \left[1 + e^{-\beta t} \sec\alpha \; \cos(\omega_1 t + \alpha) \right],
\tag{11.39}
$$

where

$$
\beta = \frac{\omega_0}{2(G+1)Q},
\tag{11.40}
$$

$$
\omega_1 = \sqrt{\frac{\omega_0^2}{G+1} - \beta^2},
\tag{11.41}
$$

and

$$
\tan\alpha = \frac{\beta}{\omega_1}.
\tag{11.42}
$$

The effect of a large gain G is clearly seen in these equations. First, the time to reach equilibrium, $1/\beta$, is substantially extended. Second, if at small gain the system is critically damped or overdamped, that is, $\omega_1 = 0$ or is imaginary, then at a larger gain, the system becomes oscillatory. Therefore, a high gain tends to cause the system to oscillate.

If somewhere in the loop has a delay τ, then the system may become absolutely unstable: The decay constant β might become negative, and the system would oscillate spontaneously. This can be shown by considering the delay τ as a perturbation to the system. Actually, by letting $RC = 0$ in the denominator of Eq. (11.35), the poles of $F(s)$ are determined by the following algebraic equation,

$$G + (1 + \tau s)\left(1 + \frac{s}{Q\omega_0} + \frac{s^2}{\omega_0^2}\right) = 0. \tag{11.43}$$

Assuming that τ is a small quantity, by differentiating Eq. (11.43) with respect to τ and s, we obtain the approximate roots of Eq. (11.43):

$$s = \left(-\beta + \frac{G\omega_0^2}{2}\tau\right) \pm i\left(\omega_1 + \frac{G\omega_0^2}{\sqrt{4Q^2 - 1}}\tau\right). \tag{11.44}$$

In other words, the real part of the root of Eq. (11.43) is increased in proportion to G and τ. When the real part of the roots becomes positive, or

$$G(G + 1)Q\omega_0\tau > 1, \tag{11.45}$$

the system becomes absolutely unstable. To improve the accuracy of imaging, a high open-loop gain G is desirable. From Eq. (11.45) it is obvious that the system would soon become unstable with a moderately high G. This situation can be partially improved by decreasing the Q of the piezo, for example, by filling the piezo tube with an elastomer.

A common method of achieving a high gain and at the same time preventing instability is to introduce a compensation circuit, for example, a large RC time constant in the feedback circuit. Sometimes, it can be as long as 1 sec. In this case, the pole given by the large RC constant dominates the response of the entire system. In other words, the $1/RC$ pole becomes the *dominant pole*. In this case, Eq. (11.35) becomes

$$Z_T(s) = \frac{1}{1 + \dfrac{RC}{G}s}\frac{1}{s}. \tag{11.46}$$

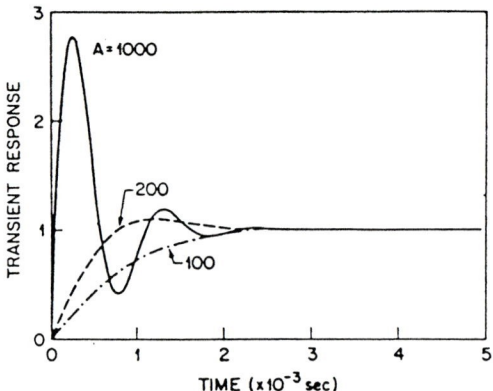

Fig. 11.7. Transient response of the STM feedback system. Three different values of the loop gain G give different results. The response is overdamped with a gain of 100, critically damped with 200, and underdamped with 1000. (After Kuk and Silverman, 1989.)

The inverse Laplace transform gives

$$z_T(t) = 1 - e^{-Gt/RC}. \tag{11.47}$$

If the gain G is sufficiently large, the response of the tip can still be sufficiently fast. Nevertheless, the other factors will show up, and oscillation will occur. Practically, the gain G is limited by the amplifiers in the circuit. The optimum condition is chosen in between, either by a more careful analysis or by measuring the actual time response of the system. An example of the measured time response is shown in Fig. 11.7.

The insertion of an RC circuit in the feedback electronics right before the high-voltage amplifier for the z piezo has some other advantages. First, much of the high-frequency noise is efficiently filtered out. Second, it facilitates the realization of the electronics for spectroscopic study, which we will discuss in the following section.

11.3 Computer interface

Computer control has been an essential part of STM ever since the very beginning. The different variations of computer interface and software are virtually unlimited. In this section, we describe the essential elements of computer interfaces.

Although in the early years the x, y scanning was made by function generators, most of the laboratory STMs and commercial STMs use software and

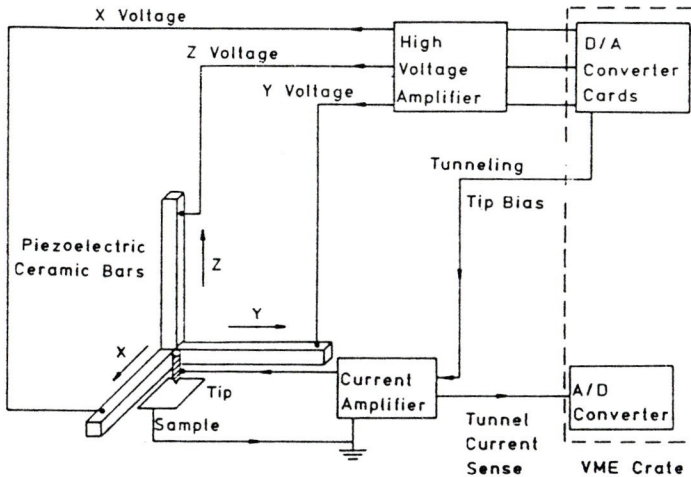

Fig. 11.8. The essential elements of a computer-controlled STM. The feedback electronics is replaced by a single-CPU computer. A Motorola 68020 microprocessor and a 68881 math coprocessor are used to perform the feedback control. A commercial VME crate is applied. The versatility of the software-controlled system facilitates the optimization of the transient response of the STM. (Reproduced from Piner and Reifenberger, 1989, with permission.)

D/A converters to generate raster scan voltages. An important point to be noted is that D/A converters have finite step sizes. A typical D/A converter has a full range of ± 10 V, with 12-bit accuracy. Each step is 20 V/4096=4.88 mV. By using it directly to drive the x, y piezo with a typical piezo constant of 60 Å/V, each step is about 0.3 Å. If amplification is installed, the step size becomes larger.

The reading of z piezo voltage is taken to the computer with an A/D converter. Using a 16-bit A/D convertor, each step is 0.305 mV. For a z piezo of 20 Å/V, each step is 0.006 Å. The whole range is 400 Å If a range of 4000 Å is needed, each step becomes 0.06 Å. To expand the range of z piezo and at the same time retain a high reading accuracy, a dc offset for the z piezo reading can be implemented.

The advantage of using a computer and A/D and D/A convertors for piezo control and reading is that for tube piezos, especially when the tip is mounted at the edge of the tube instead of at the center, the x, y scan is not linear, and there is substantial cross talk between x, y, and z. This nonlinearity and interference can be corrected by software.

The feedback electronics, an analog circuit described in the previous section, can be executed with a dedicated computer, or more precisely, by a dedicated microprocessor. Figure 11.8 shows an example of such a system, described by Piner and Reifenberger (1989). The transfer function of the feedback system is then software controlled. It is more versatile than the analog circuits. The local tunneling spectroscopy can be easily implemented on such a digital feedback system by simply holding the z piezo voltage and ramping the bias. Some of the commercial STMs use such a digital feedback system.

The bias can be implemented from the computer through a D/A converter from the computer. The typical output from a typical D/A converter, ± 10 V in range and 4.88 mV per step, is ideal for bias control and for local tunneling spectroscopy. The speed of output from a D/A convertor and the speed of reading by an A/D convertor are typically 30 kHz, which matches the speed of the current amplifier. With an additional A/D conversion for the tunneling current (the output of the current amplifier), the local tunneling spectroscopy can be implemented by the computer without additional analog electronics.

11.3.1 Automatic approaching

In the early years of STM operation, coarse approaching was conducted manually. It is a painfully difficult process. With an electrically controlled coarse advance mechanism, the coarse positioning can be performed automatically.

The process is as follows:

1. Disable the feedback loop.

2. Withdraw the z piezo to the limit.

3. Take one step forward with the coarse positioner.

4. Activate the feedback loop.

5. If a tunneling current is detected and the tip is stabilized, then stop. Otherwise repeat the loop.

The automatic approaching procedure is used in many home-made STMs and almost all commercial STMs.

CHAPTER 12
COARSE POSITIONER AND STM DESIGN

We have discussed the common elements of STMs in the previous chapters, including piezodrives, vibration isolation, and electronics. An important element of an STM is the coarse positioner, which moves the relative position of the tip versus the sample in steps exceeding the range of the piezodrive (typically a fraction of a micrometer). The initial success of the STM was partly because of the invention of the piezoelectric stepper, nicknamed the *louse*. Later, simpler stepping mechanisms were introduced. The coarse positioner largely determines the personality of the STM design. In this chapter, we discuss several important examples of the coarse positioners and the STM designs, in chronological order.

12.1 The louse

The piezoelectric stepper, nicknamed the louse, was the first successful stepper used in UHV STM (Binnig and Rohrer, 1982). A schematic of the louse is shown in Fig. 12.1. As shown, the actuating element of the louse is a piezoelectric plate (PP), which can be expanded or contracted by applying a voltage (100 to 1000 V). It is resting on three metal feet (MF), separated by high-dielectric-constant insulators (I) from the metal ground plate (GP).

Fig. 12.1. The piezoelectric stepper — the louse. It consists of a piezoelectric plate (PP), standing on three metal feet (MF), separated by high-dielectric-constant insulators (I) from three metal ground plates (GP). The feet can be clamped electrostatically to the ground plate by applying a voltage V_F. By alternatively activating the clamping voltage and the voltage on the piezo plate, the louse crawls like a creature with three legs. (Reproduced from Binnig and Rohrer, 1982, with permission.)

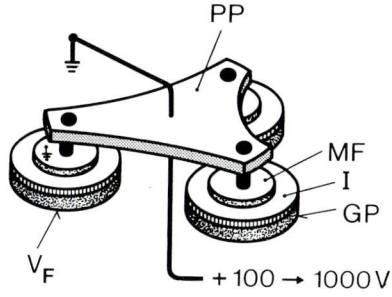

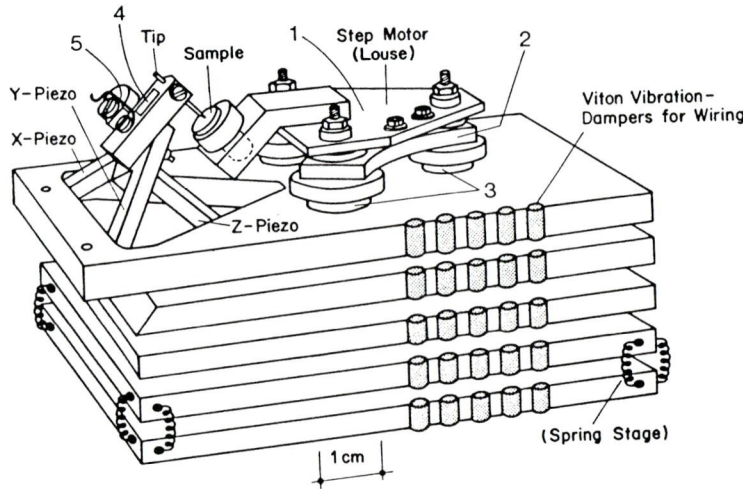

Fig. 12.2. The pocket-size STM. Vibration isolation and damping are achieved by a stack of stainless-steel plates separated by viton O rings (not shown) in between. On the top metal plate are the louse and the piezoelectric tripod. 1, A metal plate with the sample holder, with only one of the three screws tightened. 2, Piezoelectric plate. 3, Annodized aluminum feet. 4, Tip holder. 5, The current lead. Viton pieces on the edges of the metal plates are used for isolating the vibration transmitted through the wires. (After Gerber et al., 1986.)

The feet are clamped electrostatically to the ground plate by applying a voltage V_F. The motion of the louse resembles the walking of a creature with three legs. Each step consists of the following six substeps. First, loosen one of the three feet by eliminating the clamping voltage on that foot. Second, expand or contract the piezo plate (PP) to move the loose foot to a new position. Third, apply clamping voltage on that foot. Fourth, loosen another foot. Five, release the voltage on the piezoelectric plate. Six, apply clamping voltage on the other foot. The louse is then moved to a new position. By alternatively activating the three feet, the louse can be moved to any direction on the ground plate. The size of a step ranges from 100 Å to 1 μm, up to 30 steps per second. To make a louse work properly, the feet and the ground plate have to be polished and cleaned carefully. However, once working, the louse is fairly reliable.

In virtue of the small physical size of the louse, a pocket-size STM was developed and incorporated in an UHV scanning electron microscope chamber (Gerber et al., 1986); see Fig. 12.2. The actual dimensions are $10 \times 6 \times 4$ cm³. Many of the spectacular STM images were obtained by the pocket-size STM, or slightly improved versions of the original pocket-size

STM, for example, Feenstra, Thomson, and Fein (1986), Chiang, Wilson, Gerber, and Hallmark (1988). At the time being, these instruments are still working well.

The excellent performance of the pocket-size STM is largely due to the smallness and the rigidity of the piezoelectric stepper—the louse. The efficiency of vibration isolation by the stack of metal plates is not as good as the two-stage suspension-spring system. To reduce the influence of low-frequency vibration further, an additional suspension-spring stage is often applied. Feenstra, Thomson, and Fein (1986) used a two-stage suspension-spring vibration isolation system, whereas Chiang, Wilson, Gerber, and Hallmark (1988) used very long springs (longer than 25 cm) to isolate their pocket-size STM from the vibrations of their vacuum chamber.

Another advantage of the louse is that two-dimensional coarse motions can be achieved. By placing the pocket-size STM in a vacuum chamber with a scanning electron microscope (SEM), it is possible to locate interesting features on the sample surface with the tip. This enabled several authors to locate the interface of a device, for example, Muralt et al. (1987), Salmink et al. (1989), and Albrektsen et al. (1990).

12.2 Level motion-demagnifier

The piezoelectric stepper described in the previous section, the louse, is a somewhat complicated device, which requires substantial effort to make it work. In many surface-science experiments, the actual location on the sample surface does not matter. A one-dimensional stepper is sufficient. In its simplest form, a micrometer, or a fine-pitch lead screw, can make controlled steps of a few micrometers. However, it is extremely difficult for STM, where the range of the z-piezo is typically of the order of $0.1\mu m$.

Inspired by the vibration immunity of the squeezable tunneling junctions (Hansma, 1986), a simplified STM for surface-science studies using a conventional rotary feedthrough, a fine-pitch lead screw, and a mechanical lever was developed by Demuth et al. (1986, 1986a). A schematic diagram is shown in Fig. 12.3. The three piezo bars are mounted on a macor[1] piece (A). The tip is mounted near the joining point of the three piezo bars. A sharp piece of metal (I), the "foot," is also mounted on the macor block. The sharp end of the foot is manually placed about 0.5 mm horizontally from the tip, and leveled with the tip within ±0.1 mm vertically. The sample is mounted on a lever (F), which is actuated by a rotary feedthrough and a lead screw through the carriage rod (C). The lever can have two alternative pivot points, which

---

1   Macor is the trade name of a machineable glass manufactured by Corning Glass, NY 10000. It is a good insulator and compatible with ultra-high vacuum.

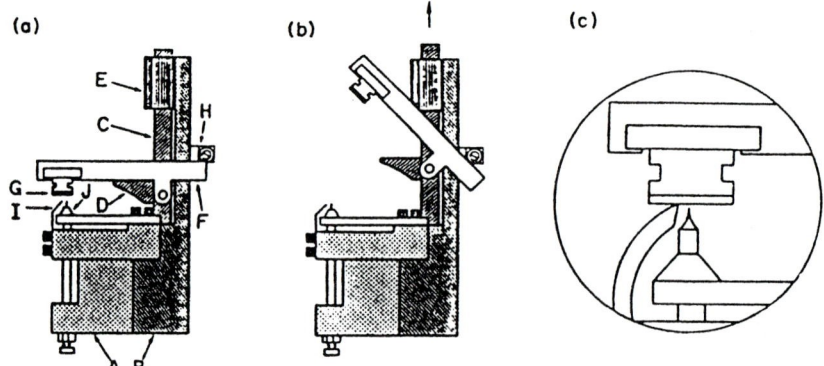

Fig. 12.3. STM with a double-action lever. Various parts are shown in (a): (A) The macor block onto which the x, y piezo bars (horizontal) and the z piezo bar (vertical) are mounted, (B) the microscope base plate, (C) carriage rod, actuated by a linear feedthrough and a lead screw, (D) stop, (E) ball bushing assembly, (F) lever, (G) sample and sample holder, (H) catch, the pivot point for coarse motion, (I) foot, the pivot point for fine motion, and (J) the probe tip assembly. (b) Shows the STM in coarse motion. Using the catch as the pivot point, the sample can be removed away from the tip. (c) Shows the STM in fine motion. Using the foot as the pivot point, the linear motion of the carriage rod is reduced by a large factor. (After Demuth et al., 1986a.)

make the lever function in two different ways. When the carriage rod is at its high position, the catch (H) is the pivot point. The sample can be moved to a vertical position far away from the tip. The sample can be resistively heated, or inspected by LEED. When the carriage rod is at a low position, the foot touches the sample surface, and the foot becomes the pivot point. Because the foot is very close to the tip, the vertical motion of the carriage rod is significantly demagnified. In the original design the distance from the foot to the carriage rod is 38 mm, whereas the distance between the foot and the tip can be adjusted to about 0.5 mm. A demagnification of 70:1 is observed. This together with the fine pitch of the lead screw (1.5 threads per mm) provides ≈ 250 Å of motion per degree of rotation of the lead screw. Also, the direct contact of the foot with the sample reduces the relative motion of the tip versus the sample, which simplifies vibration isolation.

Even with the foot, the vibration problem in such STM designs has not been resolved sufficiently. In fact, the tripod scanner has a relatively low natural vibration frequency, and the approaching mechanism is relatively bulky. In actual application, a four-to-six element metal-plate stack with viton separators is used for vibration isolation. In order to achieve atomic resolution, a spring stage, either suspension spring or compression spring, is neces-

Fig. 12.4. Single-tube STM. The tube piezo scanner is adhered inside a sturdy metal cylinder, which sits on three screws on the base plate. The two front screws make the coarse approaching. The rear screw makes fine approaching by using the two front screws as the pivot axis. The rear screw is actuated by a stepping motor for automatic approaching. The preamplifier (not shown) is mounted directly on top of the metal cylinder to eliminate the microphone effect of the coaxial cable between the tip and the input of the preamplifier. The entire unit is rigid enough that a mediocre vibration isolation device can provide atomic resolution. (After Hansma et al., 1988.)

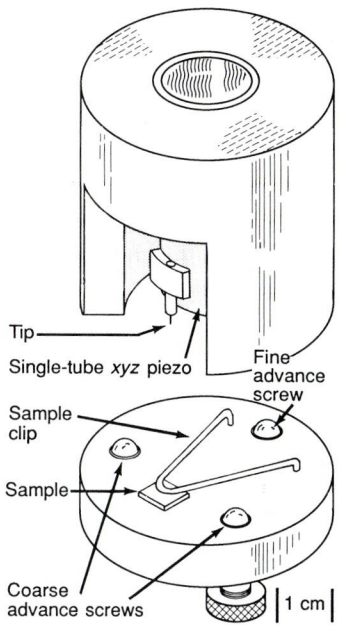

sary. For example, Boland (1990) has added a 30 cm long suspension spring and eddy-current damping system to an STM of similar design, and achieved true atomic resolution.

Because of its extremely simple structure, this type of home-built STM is routinely used in several scientific laboratories.

12.3 Single-tube STM

Once the tube scanner was invented (Binnig and Smith, 1986), it soon became the primary choice of piezo scanners in STM. Its small size and high natural resonance frequency make the mechanical design and vibration isolation much easier. Currently, most of the commercial STMs as well as home-made STMs use tube scanners.

Figure 12.4 shows an STM designed by Drake et al. (1987). The tube piezo scanner is adhered at the center of a heavy and sturdy metal cylinder. The tip is mounted on the edge of the piezo tube, similar to the original arrangement of Binnig and Smith (1986). The preamplifier, mounted on a small printed circuit board of less than 25 mm in diameter, is directly connected to the tip through a hole in the metal cup. This entire assembly constitutes the scanning head. The head is seated on three screws with polished spherical ends, which are bolted through the base plate on which the sample is mounted. The two screws near the front are coarse advance devices. By using the front screws, the tip is advanced to be very close to the sample surface. Then, fine approaching is executed through the screw at the rear. The two front screws now become the pivot axis. Because the line connecting the two front screws is very close to the end of the tip, there is a huge reduction in the mechanical motion of the rear screw. The rear screw is usually actuated by a stepping motor. Automatic approaching is easily implemented.

As expected from its structure, such an STM is very rigid. To improve its rigidity further, the polished tops of the three screws are made of permanent magnets, and the cup is made of a soft ferromangetic material. The magnetic force adds to the gravitational force to increase the natural resonance frequency of the entire unit. The lowest natural resonance frequency of the entire unit can easily be greater than 10 kHz. From the previous discussion of vibration isolation, if a single-stage vibration isolation system with a natural frequency of 10 Hz is used, the overall transfer function in the frequency range of some 10 Hz to 10 kHz is 10^{-6}. Remember that a pendulum of natural frequency 10 Hz means a length of 2.5 mm; so the entire STM unit can be made very small.

The simple single-tube STM is preferentially operated in air. With an air bag, it can be operated in controlled ambient conditions. It is also convenient to operate in liquid. A refined version of this STM has become a commercial product, Nanoscope, which is probably the most popular STM in the open market.

12.4 Spring motion demagnifier

Another STM design, also inspired by the squeezable tunneling junction (Hansma, 1986), uses two springs of very different stiffness to demagnify the linear motion of a lead screw. An example of such a design is shown in Fig. 12.5, and is a low-temperature STM. As designed by Smith and Binnig (1986), the instrument utilizes a metal reed as the sample holder. A screw at one end of the reed performs the coarse approach. Another screw, actuated from the top through a long shaft, conducts the fine approach. It compresses a helical spring pressing on the reed. The advance of the lead screw is demagnified by the ratio of the stiffness of the helical spring and the reed. The reduction ratio is easily made to exceed 100.

A disadvantage of such an STM is that there is no method to treat the tip and the sample in the cryostat. The sample and the tip have to be prepared in air, the relative position in air adjusted, then the setup immersed to a dewar. The samples to be studied are limited. However, Hess et al. (1989, 1991) have made startling progress using such an STM design in the study of superconductors. The secret of their success is to use an extremely inert superconducting material, $NbSe_2$.

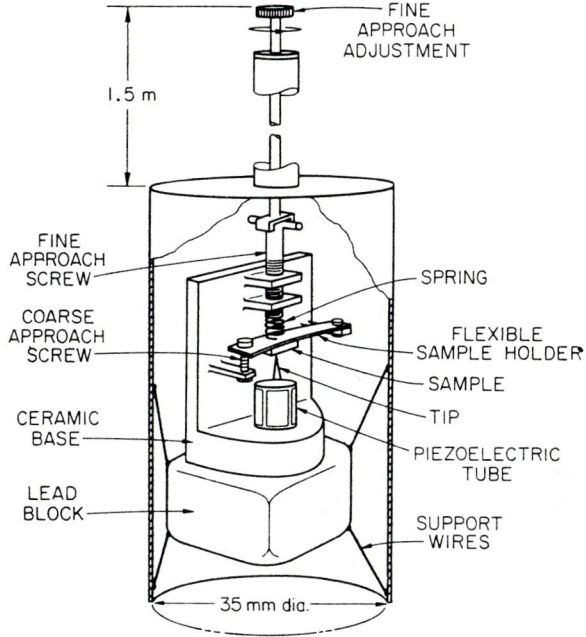

Fig. 12.5. STM with spring motion demagnifier. The design, due to Smith and Binnig, is used for low-temperature STM experiments. (Reproduced from Smith and Binnig, 1986, with permission.)

12.5 Inertial steppers

The concept of inertial steppers was developed by Pohl (1986a, 1987). A schematic of the inertial stepper is shown in Fig. 12.6. The two ends of a piezo tube, located in the middle of the device, are attached to the two support pieces. By applying a voltage to the piezo tube, the two support pieces move relative to each other. A sliding block, carrying the load, is riding on the rail at the upper support piece. By ramping the voltage to the piezo tube slowly, the sliding block follows the horizontal motion of the upper support piece. By ramping the voltage to the piezo tube quickly, the sliding block remains stationary due to inertia. Therefore, by applying an asymmetric sawtooth voltage to the piezo tube, the sliding block moves step by step to one direction or another. The step size can be controlled from a few nm to a few μm. The frequency of the step can be as high as a few kHz. Therefore, it provides

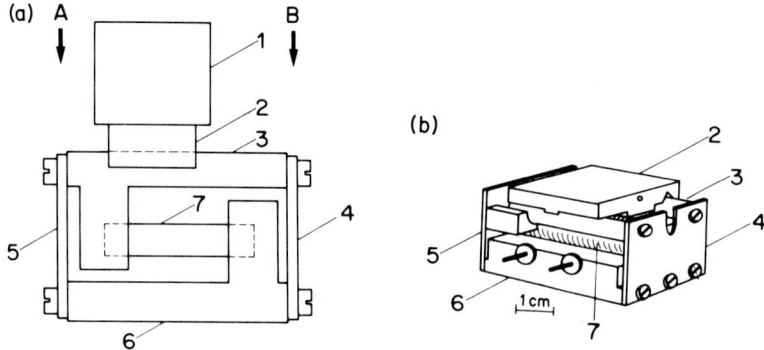

Fig. 12.6. Inertial stepper. (a) schematic, (b) actual design. A load (1) is carried by a sliding block (2), which is riding on a rail (3). The rail (3) is mounted on a mobile frame consisting of two elastic metal plates (4) and (5), a base plate (6), and a piezoelectric tube (7). By applying a sawtooth voltage on the piezo tube, the sliding block (2) moves back or forth in steps. (Reproduced from Pohl, 1987, with permission.)

both quick and coarse motion as well as fine motion well within the range of the piezodrive.

Another application of the idea of an inertial stepper is the STM with four identical tube piezoelectrical elements developed by Besocke (1987). Each of the tube piezoelectric elements has its outer metallization quartered, similar to that of Binnig and Smith (1986). The center piezo tube is used as the scanner, whereas the three outer tubes are used for sample manipulation.

The carriers can move the sample in any horizontal direction through a sequence of two steps. First, the carriers moves quickly downwards, to the right, then upwards. During this step, the absolute position of the sample is frozen by its inertia. Second, the carriers then move slowly to the left, recovering their original position. The sample follows the motion of the carriers because of friction. By repeating these two steps, the sample can be moved in any direction, by steps of 50 to 1500 Å, up to a few millimeters.

The concept of inertial steppers of Pohl (1986a, 1987) has been applied by several other authors to build very compact STMs. An example is the concentric-tube STM of Lyding et al. (1988), as shown in Fig. 12.7. This design utilizes two concentric piezoelectric tubes, glued onto a common base. The inner tube (1.27 cm long with 0.635 cm outer diameter) supports and scans the tunneling tip, located along the axis of the tube. The outer tube (1.27 cm long with 0.953 cm outer diameter) supports a collar that holds the quartz sample mounting tube.

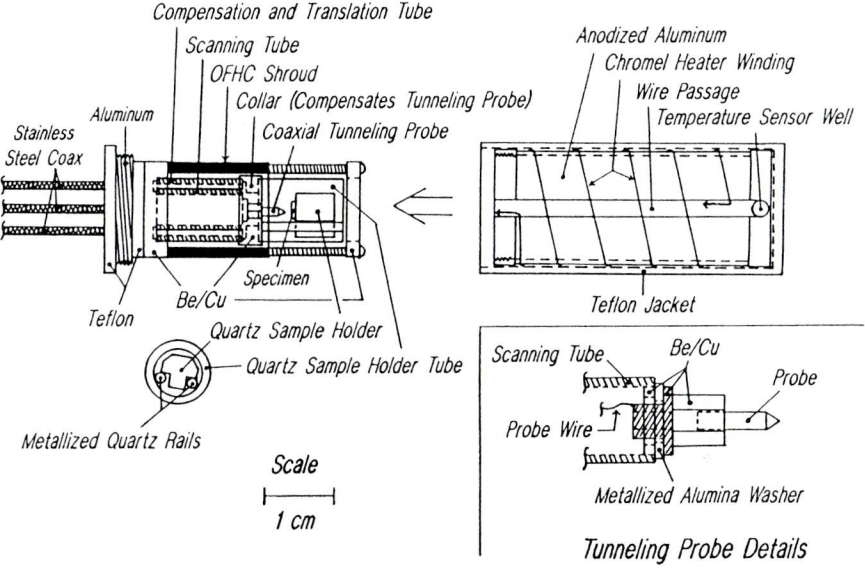

Fig. 12.7. The concentric-tube STM. The inner piezoelectric tube acts as the scanner. The outer piezoelectric tube acts as an inertial walker. (Reproduced from Lyding et al., 1988, with permission.)

The outer piezoelectric tube functions as the inertial walker to move the sample holder back or forth with submicrometer steps (Pohl, 1986a, 1987). For example, to translate the sample toward the tip, a relatively slow voltage ramp is applied to the outer piezoelectric tube, causing it to contract. At the end of the ramp, the voltage is rapidly returned to its initial value, causing the tube to return rapidly to its initial position. Due to inertia, the sample holder, which slides on the rails of the quartz tube, cannot follow the rapid motion of the rails. Thus the sample holder has been translated toward the tip by one "step." By adjusting the amplitude of the ramp, step sizes ranging from 1 μm to 5 Å are readily achieved. This stepping process can be repeated rapidly (up to several kHz) resulting in rapid translation (up to 1 mm/sec). By reversing the direction of the ramp, the sample holder can be stepped away from the tip.

By carefully selecting the thermal expansion coefficients of the materials, this design can compensate the thermal drift of the STM, which makes it stable even if the temperature is varying.

12.6 The Inchworm®

Translational motions of very small steps can be utilized by a commercial device, the Inchworm®.[10]   Figure 12.8 shows the principle of the walking mechanism of the Inchworm®.   An alumina shaft is slid inside a PZT sleeve, which has three sections.   The three sections of the PZT sleeve are numbered 1 through 3, respectively.   By sequentially activating the three sections, axial motions in small steps are made.

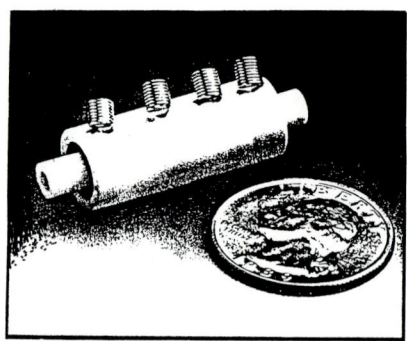

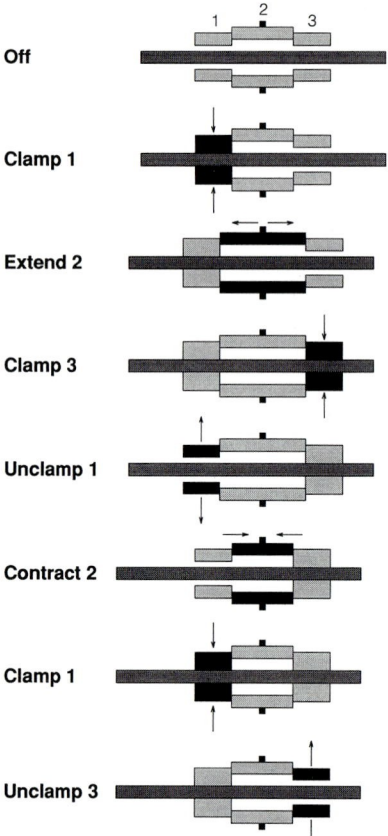

Fig. 12.8. The Inchworm®. It consists of an alumina shaft, slid inside a piezoelectric tube, which consists of three sections. Top, a photograph of the MicroInchworm® Motor, actual size. Right, walking mechanism: (1) Clamp the left-hand side by applying a voltage to section 1. (2) Extend section 2. (3) Clamp the right-hand end. (4) Unclamp section 1. (5) Contract the center section. (6) Clamp section 1. (7) Unclamp section 3. (Courtesy of Burleigh Instruments, Inc.)

---

10 Inchworm® is a registered trademark of Burleigh Instruments, Inc., Burleigh Park, Fishers, New York 14453.

The outer two PZT elements, numbered 1 and 3, act as clamps. When a voltage is applied to one of them, its diameter shrinks to grip the shaft tightly. The center section does not contact with the shaft, but the length can change according to the applied voltage. The function of the Inchworm® is illustrated in Fig. 12.8.

Figure 12.9 shows an STM design utilizing an Inchworm®. The entire STM, including a fixed sample mounting block and a tube scanner attached to the center shaft of an Inchworm®, is on a thick base plate. The base plate is vibration isolated with four suspension springs from the flange. For sample loading and tip loading, the base plate can be locked down to the frame, which is attached to the flange.

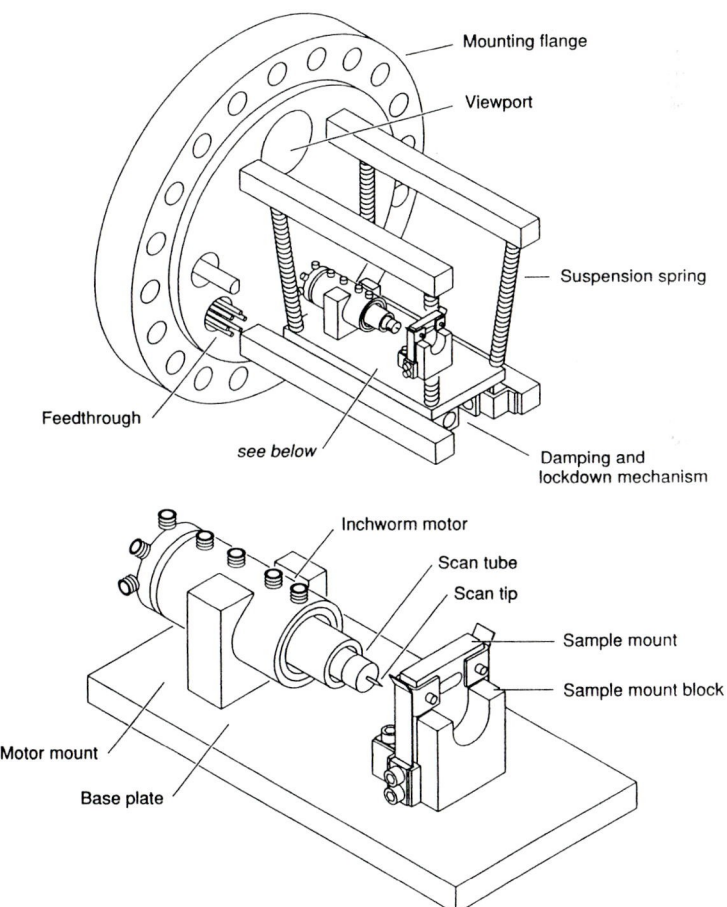

Fig. 12.9. An STM with an Inchworm® as coarse positioner. (Courtesy of Burleigh Instruments, Inc.)

CHAPTER 13
TIP TREATMENT

13.1 Introduction

The importance of tip treatment in STM was recognized by Binnig and Rohrer from the very beginning of their experimentation (Binnig and Rohrer, 1982). In order to understand the observed resolution, they realized that if the tip would be considered as a piece of continuous metal, the radius must be smaller than 10 Å. The number does not appear very meaningful for a tip radius, but it is nevertheless clear that conventional field-emission tips of radius $\approx$ 100–1000 Å would not provide the resolution they observed. In the first set of STM experiments, the tips were mechanically ground from Mo or W wire of about 1 mm in diameter. Scanning electron micrographs showed an overall tip radius of <1 μm, but the rough grinding process created a few rather sharp minitips. The extreme sensitivity of the tunneling current versus distance then selects the minitip closest to the sample surface for tunneling. They also reported (Binnig and Rohrer, 1982) several in situ tip sharpening procedures: By gently touching the tip with sample surface, the resolution was often improved, and the tips thus formed were quite stable. By exposing the tip to high electric fields, of the order of 10^8 V/cm, the tips are often sharpened. Even now, some 10 years after, the problem of obtaining reproducible tips with atomic resolution is not fully resolved.

In the STM literature, there are many scattered discussions of the problem of tip treatment. A brief summary of the facts found in the STM literature is as follows:

1. A tungsten tip, prepared by electrochemical etching, with a perfectly smooth end of very small radius observed by SEM or TEM, would not provide atomic resolution immediately.

2. Atomic resolution might happen spontaneously by repeated tunneling and scanning for an unpredictable time duration.

3. A crashed tip often recovers to resume atomic resolution, unexpectedly and spontaneously.

4. During a single scan, the tip often undergoes unexpected and spontaneous changes that may dramatically alter the look and the resolution from one half of an STM image to another half.

5. Various in situ and ex situ tip sharpening procedures were demonstrated. These procedures might provides atomic resolution, but often makes the tip end looks even worse under SEM or TEM.

6. The tip that provides atomic resolution often gives unpredictable and nonreproducible tunneling I/V curves.

7. A tip mechanically cut out from a Pt–Ir wire often works even if it looks like a warthog under SEM or TEM.

In this chapter, we describe various experimental methods of tip preparation and treatment. Most of them are found empirically. A thorough understanding of these procedures is still lacking. Some of the explanations are tentative. Tip preparation and characterization is one of the central experimental problems in STM, and we can certainly expect that a lot more development and understanding will be achieved in the near future.

13.2 Electrochemical tip etching

The art of making sharp tips using electrochemical etching was developed in the 1950s for preparing **samples** for field ion microscopy (FIM) and field electron spectroscopy (FES). A description of various tip-etching procedures can be found in Section 3.1.2 in the book of Tsong (1990).

The preferred method for preparing STM tips is the dc dropoff method. The basic setup is shown in Fig. 13.1. It consists of a beaker containing an electrolyte, typically 2M aqueous solution of NaOH. A piece of W wire, mounted on a micrometer, is placed near the center of the beaker. The height of the W wire relative to the surface of the electrolyte can then be adjusted. The cathode, or counterelectrode, is a piece of stainless steel or platinum placed in the beaker. The shape and location of the cathode has very little effect on the etching process, which can be chosen for convenience. A positive voltage, 4 V to 12 V, is applied to the wire, which is the anode. Etching occurs at the air–electrolyte interface. The overall electrochemical reaction is (Ibe et al., 1990):

cathode $6H_2O + 6e^- \rightarrow 3H_2(g) + 6OH^-$ $\qquad$ SRP $= -2.48$ V

anode $\quad W(s) + 8OH^- \rightarrow WO_4^{2-} + 4H_2O + 6e^-$ $\qquad$ SOP $= +1.05$ V

$\rule{8cm}{0.4pt}$

$\qquad W(s) + 2OH^- + 2H_2O \rightarrow WO_4^{2-} + 3H_2(g)$ $\quad E^0 = -1.43$ V

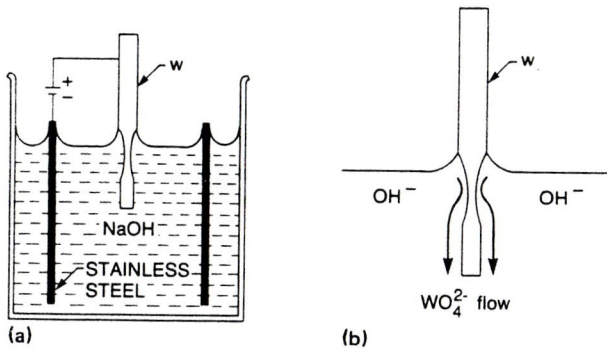

Fig. 13.1. Electrochemical etching of tungsten tips. (a) A tungsten wire, typically 0.5 mm in diameter, is vertically inserted in a solution of 1N NaOH. A counterelectrode, usually a piece of platinum or stainless steel, is kept at a negative potential relative to the tungsten wire. (b) A schematic illustration of the etching mechanism, showing the "flow" of the tungstate anion down the sides of the wire in solution. (Reproduced from Ibe et al., 1990, with permission.)

The etching takes a few minutes. When the neck of the wire near the interface becomes thin enough, the weight of the wire in electrolyte fractures the neck. The lower half of the wire drops off. Actually, this procedures makes two tips at the same time. The sudden rupture leaves a ragged edge of work-hardened asperities at the very end of the tip, the detailed structure of which is usually irregular, as shown in Fig. 13.3. To remove the residual NaOH from the tip surface, a thorough rinsing with deionized water and pure alcohol is necessary.

Several parameters affecting the etching process have been studied by Ibe et al. (1990): first, the cell potential, or the voltage between the electrodes. It was found experimentally that the etching rate is virtually flat for a voltage between 4 and 12 V. Below 4 V, the etching rate drops significantly. Virtually any voltage between 4 and 12 V is adequate, although a voltage close to 4 V may reduce the oxide thickness. Second, the shape of the meniscus affects the aspect ratio and the overall shape of the tip. If the position of the meniscus drops too much during the etching process, a readjustment of the wire height is helpful in obtaining the desired tip shape. Third, the electrolyte concentration. The lower the concentration, the slower the etching process. Because the OH^- is consumed in the reaction, the NaOH solution should be changed after a few etching processes. Fourth, the length of the wire in solution. It affects the radius of curvature of the tip end. Since the dropoff occurs when the narrow neck no longer supports the weight of the wire in solution, a shorter wire in the liquid is favorable for generating sharper tips. For the 0.25 mm diameter W wire, the optimum wire length in the liquid is found to be 1–3 mm.

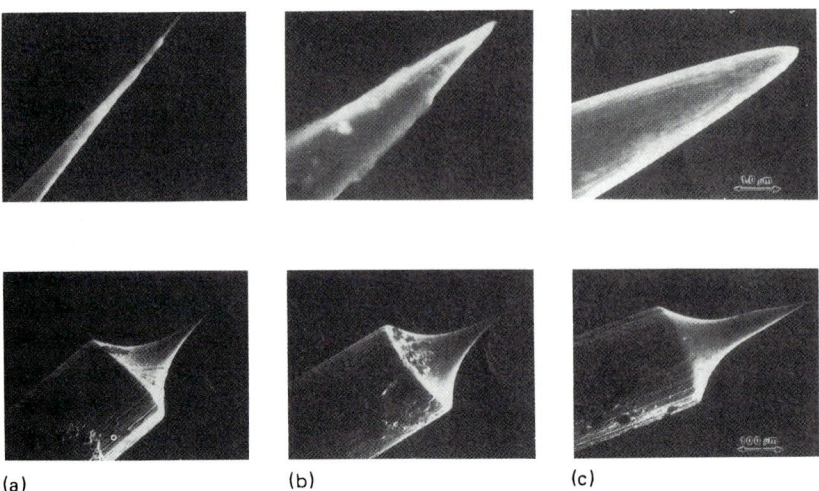

(a)   (b)   (c)

Fig. 13.2. Dependence of tip radius of curvature with cutoff time. Scanning electron micrographs of tips with different etching-current cutoff time. (a) 600 ns, with an average radius of curvature 32 nm. (b) 140 ms, with an average radius of curvature 58 nm. (c) 640 ms, with an average radius of curvature 100nm. (reproduced from Ibe et al., 1990, with permission.)

The most important parameter that affects the final shape of the tip end is the time for the etching current to cut off after the lower part drops off. The shorter the cutoff time, the etching current, the sharper the tip end, as shown in Fig. 13.2. To shorten the cutoff time, a simple electronic circuit is helpful. When the lower part of the wire drops off, the etching current suddenly drops. The electronic circuit senses the drop of the etching current and turns off the current completely through an electronic switch. An example of such a circuit is given in detail by Ibe et al. (1990).

Various improved methods for making STM tips linked with SEM studies have been discussed in the literature, for example, Lemke (1990), Melmed (1991), and references therein. However, since the STM resolution does not have a direct correlation with the look of the tip under SEM, the simple dc dropoff method, as described here, is usually sufficient. From the experience of the author, two simple improvements can be helpful. The first is to install an insulating piece between the cathode and anode across the liquid surface to prevent the hydrogen bubbling on the cathode from perturbing the meniscus near the anode. The second is to save the dropped piece as the tip, which might be better than the upper one.

The exposure of the tip to the electrolyte and air results in the formation of a surface oxide (WO_3). This oxide coating has to be removed before tunneling can occur. We will describe various methods for removing the oxide in the following.

To etch Pt–Ir tips, a solution containing 3 M NaCN and 1 M NaOH is used (Nagahara et al., 1989). A circular Ni foil is used as the counterelectrode. The etching current, which depends on the area of immersed wire and the applied voltage, is adjusted to an initial value of 0.5 A using a bias of 20 V ac by varying the immersed depth of the Pt–Ir wire with a micrometer. The etching proceeds faster at the air–solution interface, and a narrow neck is formed. The part of the wire in the liquid breaks off, and two tips are formed. The tips should be thoroughly rinsed, similar to the case of W tips.

13.3 Ex situ tip treatments

The W tips generated by electrochemical etching are seldomly applicable immediately. First, immediately after etching, the tip is covered with a dense oxide layer, and often contaminated with sodium compounds from the etchant, as well as organic molecules. Therefore, procedures to remove the oxides and various contaminants must be executed before beginning STM experiments. Next, the arrangement of the apex atoms may not generate the electronic states required to generate atomic resolution, or to generate reproducible tunneling spectra. Fig. 13.3 shows a typical FIM image of an as-etched W tip from a single-crystal W wire with (111) orientation. Although, on a large scale, the

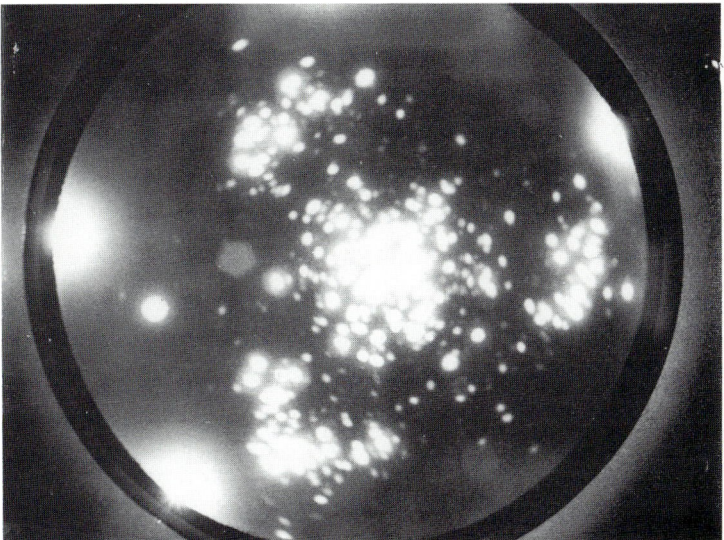

Fig. 13.3. FIM image of a W tip immediately after etching. The tip is etched from a single-crystal W wire with (111) orientation. The threefold symmetry is visible on a large scale. Locally, severe dislocations are observed. (Courtesy of U. Staufer.)

threefold feature is preserved, locally, serious disorder is always present. This is because immediately before the lower part of the W wire drops off, the narrowest part of the wire experiences an enormously large stress. A plastic deformation occurs, which eventually causes the wire to rupture. The very end of the tip is often seriously distorted. The distortion does not mean that atomic resolution is impossible. If by accident the arrangement of the last few atoms are in a favorable configuration to provide a protruding local orbital, atomic resolution is immediate. However, the chances are slim. Tip sharpening procedures are often needed. Many tip-sharpening procedures have been reported. Some of them are in situ, that is, executed under nearly tunneling conditions, which we will discuss later. In this section, several ex situ tip-treatment procedures are described.

13.3.1 Annealing

The purpose of tip annealing is to remove the contaminants and oxides without causing blunting. Because tungsten has a very high melting temperature (3410°C), the temperature-and-time process window is wide.

The removal of tungsten oxide is based on the following mechanism. On tungsten surfaces, the stable oxide is WO_3. At high temperature, the following reaction takes place:

$$2WO_3 + W \rightarrow 3WO_2\uparrow. \tag{13.1}$$

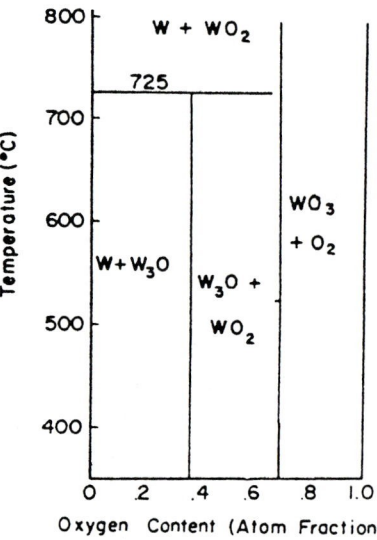

Fig. 13.4. The phase diagram of the W–O system. At lower temperatures, the W surface is always covered with some kind of oxide. Above 725°C, at relatively low oxygen pressure, only W and WO_2 are present. Because WO_2 is volatile, a metal W surface is generated. (After Promisel, 1964.)

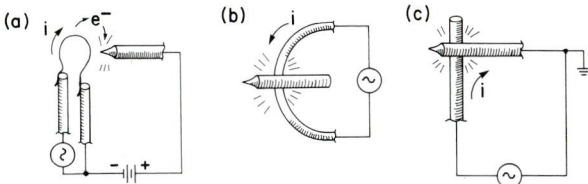

Fig. 13.5. Tip annealing methods. (a) Electron bombardment. A filament, biased negatively with regard to the tip, emits electrons to heat up the tip. (b) Resistive heating by a W filament. The tip is spotwelded to the filament. After heating, the tip is removed from the chamber and detached from the filament, and put into the vacuum chamber quickly. (c) Using the tip shank as the heating element. The tip is made in touch with a thicker W wire, which is connected to a power supply. Current flows through the tip shank to the ground.

While WO_3 has very low vapor pressure even at its melting point (1473°C), WO_2 sublimes at about 800°C. Figure 13.4 is the phase diagram of the W–O system. As shown, with a high oxygen content, at low temperature, the mixture of tungsten oxides are stable. At $T > 725$°C, only metal tungsten and WO_2 are present. Because WO_2 is volatile, as well as the organic substances, a clean metal surface is generated. Several methods for annealing the tips are shown in Fig. 13.5.

13.3.2 Field evaporation and controlled deposition

Fink (1986) conducted a systematic study of the methods for making tips with a well-defined atomic arrangement near the apex, The motivation is twofold. First, the lateral resolution of STM is determined by the atomic arrangement and its electronic structure. Second, this study is essential for understanding the tunneling between two microscopic objects—the tip and the sample. The atomic structure of both microscopic objects will eventually be known.

The methods used by Fink (1986) originated from field ion microscopy (FIM), described in Section 1.8.1. Tips made of single-crystal W(111) wires are chosen as the starting material. By controlling the field intensity at the tip apex, the most protruding W atoms are stripped off, leaving a well-defined tip shape. This process is known as *field evaporation* (Müller and Tsong 1969, Tsong 1990). By carefully controlling the field intensity, a well-defined pyramid is formed (see Fig. 13.6). A single W atom is located at the apex. By carefully controlling the conditions, the first atom can be removed. Using well-controlled vacuum evaporation, a tip atom of different and known chemical identity can be put on top of the three-atom platform. The electronic con-

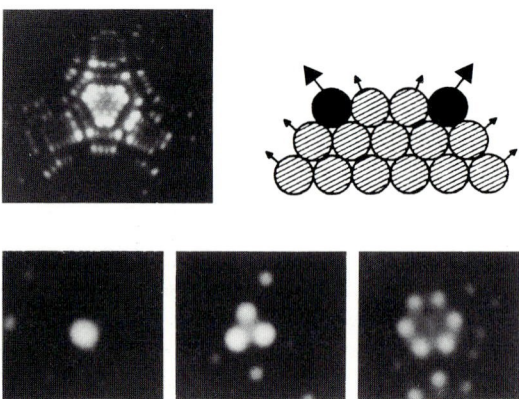

Fig. 13.6. Tip formation by field evaporation. Top left, FIM image of a (111)-oriented W tip; the (111) apex plane contains 18 atoms. Top right, the field evaporation process. Bottom, tip with pyramidal apex with one, three, and seven W atoms at the apex plane. (Reproduced from Fink, 1986, with permission.)

figuration of the tip can be optimized by choosing the right tip atom, to provide the highest lateral resolution.

13.3.3 Annealing with a field

Vu Thien Binh (1988, 1988a) reported that by heating the tip in the presence of an external electrical field, the apex of the tip will evolve into a pseudostationary profile that is defined by the competitive actions between surface diffusion and evaporation. As shown in Fig. 13.7, at high temperature, about 3200 K, the diffusion process dominates, and the surface tension favors a spherical end. At lower temperatures, about 1600–1800 K, the directional effect of the external field dominates and tips with sharpened ends can be formed. By conducting this process in an oxygen atmosphere, the corrosion process of forming volatile tungsten oxides better results. Because the process is fairly reproducible under identical temperature and external field intensity conditions, the need for extra heavy control devices (such as an FIM) is eliminated. By using such a procedure, even a crashed tip can be recovered. The effectiveness and reproducibility of this process have been demonstrated by FIM imaging.

As shown in Fig. 13.7, at high temperatures, about 3200 K, the diffusion process dominates, and the surface tension favors a spherical end. At lower temperatures, about 1600–1800 K, the directional effect of the external field dominates. By conducting this process in an oxygen atmosphere, a nanotip with a single W atom at the apex is routinely produced.

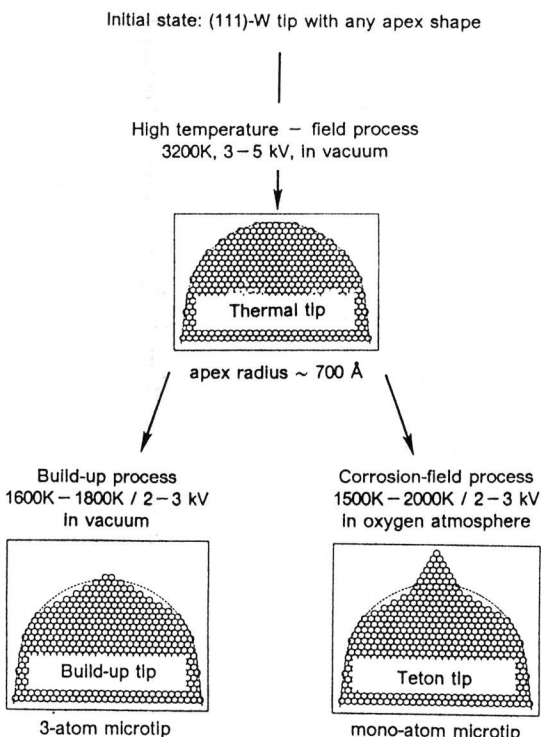

Fig. 13.7. Tip treatment by annealing in a field. Top: At a temperature close to the melting point of tungsten, the shape of the tip is basically determined by diffusion process. A rather round-shaped thermal tip is formed. Bottom left, at lower temperature, the directional effect of the field dominates. A built-up tip is formed. Bottom right, in an oxygen atmosphere, the corrosion process further generates a nanotip. After Binh (1988a).

13.3.4 Atomic metallic ion emission

Binh and Garcia (1991, 1992) reported that, at temperatures around one third of the bulk melting temperature, by applying an even higher electric field to th tip, the metal ions move to the protrusions and emit from the ends. Sharp, pyramidal nanotips, often ending with a single atom, can be reproducibly generated. The experiment was performed on W, Au and Fe tips. In particular, for W tips, the conditions to observe metal ion emission are: electrical field, 4–6 kV, or 1.2–1.8 eV/Å; and temperature, 1200–1500 K. This new phenomenon is termed the atomic metallic ion emission (AMIE).

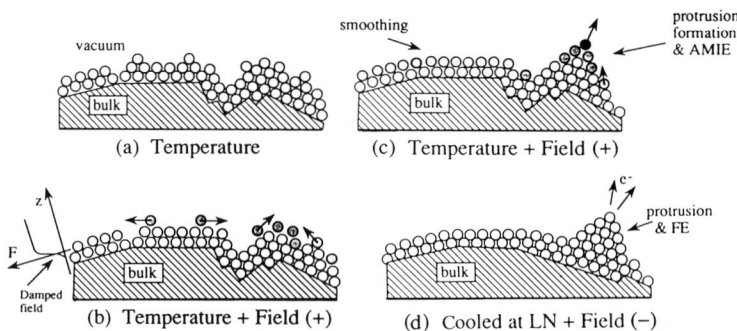

Fig. 13.8. Atomic metallic ion emission and nanotip formation. (a) At high temper-ature, the atoms on a W tip becomes mobile. The tip surface is macroscopically flat but microscopically rough. (b) By applying a high field (1.2–1.8 V/Å,), the W atoms move to the protrusions. (c) The apex atom has the highest probability to be ionized and leave the tip. The W ions form an image of the tip on the fluorescence screen. (d) A well-defined pyramidal protrusion, often ended with a single atom, is formed. By cooling down the tip and reversing the bias, a field-emission image is observed on the fluorescence screen. The patterns are almost identical. (Reproduced from Binh and Garcia, 1992, with permission.)

The mechanism of the AMIE is shown schematically in Fig. 13.8. By applying of a high temperature alone, the surface atoms become mobile and a equilibrium tip surface is formed. Macroscopically, the tip surface is smooth. Microscopically, it is rough. By further applying a high electrical field, the surface atoms become even more mobile, with a diffusion coefficient near that of a liquid. The metal atoms near the protrusions experience a greater force than those near the depressions. They move to the end of the protrusion, and make it even sharper. The single atom at the very end of the protrusion expe-riences the greatest force (Atlan et al., 1992). The apex atom is then ionized and accelerated by the field. The nearby surface metal atoms move to the end of the protrusion and are then ionized. Therefore, a continuous flux of positive metal ions, with an origin at the apex of the protrusion, is created. By fast cooling, the nanotip is fixed. By applying a negative voltage, it then becomes an electron source with atomic dimension.

The AMIE is different from the standard field ionization or evaporation. The emitted positive ion beams, with a flux of approximately 10^5 ions/sec, are obtained under ultra-high vacuum conditions ($\approx 10^{-11}$ Torr). Due to the ultra-high vacuum, the observed ion beams cannot be due to the impurities in the vacuum chamber. Also, the beam is very difficult to deflect by an external magnetic field, indicating that they consist of heavy ions. Furthermore, the emitted ion beam is highly collimated, and form a sharp image of the tip structure on the fluorescent screen. By comparing the metal ion image with FIM and FEM, it is confirmed that the metal ions come from the single-atom protrusions on the tip apex. The FIM and FEM studies also confirmed that most of the protrusions generated through this process ended with a single metal atom.

The tips produced through AMIE are highly reproducible. Apart from the highly collimated electron or ion beams, tips ending with a single metal atom have other reproducible characteristics, in particular, its field emission spectrum (FES) deviates dramatically from the well-known free-electron-metal behavior, Eq. (4.20). The peak structure in the FES and its implications to STM and STS (Binh et al. 1992) is discussed in Section 14.5. As these deviations are characteristic to the FES from single-atom nanotips, they can be used as signatures for attesting the presence of such atomic structures.

13.4 In situ tip treatments

The tip treatment can be done during actual tunneling. Often, these in situ tip treatments take a few seconds to complete. The effect of the tip treatment process can be verified by actual imaging immediately. If one action is not successful, another action can proceed immediately—it takes a few more seconds. As we have mentioned at the beginning of this chapter, these methods was already used at the birth of the STM by the inventors in their first set of experiments (Binnig and Rohrer, 1982).

13.4.1 High-field treatment

A detailed description of the high-field treatment was first presented by Wintterlin et al. (1989). It starts with a clean W tip made through electrochemical etching, and a clean Al(111) surface. The tunneling conditions are -500 mV at the sample, with a current set point 1 nA (see Fig. 13.9). The images produced with this tip do not show atomic resolution immediately. The tip sharpening is performed by suddenly raising the bias to -7.5 V (at the sample), and leaving it at this voltage for approximately four scan lines. The tip responds to the voltage jump by a sudden withdrawal by $\approx$ 30 Å. This is much more than the 2–4 Å that could be expected from the constant current-condition. While the bias is kept at -7.5 V, the scan lines are strongly disturbed. Subsequently the bias voltage is reduced to its initial value of -500 mV again. Now the tip does not return to its former z position but remains displaced from that by about 25 Å. It is obvious that the tip actually gets longer by about 25 Å. This process turns out to be completely reproducible and in most cases results in tips achieving atomic resolution.

At beginning, the mechanism of such tip-sharpening process was not well understood. Two hypotheses were proposed: it is either a restructuring of

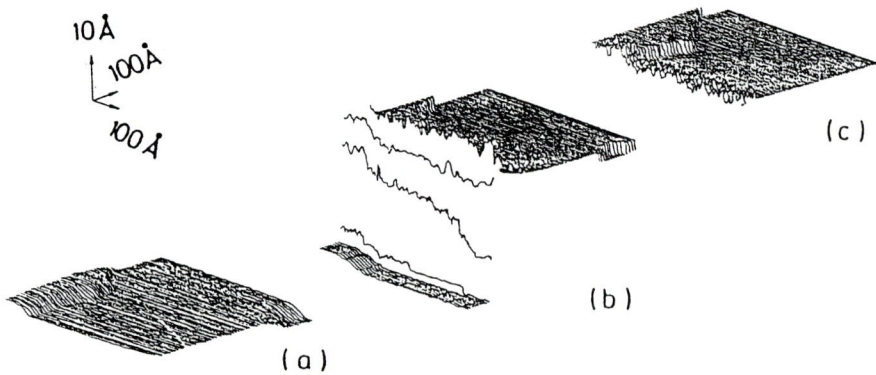

Fig. 13.9. In situ tip sharpening by electrical field. (a) Tunneling at -500 mV and 1 nA with no atomic resolution. (b) By suddenly raising the bias to -7.5 V for several scan lines, the tip end is elongated. (c) A tip with atomic resolution is formed. (Reproduced from Wintterlin et al., 1989, with permission.)

the tip itself or a transfer of material from the sample to the tip. If the latter is the correct mechanism, its result should depend on the sample material. Later, the same tip sharpening procedure was successfully applied on Ru(0001), Ni(100), NiAi(111), and Au(111), indicating that this phenomenon is not specific for certain surfaces (Behm, 1990). Therefore, there must be a restructuring of the tip, that is, the W atoms move from the shank surface to the apex, as shown in Fig. 13.10.

The magnitude and the direction of the tip-treatment voltage is consistent with the tip-restructuring hypothesis. Actually, the directional walk of W atoms on W surfaces under a nonuniform electric field was observed using FIM by Tsong and Kellogg (1975), studied in detail by Wang and Tsong (1982), as summarized in Tsong's book (1990). Near the end of a W tip, the

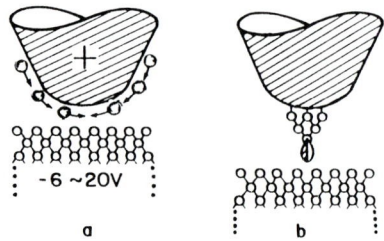

Fig. 13.10. Mechanism of tip sharpening by an electrical field. (a) W atoms on the tip shank walk to the tip apex due to the nonuniform electrical field. (b) A nanotip is formed. (Reproduced from Chen, 1991, with permission.)

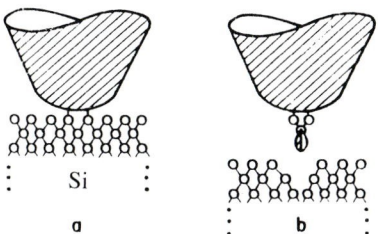

Fig. 13.11. Mechanism of tip sharpening by controlled collision. (a) The W tip picks up a Si cluster from the Si surface. (b) A Si cap forms at the apex of the tip, providing a p_z dangling bond. (Reproduced from Chen, 1991, with permission.)

electrical field is highly nonuniform. The W atoms on the surface, polarized by the high field, are attracted to the apex, where the field intensity is the highest. The topmost atoms are then ionized and stripped off by the electrical field. The experimental demonstration of the atomic metallic ion emission, as discussed in the previous section, further confirms the hypothesis of tip restructuring (Binh and Garcia, 1991, 1992).

13.4.2 Controlled collision

Tip sharpening by controlled collision (Fig. 13.11) was also used by Binnig and Rohrer in their very first experiments with Si(111)-7 $\times$ 7 (1982, 1987). Demuth et al. (1988) provided experimental evidence that during a mild collision of a W tip with a Si surface, the W tip picks up a Si cluster. The tip then provides atomic resolution, and a crater is left on the Si surface. The p_z dangling-bond state on the Si cluster is apparently the origin of the observed atomic resolution.

CHAPTER 14

SCANNING TUNNELING SPECTROSCOPY

By interrupting the feedback loop and keeping the tunneling gap constant and applying a voltage ramp on the tunneling junction, the tunneling current as a function of bias provides information that is the convolution of the tip DOS and the sample DOS. It is an extension of the tunneling junction experiment (Giaever, 1960, 1960a; Bardeen, 1960), but with a richer information content: While the tip is scanning over the sample surface, a map of tunneling spectra is generated. Actually, the original idea of building the STM was to perform tunneling spectroscopy locally on an area less than 100 Å in diameter (Binnig and Rohrer, 1987). A concrete suggestion of a scanning tunneling spectroscopy (STS) experiment was made by Selloni et al. (1985) on graphite. The first demonstration of STS was made by Feenstra and co-workers (Stroscio et al., 1986, 1987a; Feenstra et al., 1987a).

14.1 Electronics for scanning tunneling spectroscopy

In this section, we discuss a typical circuit and the methods for such measurements. Although a commercial sample-and-hold device can be used to hold the z voltage, the accuracy and holding time of those commercial integrated circuits are not good enough. In the feedback circuit shown in Chapter 11, there is a natural point where a custom sample-and-hold device can be easily inserted, as shown in Fig. 14.1. The integration capacitor C is usually of the order of 10 μF and is located right before the amplifier for the z piezo. An op-amp with FET input stage has a typical leakage current I_0 of 1–10 pA. Assuming $I_0 = 10^{-11}$ A and $C = 10^{-5}$ F, the time for the voltage on the capacitor to slip 1 mV is

$$\Delta t = 0.001 \frac{C}{I_0} \approx 10^3 \text{ sec.} \tag{14.1}$$

Therefore, leakage due to the amplifier after the capacitance is not a problem. The problem is the switching device before the capacitor. Feenstra et al. (1987a) used a reed relay as the switching device. The open-circuit resistance of a reed relay is larger than 10^{12} Ω. The leakage current is even smaller than the input of the FET op-amp. (Experience has shown that the printed circuit

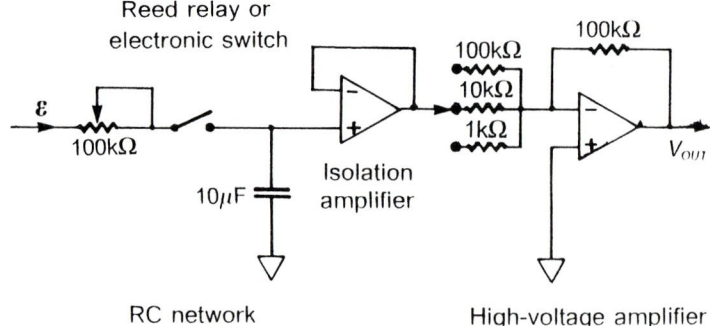

Fig. 14.1. Electronics for local tunneling spectroscopy. By using an op-amp with FET input stage as the isolation amplifier to the high-voltage amplifier for the z piezo, the holding time on the capacitor can be as long as 100 sec. The values of R and C show typical ranges.

board containing those parts must be cleaned meticulously after soldering to avoid any unexpected leakage.) Therefore, if the mechanical structure of the STM is sufficiently stable, the position can be held for several seconds at virtually the same voltage. However, the reed relay has a switching time of a few milliseconds. To obtain one I/V curve, at least 10 msec is required. Therefore, this circuit is suitable for measuring tunneling spectra at selected points with a long holding time (up to a few seconds). Another choice is to use an electronic switch. The switching time is of the order of a few microseconds. However, the leakage current for a typical analog switch (AD7510) is 1–10 nA. An estimation similar to Eq. (14.1) gives a holding time of 1 sec. In most cases this is enough. The reason is as follows. The typical sampling rate for I/V data can be estimated from the sampling frequency of the A/D converter and the time delay of the current amplifier, which is about 50 kHz. At room temperature, the thermal broadening is $4k_BT \approx 0.1$ V. A 100-point spectrum contains a sufficient amount of information. Even if a few repeated ramps have to be executed for making an average, a few milliseconds at a point is sufficient. Therefore, if tunneling spectra are taken from every point of the image, or a substantial number of points, the electronic switch is the appropriate choice as the input switching device for the feedback capacitor.

14.2 Nature of the observed tunneling spectra

In planning and interpreting STS experiments, it is essential to understand the nature of the local tunneling spectra obtained by an STM.

In Chapter 2, we have obtained the general expression for the tunneling current under a bias V:

$$I = \frac{4\pi e}{\hbar} \int_{-\infty}^{\infty} \left[f(E_F - eV + \epsilon) - f(E_F + \epsilon) \right]$$

$$\times \rho_s(E_F - eV + \epsilon) \, \rho_T(E_F + \epsilon) \left| M \right|^2 d\epsilon, \tag{14.2}$$

where $f(E) = \{1 + \exp[(E - E_F)/k_B T]\}^{-1}$ is the Fermi distribution function, $\rho_s(E_F)$ and $\rho_T(E_F)$ are the densities of states (DOS) of the two electrodes. The tunneling matrix element, M, is a surface integral over a separation surface between the tip and the sample,

$$M_{\mu\nu} = -\frac{\hbar^2}{2m} \int_{\Sigma} (\chi^*_\nu \nabla \psi_\mu - \psi_\mu \nabla \chi^*_\nu) \bullet d\mathbf{S}. \tag{14.3}$$

where ψ is the sample wavefunction, modified by the potential of the tip; and χ is the tip wavefunction, modified by the potential of the sample.

In the following, we discuss some general features of the tunneling spectra in terms of Equations (14.2) and (14.3).

14.2.1 The role of tip DOS

The tip DOS, $\rho_T(\epsilon)$, and the sample DOS, $\rho_s(\epsilon)$, appear in a symmetric way in determining the tunneling current. In the simplest case, when the temperature is low and the tunneling matrix element is a constant, the tunneling current is a convolution of the tip DOS and the sample DOS over an energy range eV,

$$I \propto \int_0^{eV} \rho_s(E_F - eV + \epsilon) \, \rho_T(E_F + \epsilon) \, d\epsilon. \tag{14.4}$$

The usual goal of the STS experiment is to probe the DOS distribution of a sample surface. Eq. (14.4) means that this measurement is meaningful only when the tip DOS as a function of energy over the range of measurement is known a priori.   Otherwise, the sample DOS does not have a definitive relation to the tunneling spectrum. If the tip DOS is a constant, then Eq. (14.4) implies

$$\frac{dI}{dV} \propto \rho_s(E_F - eV + \epsilon). \tag{14.5}$$

In other words, with a flat-DOS tip, the dynamic tunneling conductance is proportional to the sample DOS. This simple situation requires a special procedure to treat the tip. On the other hand, a reproducible sample surface is easier to prepare than a reproducible tip. In this case, the sample DOS as a function of energy in the range of measurement can be determined by first-principles calculations or independent experiments. The tunneling spectra thus provide information on the tip DOS, or the tip electronic structure.

Several authors have reported that the simultaneously acquired tunneling spectra with an STM experiment contain information about the electronic states of the tip. For example, Klitsner, Becker, and Vickers (1990) observed a case in which two microtips on one tip body produced two different tunneling spectra at the same spot on the same sample surface. The two independently acquired tunneling spectra contained information about the electronic structures of the two microtips. Pelz (1991) reported in detail a case where the tip electronic state changed during a single scan, which dramatically altered the local tunneling spectra, which we will discuss later on.

14.2.2 Uncertainty-principle considerations

As we have shown in the previous subsection, the tip DOS and the sample DOS are equally important in deterring the observed tunneling spectra. This is a basic fact established in the theory and practice of the tunneling junctions (Giaever, 1960, 1960a; Bardeen, 1960). However, there is an important difference between the tunneling junctions and the STS. In the tunneling junction experiments, the (x,y) dimensions are very large on the microscopic scale, which is in the order of a fraction of a millimeter. The observed tunneling spectra reflect the two-dimensional band structures of both electrodes. In STS, the area through which the tunneling current passes is of the order of 1 nanometer or less, as shown in Fig. 14.2. The local modification of the sample states by the tip may have a substantial impact in the observed tunneling spectra.

On perfect crystalline surfaces, the unperturbed electronic structure is determined by the energy band structure of the surface Bloch waves. This is a consequence of the two-dimensional translational symmetry of the surface. The presence of the tip breaks the translational symmetry of the surface, and the surface electronic structure of the sample is perturbed.

This effect can be illustrated by Fig. 14.2. The effective range of local modification of the sample states is determined by the effective lateral dimension l_{eff} of the tip wavefunction, which also determines the lateral resolution. In analogy with the analytic result for the hydrogen molecular ion problem, the local modification makes the amplitude of the sample wavefunction increase by a factor $\exp(-\tfrac{1}{2}) \approx 1.213$, which is equivalent to inducing a localized state of radius $r \approx l_{\text{eff}}/2$ superimposed on the unperturbed state of the solid surface. The local density of that state is about $(4/e - 1) \approx 0.47$ times the local electron density of the original state in the middle of the gap. This superimposed local state cannot be formed by Bloch states with the same energy eigenvalue. Because of dispersion (that is, the finite value of $\partial E/\partial k_x$ and

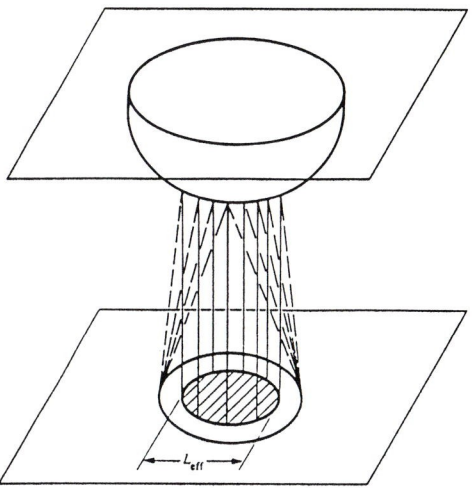

Fig. 14.2. The "illuminated area" of the tip current. The tunneling current of a tip with atomic resolution is concentrated in an area with a small diameter. On one hand, the diameter determines the lateral resolution. On the other hand, the diameter determines the lateral extent of the modification of the sample wavefunction due to the presence of the tip. The higher the resolution, the more severe the modifications is. (Reproduced from Garcia, 1986, with permission.)

$\partial E/\partial k_y$), that tip-induced local state is a superposition of Bloch states with a range of energy values. The following is an estimation based on the uncertainty principle:

As we showed in Section 8.1, for simple metals, the dispersion relation is

$$E = E_0 + \frac{\hbar^2 k^2}{2m*} \, , \tag{14.6}$$

where $m*$ is the effective mass and E_0 is the bottom of the valance band. The Fermi wave vector is

$$k_F = \frac{\sqrt{2m*(E_F - E_0)}}{\hbar} \, . \tag{14.7}$$

It follows from the uncertainty relation

$$\Delta x \Delta k \geq \frac{1}{2} \tag{14.8}$$

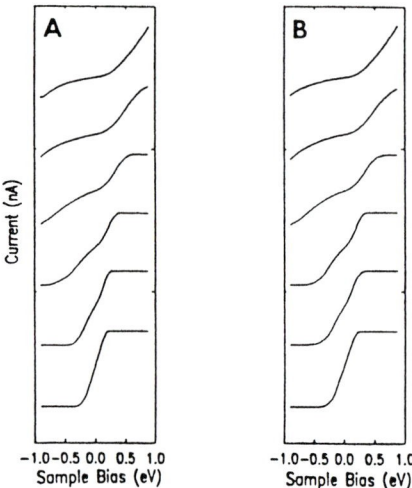

Fig. 14.3. Dependence of local tunneling spectra with distance. By moving a tip with atomic resolution closer and closer to the sample surface, the features in the tunneling spectra is broadened. At the contact point, the tunneling spectrum becomes featureless. This phenomenon demonstrates the effect of local modification of the sample wavefunction and the consequence of uncertainty principle in scanning tunneling spectroscopy. (After Avouris et al., 1991b.)

that

$$\Delta E \geq \frac{\hbar^2 k_F}{2m^* \Delta x} = 0.47 \frac{E_F - E_0}{r k_F} .\tag{14.9}$$

The factor 0.47 is the relative intensity of the tip-induced local states, estimated from an analogy to the hydrogen molecular ion problem. For example, for $(E_F - E_0) = 4$ eV and $\kappa_F = 1$ Å^{-1}, when atomic resolution is achieved, that is, $r \leq 2$ Å, the energy resolution is about 0.2 eV. Even for the reconstructed surfaces, such as Si(111) 7×7, when the adatoms are barely resolved, that is, $r \approx 4$ Å, the energy resolution is 0.12 eV.

An immediate consequence of this argument is that the observed tunneling spectra should exhibit a dependence on the tip–sample distance. At large tip–sample distance, the details of the tunneling spectra should be observed. By reducing the distance, the fine features in the tunneling spectra would disappear gradually. This was indeed observed experimentally, as shown in Fig. 14.3 (Avouris et al. 1991). The experiment was performed on a Si(111) surface. As shown, at large distances (the curves at the top), the band gap is clearly observed in the tunneling spectrum. By gradually pushing the tip into the sample surface (the curves in the middle ofFig. 14.3). the semi-conducting behavior gradually disappears. At very short distances (the bottom curves), the bend gap disappears completely. A metallic characteristics is observed. In other words, the closer the tip–sample distance, the higher the resolution in topography, and the more the features in the tunneling spectra are blurred. At mechanical contact, the tunneling spectrum is virtually flat.

14.2.3 Effect of finite temperature

The Fermi-distribution factor in Eq. (14.2), imposes another limit on spectroscopic resolution. At room temperature, $k_B T \approx 0.026$ eV. The spread of the energy distribution of the sample is $2k_B T \approx 0.052$ eV. The spread of the energy distribution of the tip is also $2k_B T \approx 0.052$ eV. The total deviation is $\Delta E \approx 4k_B T \approx 0.1$ eV.

To make high-resolution STS, in millivolts, two conditions must be met. First, the experiment must be conducted at cryogenic temperatures. Second, the spatial resolution must be sacrificed. In the classical tunneling junction experiments of Giaever (1960), to observe the energy gap of superconducting aluminum, the temperature was dropped to 1.0 K, and the dimension of the tunneling junction was about 0.1 mm. The tunneling spectra of $NbSe_2$ reported by Hess et al. (1989, 1990) were obtained at a few degrees kelvin and with a spatial resolution of about 1000 Å.

14.3 Tip treatment for spectroscopy studies

As we showed in the previous section, in order to obtain reproducible tunneling spectra, the STM tip must have reproducible DOS, preferably a flat DOS, that is, with a free-electron-metal behavior. However, the tips freshly made by mechanical or electrochemical methods, especially those providing atomic resolution, often show nonreproducible tunneling spectra. The DOS of such tips is often highly structured. To obtain reproducible STS data, a special and reproducible tip treatment procedure is required.

Feenstra et al. (1987a) developed an in-situ "tip DOS flattening" proce-dure. This procedure uses field emission current to locally heat up the end of the tip, which eventually causes a local melting and recrystallization. Details of this process is as follows:

1. Start with an electrochemically etched W tip. The cleaning or sharpening is not critical.

2. The sample, or the anode, can be any conducting material. Because the process will cause local damage to the anode, it is better to use a scratch sample or an unused area on the sample.

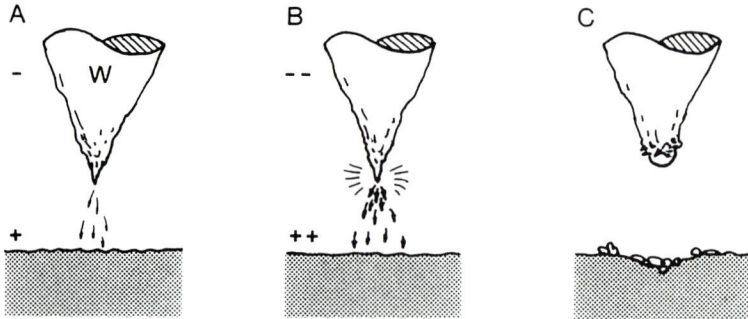

Fig. 14.4. Tip treatment for tunneling spectroscopy. (A) By applying a relatively large positive bias on the sample, a sharp tip generates a field-emission current. (B) When the field-emission current is very high, the tip end melts. (C) The tip end recrystallizes to form facets with low surface energy. In the case of tungsten, the W(110) facets are preferred. Its surface DOS resembles a free electron metal. (After Feenstra et al., 1987a.)

3. Apply a relatively high voltage (for example +10 volts) to the sample, then use the coarse positioning mechanism and z piezo to bring the tip and sample into tunneling distance. The actual value of the tunneling current is not important. The use of a relatively high bias is to avoid tip crashing due to the oxide layer on the tip surface.

4. Pull the tip back by about 1000 Å.

5. Increase the bias slowly up to some 100 V until a field emission current is observed.

6. Either by further increasing the bias voltage or by reducing the tip–sample distance, carefully ramp up the field-emission current.

7. At a current of the order of 1 to 2 µA, a catastrophic decrease of the field-emission current is observed. In a small fraction of a second, the field emission current drops to zero.

Schematically, the steps in this process are shown in Fig. 14.4. At the beginning, the local radius of the tip end is small. Field emission can be easily established. A high current though the tip end then causes local melting. The local curvature at the end of the tip suddenly decreases. The field emission current is then reduced dramatically. The tip end recrystallizes to have a relatively large radius. Feenstra et al. (1987a) observed that the tips prepared in this way always provide reproducible tunneling spectra, although atomic-resolution topographic images are generally not observed.

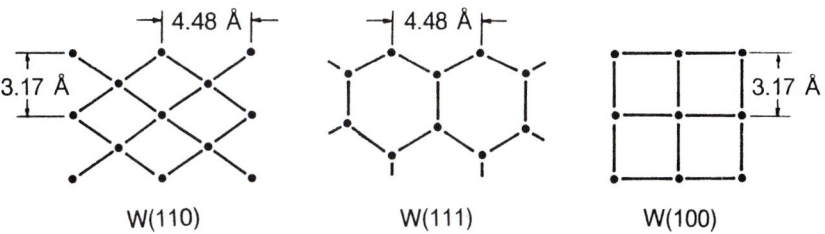

Fig. 14.5. Structure of some low-Miller-index W surfaces. W(110), 1.41×10^{15} atoms per cm^2. W(111), 1.15×10^{15} atoms per cm^2. W(100), 9.98×10^{14} atoms per cm^2.

A possible interpretation of this process is as follows. As the tungsten tip recrystallizes, it always tends to reduce the surface energy by forming facets with the highest density of surface atoms. For tungsten, a bcc metal, the (110) facet has by far the highest density of atoms, 1.41×10^{15} cm^{-2}, followed by the (111) facet, 1.15×10^{15} cm^{-2}. The (100) facet, with a surface density of atoms 9.98×10^{14} cm^{-2}, is unlikely to be formed when recrystallized, as shown in Fig. 14.5. The (112) facet is nearly impossible to form spontaneously. Both field-emission spectroscopy (Swanson and Crouser 1967) and photoemission spectroscopy (Willis and Feuerbacher, 1975; Drube et al., 1986; Gaylord et al., 1989) show that near the Fermi level, the DOS of the W(110) and W(111) surfaces is similar to free-electron metals, in contrast to the W(100) and W(112) surfaces, where localized d_z surface states exhibit highly structured surface densities of states near the Fermi level.

14.4 The Feenstra parameter

One of the major problems in doing tunneling spectroscopy with STM is that the tunneling current exhibits a dramatic dependence on tip–sample separation

$$I = I_0 \, e^{-2\kappa z}, \qquad (14.10)$$

which means that the tunneling current varies approximately one order of magnitude per Å. Therefore, the measured spectroscopic data need a *normalization*. Similar to most spectroscopic methods, the interest lies in the variation of an intensity quantity with respect to energy, rather than its absolute value. Often, an "arbitrary unit" is used on the ordinate. For metals and narrow-band-gap semiconductors with metallic surface states, Feenstra et al. (1987) proposed a parameter, the normalized dynamic conductance,

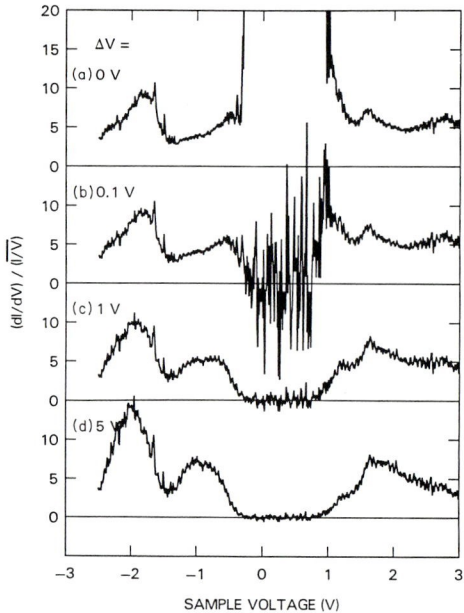

Fig. 14.6. Effect of averaging to the normalized tunneling conductivity. Normalization of the conductivity spectrum by broadening. Even with a small ΔV, the noise is substantially suppressed. (Reproduced from Mårtensson and Feenstra, 1989, with permission.)

$$g_N(V) \equiv \frac{d \ln I}{d \ln V}, \tag{14.11}$$

which is a dimensionless quantity. In the immediate vicinity of $V = 0$, the condition of being metallic implies $I = SV$, where S is a constant. From Eq. (14.11), it follows that $g_N(0) = 1$.

Another possible definition of the relative dynamic conductance is

$$g(V) \equiv \frac{dI}{dV} \left(\frac{dI}{dV} \right)^{-1}_{V=0}, \tag{14.12}$$

which is also dimensionless and has the condition $g(0) = 1$. If the values of one quantity are known in an interval $[0, V]$, then the values of the other one on the same interval can be derived. Actually, it is easy to show that these two dimensionless quantities are related by the following equations:

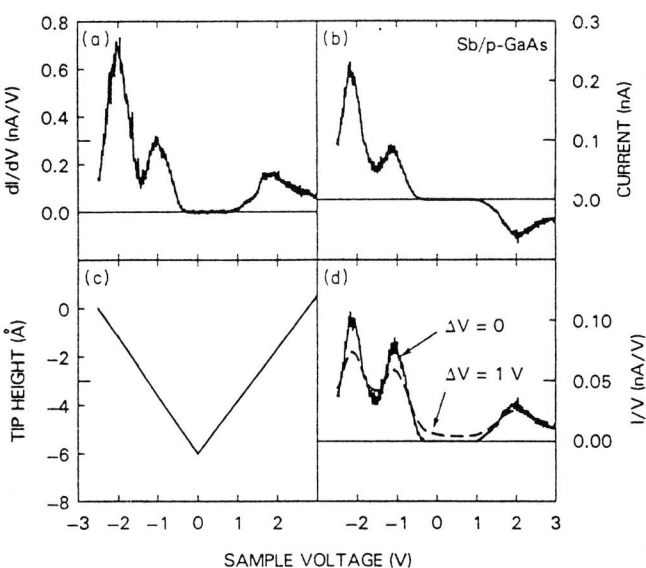

Fig. 14.7. Tunneling spectra with the varying-gap method. Raw data for (a) the differential conductivity, and (b) the current, as a function of bias voltage (at the sample). The applied variation in tip–sample separation is shown in (c). The total conductivity I/V is shown in (d), with no broadening (solid curve) and with broadening of $\Delta V = 1$ V (dashed curve). (Reproduced from Mårtensson and Feenstra, 1989, with permission.)

$$g_N(V) = V g(V) \left(\int_0^V g(u)\, du \right)^{-1}, \tag{14.13}$$

and

$$g(V) = \frac{V_0\, g_N(V)}{V}\ \exp\left(\int_{V_0}^V g_N(u)\, \frac{du}{u} \right), \tag{14.14}$$

where V_0 is a small positive quantity with the condition,

$$\text{when } |V| < V_0, \quad g_N(V) \approx 1. \tag{14.15}$$

For semiconductors, in the vicinity of $V = 0$, the conductance is very small, and the relative accuracy is low. A common practice to deal with this situation is to replace the inaccurate section by a small constant using interpolation, which will not affect the overall features of the spectrum. Feenstra's dimensionless parameter is widely used in the STS literature for presenting measured data.

However, for surfaces with large band gaps such as GaAs($1\bar{1}0$), the quantity in Eq. (14.11) diverges at the band edge. It is evident from the form of Eq. (14.11) that at the band edge $I = 0$, but $dI/dV \neq 0$. Two methods for dealing with STS of wide-gap semiconductors are further proposed by Feenstra and co-workers. The first method is a data processing procedure for eliminating the divergence in $g_N(V)$ (Mårtensson and Feenstra, 1989). Take the average of the measured I/V with an envelop function, for example,

$$\overline{I/V} \equiv \int_{-\infty}^{\infty} \frac{I}{V} \exp\left(\frac{|V' - V|}{\Delta V} \right) dV', \qquad (14.16)$$

where ΔV is a constant. An example of its application is shown in Fig. 14.6. Even with a small averaging parameter ΔV, the divergence is substantially suppressed.

The second method concerns improving the measuring procedure by varying the tip–sample separation while varying the bias. For semiconductors, the tunneling current increases dramatically with the magnitude of bias. With a very stable mechanical system and a very long holding time for the z position, the relative gap width can be well controlled (Shih et al., 1990, 1991). To proceed, the bias is first set at a relatively large value, typically ± 2 V, to induce a sizeable tunneling current. The tip is stabilized by the feedback loop. Then the feedback is disabled, and a v-shaped voltage is added to the z piezo voltage, synchronized with the bias ramp. The tunneling current as a function of the bias is recorded. An example of the varying-gap method for tunneling spectroscopy is shown in Fig. 14.7. Even if there is no averaging or broadening, the observed spectrum shows very little noise. This method is more natural than the averaging method in Eq. (14.16).

14.5 Ex situ determination of the tip DOS

In general, the tip DOS does not resemble that of a free-electron metal. To have meaningful STS measurements, the tip DOS must be determined independently. A well-established experimental method for determining the energy spectrum a sharp metal tip is field emission spectroscopy (FES), which was described in Section 4.4. For free-electron metal tips, the FES is described by the Young formula, Eq. (4.20). A deviation from the Young formula indicates a deviation from a free-electron metal behavior.

In Section 13.3.4, a method of generating sharp tips ended with single atom was described (Binh and Garcia, 1991, 1992). Such tips are necessary for

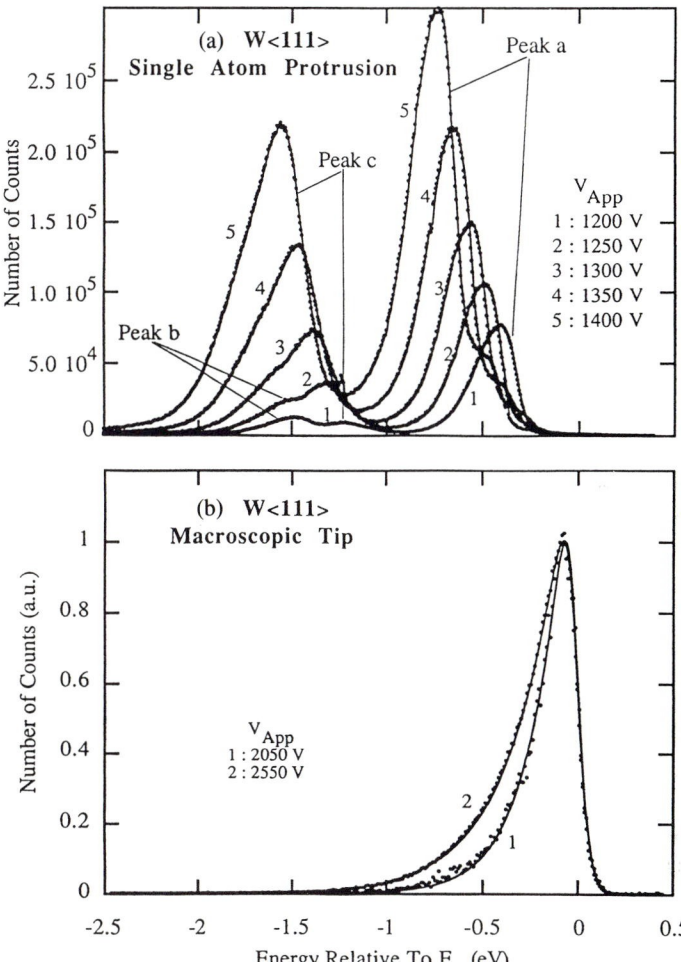

Fig. 14.8. Experimental field-emission spectra of a W tip with a single protruded atom. (a) The FES of a tip with single-atom protrusion. Well-separated peaks are observed. The position of the peaks vary with the applied voltage. (b) The FES of a macroscopic tip, after the single-atom protrusion is destroyed by heating. The FES fits accurately with the Young formula for free-electron metals, Eq. (4.20). The edge of the peak is always at the Fermi level, independent of the applied voltage. (Reproduced from Binh et al., 1992, with permission.)

achieving atomic resolution in STM. A field-emission-spectrum study of the single-atom tips (Binh, Purcell, Garcia, and Doglioni, 1992) shows a dramatic deviation from the free-electron-metal behavior, as shown in Fig. 14.8(a). For comparison, also included is the field-emission spectrum of a macroscopic tip, which is prepared by heating the single-atom tip to destroy the protrusion, Fig. 14.8(b). Several salient features are discerned from the experimental observations:

1. The spectra of the single-atom tips are composed solely of well-separated peaks. An example is shown in Fig. 14.8(a). Here, three peaks are observed. The actual position and intensity of the peaks depends on the details of the atomic structure near the apex atom. For comparison, with a macroscopic tip, only one peak is observed right below the Fermi level, see Fig. 14.8(b).

2. The positions of the peaks shift with applied voltage, as shown in Fig. 14.8(a). For comparison, the edge of the single peak for a free-electron metal tip does not change with the applied voltage, see Fig. 14.8(b).

3. Although the FES of the macroscopic tip fits well with the Young formula, Eq. (4.20), none of the peaks in the FES of the single-atom tip fits it.

These features of the FES of the single-atom tips strongly suggest that the electrons do not tunnel directly from the Fermi sea to the vacuum. Rather, they come solely from the localized states of the protruding atom.

The existence of the localized states due to the protruding atom are important for interpreting atomic resolution in STM (see Chapter 5 and Chapter 6). They are equally important for the correct interpretation of STS data. Especially, when atomic resolution is achieved, the tip DOS is usually highly structured. This fact must be considered carefully in the interpretation of the simultaneously acquired tunneling spectra.

14.6 In situ determination of the tip DOS

In the topographic mode of STM, the major objective is to achieve high resolution. The flat-DOS tip is not appropriate for this application. The energy spectra of the tips that allow atomic resolution are usually highly structured. Moreover, direct experimental observation shows that the tip states change frequently and *spontaneously*. Actually, the tip states are very sensitive to tip structure. An addition, removal, or displacement of a single atom near the apex would change the tip state dramatically. During scanning, the tip often picks up an atom from the sample, or vice versa. The *spontaneous switching of STM resolution,* resulting from a *spontaneous tip restructuring,* is a manifestation of the tip-state effect (see Fig. 14.9). In order to achieve a better understanding of the STM imaging mechanism, an in situ characterization of the tip electronic structure is useful.

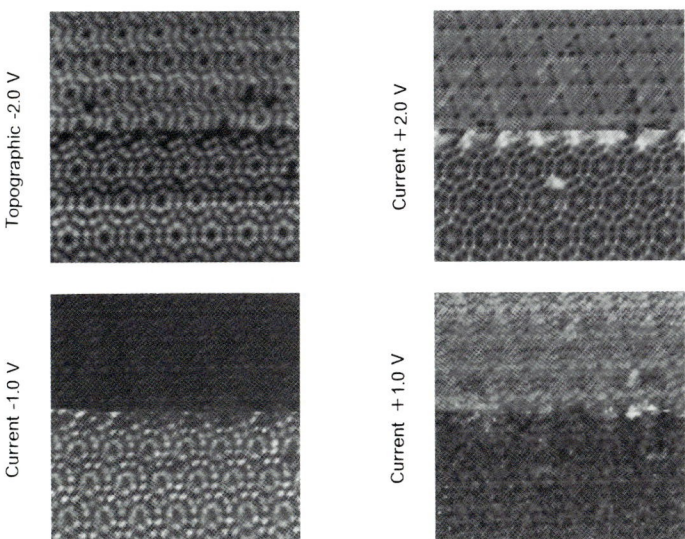

Fig. 14.9. A recorded case of spontaneous tip restructuring. (a) The tip restructured at the middle of a scan. The adatoms of the Si(111)-7×7 are resolved in both halves of the topographic image. (b)–(d) The current images at different biases changed dramatically, indicating that the electronic states of the tip have undergone a dramatic change. (Courtesy of J. Pelz, see also Petz, 1991).

The following method, based on the modified Bardeen approach, allows the electronic states of the tip to be characterized. If the energy scale of feature in the DOS is larger than $k_B T$, the tunneling current is

$$I = \text{const.} \times \int_{E_F}^{E_F + eV} \rho_S(E)\, \rho_T(E - eV)\, |M(E, E - eV)|^2 \, dE, \quad (14.17)$$

where $\rho_T(E)$ and $\rho_S(E)$ are the tip DOS and the sample DOS, respectively. The tunneling matrix element M is expressed as the modified Bardeen integral of the sample wavefunction and tip wavefunction on a separating surface in the gap. Here, we are more interested in finding sharp structures in the tip DOS. To simplify the procedure, we treat the tunneling matrix element as a constant.

From Eq. (14.17), we observe that for $V > 0$, the electrons, from the occupied states in the tip, tunnel to the empty states of the sample. The tunneling current has no relation to the empty states of the tip. On the other hand, if $V < 0$, the tunneling current is determined by the occupied states of the sample and the empty states of the tip. Therefore, the determinations of the empty-state tip DOS and the occupied-state tip DOS are completely independent problems. In the following, we present the method for the determination of the DOS of occupied tip states, that is, for the case of $V > 0$. The same method is valid for the case $V < 0$.

The relative DOS for the sample and the tip can be defined as follows. For the case $V > 0$, by defining a parameter $u = (E - E_F)/e$, a convenient form for the relative sample DOS is:

$$\sigma(u) = \rho_S(E)/\rho_S(E_F). \qquad (14.18)$$

For the tip, we define a parameter $u = (E_F - E)/e$, and let

$$\tau(u) = \rho_T(E)/\rho_T(E_F). \qquad (14.19)$$

Obviously, because of the condition $V > 0$, the arguments u in both Equations (14.18) and (14.19) are always positive. (For the case of $V < 0$, simply exchange the definitions of u.) Clearly, we have $\sigma(0) = 1$ and $\tau(0) = 1$. By differentiating Eq. (14.17), we obtain a pair of symmetric equations:

$$\sigma(V) = g(V) + \int_0^V \tau'(V - u)\,\sigma(u)\,du, \qquad (14.20)$$

$$\tau(V) = g(V) + \int_0^V \sigma'(V - u)\,\tau(u)\,du. \qquad (14.21)$$

These equations are the standard form of a well-studied class of integral equations, the *Volterra equation of the second kind* (see, for example, Brunner and van der Houwen, 1986). Before discussing the numerical method, we draw a few simple conclusions from those equations. Using a free-electron-metal tip (that is, if in the entire energy range of interest),

$$\tau'(V) = 0, \qquad (14.22)$$

then from Eq. (14.21),

$$g(V) = \sigma(V). \qquad (14.23)$$

In this case, the dynamic conductance equals the sample DOS up to a constant factor. Now, we measure the tunneling spectrum on the same sample using another tip of unknown DOS, and find a new dynamic conductance as a function of bias voltage, $g(V)$. By solving Eq. (14.21), we obtain the relative DOS of the unknown tip. [Similarly, if the sample is a free electron metal, that is,

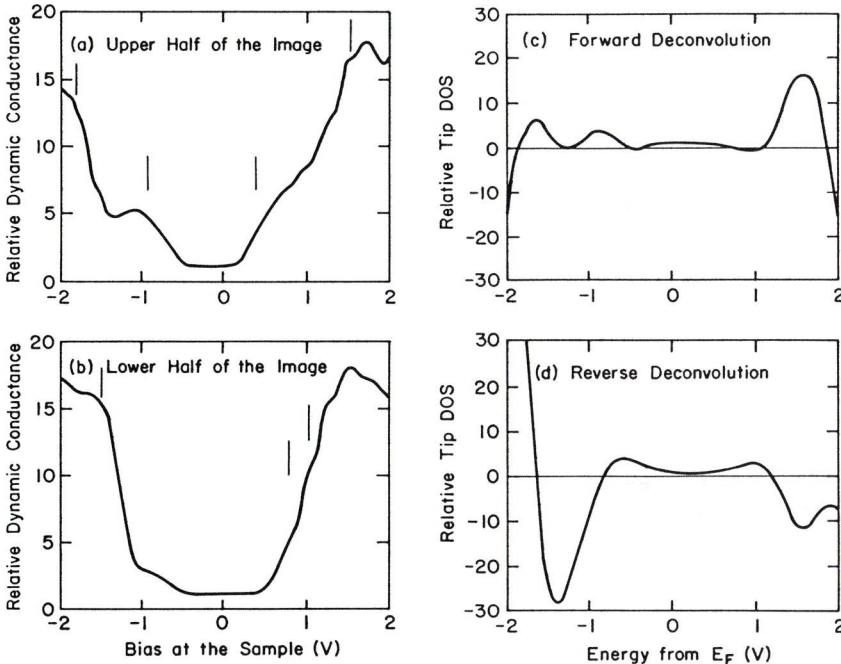

Fig. 14.10. Deconvolution of the tunneling spectra. (a) and (b) The tunneling spectra from the Si surface image in Fig. 14.9. The observed tunneling spectra changed dramatically from the upper half to the lower half of the image. (c) and (d) Results of deconvolution. It indicates that during the scan, the tip picked up a Si cluster, and the tip DOS resembles that of a Si surface. (Reproduced from Chen, 1992b, with permission.)

$\sigma'(V) = 0$, then the dynamic conductance equals the tip DOS up to a constant factor.]

In the following, we show that for application to the STM problem, a simple numerical procedure is sufficiently accurate for the solution of the relevant Volterra equation. By dividing the range of integration $[0, V]$ into N segments of equal length $h = V/N$, using the trapezoidal rule, the integral can be approximated by a sum,

$$\int_0^V f(u)\,du \approx h\{\tfrac{1}{2}f(0) + f(h) + \ldots$$
$$+ f[(N-1)h] + \tfrac{1}{2}f(Nh)\}.$$

(14.24)

Now, we replace the integral in Eq. (14.20) with such a sum. If the value of $\tau(V)$ is known up to $V = (n-1)h$, then

$$
\tau(nh) = \left[1 - \tfrac{1}{2}h\sigma'(0)\right]^{-1} \{g(nh) + \tfrac{1}{2}h\sigma'(nh)\tau(0) + \ldots
$$
$$
+ h\sigma'(2h)\tau\left[(n-2)h\right] + h\sigma'(h)\tau\left[(n-1)h\right]\}.
$$
(14.25)

This method is known as the *marching method.* The accuracy of the procedure and the correctness of the program can be verified by testing it with analytically soluble Volterra equations, for example, the test problems with nonsingular convolution kernels listed on pp. 505–507 of Brunner and van der Houwen's book (1986).

As an example, we analyze the tunneling spectra published by Pelz (1991), where the effect of spontaneous tip restructuring on the observed tunneling spectra is recorded: During a single scan, the observed image suddenly changes. Although the atomic structure on the Si(111)-7×7 surface is resolved on both parts of the image, the normalized dynamic conductance changes dramatically. Using Eq. (14.19), the corresponding relative dynamic conductance is obtained. There is no a priori reason to determine which tip state is more similar to a free-electron metal. By assuming that the tip DOS before restructuring (upper half of the image) is nearly flat, the deconvolution procedure produces a tip DOS after restructuring (lower half of the image), as shown in Fig. 14.10(c). A reversed deconvolution generates Fig. 14.10(d). However, in Fig. 14.10(d), a large portion of the obtained tip DOS is negative, which is unphysical. Therefore, the tip DOS before tip restructuring is more likely to be flat. There is still no consensus of what the accurate tunneling spectrum of the averaged Si(111)-7×7 surface is. Nonetheless, the general feature of Fig. 14.10(a) is consistent with the tunneling spectra most authors reported, whereas the general feature of Fig. 14.10(b) is not. As shown in Fig. 14.10(c), the tip DOS after restructuring has two peaks below the Fermi level, and one peak above the Fermi level. This feature is similar to the observed tunneling spectrum of Si(111)-7×7. Therefore, a likely explanation is that the tip picks up a small Si cluster, and the tip DOS becomes similar to that of silicon.

CHAPTER 15
ATOMIC FORCE MICROSCOPY

15.1 Introduction

The STM resolves individual atoms on conducting surfaces. For resolving individual atoms on insulating surfaces, Binnig, Quate, and Gerber (1986) introduced a method similar to the STM, atomic force microscopy (AFM). In AFM, the tip of a flexible force-sensing cantilever stylus is raster scanned over the surface of the sample. The force acting between the cantilever and the sample surface causes minute deflections of the cantilever. The deflections are detected and utilized as the feedback signal. By keeping the force constant, a topographic image of *constant force* is obtained. A schematic diagram of an AFM is shown in Fig. 15.1.

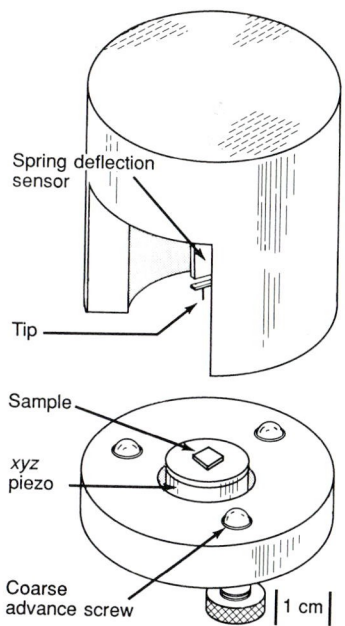

Fig. 15.1. The atomic force microscope (AFM). The sample is mounted on top of a tube scanner. Using the coarse and fine advance screws, the tip, mounted on a flexible spring (cantilever), is brought into gentle contact with the sample surface. By actuating the tube scanner, the tip is raster scanning the sample surface. The force between the tip and the sample is detected by the deflection sensor. The signal, after amplification, is compared with a reference value. The difference signal is again amplified to drive a feedback circuit. A constant-force topographic image of the surface structure is obtained. (Reproduced from Hansma et al., 1988, with permission.)

The force microscope, in general, has several modes of operation. In the repulsive-force or contact mode, the force is of the order of 1–10 eV/Å, or 10^{-9}–10^{-8} newton, and individual atoms can be imaged. In the attractive-force or noncontact mode, the van der Waals force, the exchange force, the electrostatic force, or magnetic force is detected. The latter does not provide atomic resolution, but important information about the surface is obtained. Those modes comprise different fields in force microscopy, such as electric force microscopy and magnetic force microscopy (Sarid, 1991). Owing to the limited space, we will concentrate on atomic force microscopy, which is STM's next of kin.

The AFM has a number of elements common to STM: the piezoelectrc scanner for actuating the raster scan and z positioning, the feedback electronics, vibration isolation system, coarse positioning mechanism, and the computer control system. The major difference is that the tunneling tip is replaced by a mechanical tip, and the detection of the minute tunneling current is replaced by the detection of the minute deflection of the cantilever.

In order to achieve sufficient sensitivity for atomic resolution, the cantilever has to satisfy several requirements (Albrecht et al., 1990).

First, the cantilever must be flexible yet resilient, with a force constant from 10^{-2} to 10^2 N/m. Therefore, a change of force of a small fraction of a nanonewton (nN) can be detected.

Second, the resonance frequency of the cantilever must be high enough to follow the contour of the surface. In a typical application, the frequency of the corrugation signal during a scan is up to a few kHz. Therefore, the natural frequency of the cantilever must be greater than 10 kHz.

Third, as a direct consequence of the conditions listed, the cantilever must be very small. Actually, from Eq. (F.32), the resonance frequency f of a cantilever is related to its length L, cross-sectional area S, and density ρ as

$$f^2 = \frac{0.314EI}{L^4 \rho S}.\tag{15.1}$$

On the other hand, from Eq. (F.23), the force constant K of a beam is

$$K = \frac{3EI}{L^3}.\tag{15.2}$$

Eliminating Young's modulus E and moment of inertia I from Equations (15.1) and (15.2), we find

$$K = 9.57\rho LSf^2 = 9.57Mf^2,\tag{15.3}$$

where M is the mass of the cantilever. Eq. (15.3) is very similar to the equation of a spring–mass oscillator, except the factor $(2\pi)^2$ is replaced by 9.57. If we need $K < 1$ N/m, and $f > 10^4$ Hz, then the mass of the cantilever must be much smaller than 1 μg. In other words, the dimensions of the cantilever must be in the micrometer range.

Fourth, in the vertical and horizontal directions, the stiffness should be very different. When the AFM is operated in the repulsive-force mode, frictional forces can cause appreciable image artifacts. Choosing an appropriate geometry for the shape of the lever can yield substantial lateral stiffness, thus minimizing the disturbing artifacts.

Fifth, when optical beam deflection is used to measure cantilever deflection, the sensitivity is inversely proportional to the length of the cantilever (see the following). If the length of the cantilever is on the order of 100 μm, the length of the "optical lever" can be as short as 1 cm for subangstrom resolution with an inexpensive position-sensitive detector.

Finally, a sharp protruding tip must be formed at the end of the cantilever to provide a well-defined interaction with the sample surface, presumably with a single atom at the apex. The slope of the tip should be as steep as possible, and as smooth as possible. In other words, a high aspect ratio is preferred.

In the following section, a typical method of fabricating the cantilevers and their typical characteristics is described.

15.2 Tip and cantilever

In the early years of AFM operation, the cantilevers were cut from a metal foil, and the tips were made from crushed diamond particles, picked up by a piece of eyebrow hair, and painstakingly glued manually on the cantilevers. This situation has changed completely since the methods for mass production of cantilevers with integrated tips were developed. A review of various methods for making cantilevers using standard microfabrication techniques was published by Albrecht et al. (1990), and an improved method is described by Akamine et al. (1990). Those AFM cantilevers with integrated tips are now available commercially.[14]

In this section, the method of making cantilevers with integrated tips of Si_3N_4 is briefly described (Albrecht et al., 1990). The process starts with a Si(100) wafer with a thermally grown SiO_2 layer, as shown in Fig. 15.2. The steps are as follows:

1. Using a photolithographic method, etch a square opening on the SiO_2 film.

---

14 For example, Park Scientific Instruments, 1171 Borregas Ave., Sunnyvale, California 94089; Digital Instruments, Inc., 6780 Cortona Drive, Santa Barbara, CA 93117.

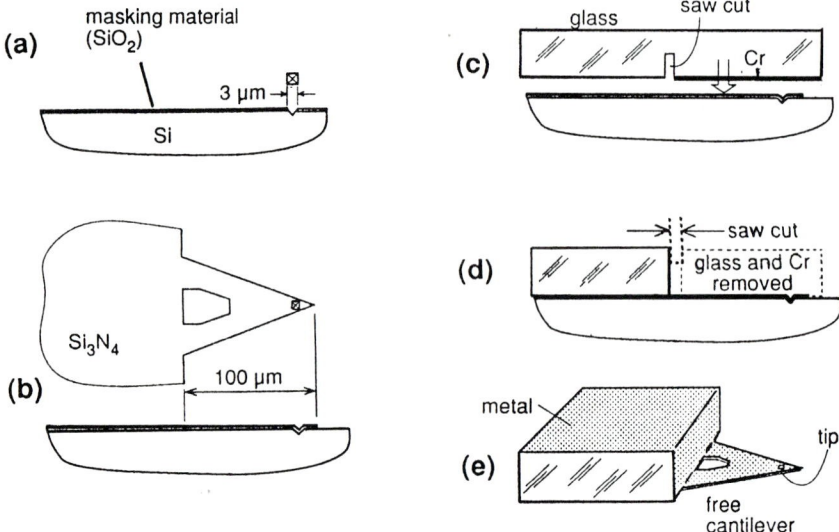

Fig. 15.2. Fabrication of silicon nitride microcantilevers with integrated tips. (a) A pyramidal pit is etched in the surface of a Si(100) wafer using anisotropic etching. (b) A Si_3N_4 film is deposited over the surface and conforms to the shape of the pyramidal pit, and patterned into the shape of a cantilever. (c) A glass plate is prepared with a saw cut and a Cr bond-inhibiting region. The glass is then anodically bonded to the annealed nitride surface. (d) A second saw cut releases the bond-inhibited part of the glass plate, exposing the cantilever. (e) All Si is etched away, leaving the Si_3N_4 microcantilever attached to the edge of a glass block. The back side of the cantilever is coated with metal (Au) for the deflection detector. (Reproduced from of Albrecht et al., 1990, with permission.)

2. Use KOH solution to etch the part of the silicon wafer exposed through the square opening. The etch self-terminates at the Si(111) planes, and a pyramidal pit is formed.

3. Remove the SiO_2 protection layer.

4. Deposit Si_3N_4 on the wafer, to form the shape of the cantilever using a lithographic method.

5. Attach a piece of glass as the carrying substrate using anodic bonding. The area with the cantilever is protected by a Cr layer on the glass.

6. Remove the unwanted part of glass and all the remaining Si.

7. The wafer is diced into pieces. Each piece is a small glass block with several cantilevers attached to its edges.

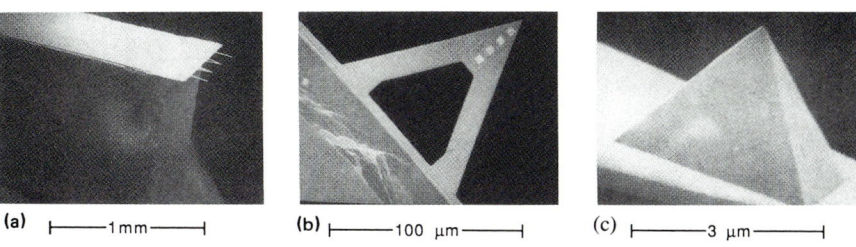

Fig. 15.3. Microcantilever for atomic-force microscopy. (a) A glass substrate with four cantilevers. (b) One of the cantilevers. (c) Close-up view of the tip. (After Albrecht et al. 1990.)

Figure 15.3 is a micrograph of a tip made with this process. The tips made through a similar process have been used commercially for the following reasons. First, the cantilever material, silicon nitride, is robust and inert. Second, the tip sidewalls are extremely smooth and have a slope of 55°, which facilitates low friction sliding over rough surfaces. In addition, since the shape of the tip is well defined, the effects of tip morphology on the image can be understood and taken into account. The typical force constant of such mass-produced cantilevers is 0.0006 to 2 N/m, and the typical resonance frequency is 3 kHz to 120 kHz (Albrecht et al., 1990).

15.3 Deflection detection methods

Many different methods have been developed for detecting the minute deflection of the cantilever (Sarid, 1991). In this section, we present several important ones, including vacuum tunneling (Binnig, Quate, and Gerber, 1986), mechanical resonance (Dürig, Gimzewski, and Pohl 1986), optical interferometry (Martin et al., 1988; Erlandson et al., 1988), and optical beam deflection (Meyer and Amer, 1988).

15.3.1 Vacuum tunneling

The tunneling current between two metal electrodes separated by a vacuum gap varies about one order of magnitude per Å. Therefore, vacuum tunneling provides an extremely sensitive method for detecting minute displacements. The first AFM, demonstrated by Binnig, Quate, and Gerber (1986), utilized vacuum tunneling to detect the cantilever deflection.

Fig. 15.4. Detection of cantilever deflection by vacuum tunneling. (Binnig, Quate, and Gerber, 1986.) The cantilever with tip 1 is sandwiched between the sample tip 2, the tunneling tip. The deflection of the cantilever changes the tunneling gap, and thus the tunneling current. (Reproduced from Yamada et al., 1988, with permission.)

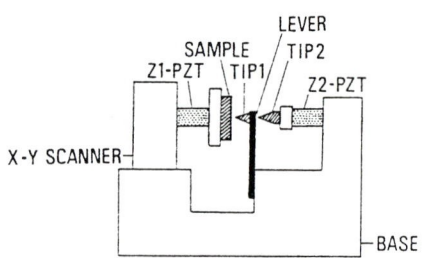

A schematic diagram of an AFM with a tunneling deflection detector is shown in Fig. 15.4. A soft and resilient lever with a tip is sandwiched between the sample and another tip, tip 2, for tunneling. In addition to the x, y, and z_1 scanner for the sample, another piezoelectric element z_2 is used to control the spacing between the tunneling tip (tip 2) and the back of the lever. Using a coarse positioning mechanism (not shown) and the z piezos, the force-sensing tip (tip 1) and the sample surface are brought into a gentle touch, and the tunneling tip is brought to within the detection distance of vacuum tunneling to the back of the lever. While making a raster scan with the sample, the tunneling current between tip 2 and the lever is compared with a reference current, and the difference is acting as the feedback signal for the z_1 piezo. Actually, by requiring the tunneling current to be a constant, the deflection of the cantilever and thus the force acting on the lever are kept constant.

15.3.2 Mechanical resonance

In the second method (Düring, Gimzewski, and Pohl, 1986), the sample is mounted on the cantilever (e.g., a layer of material vacuum deposited on the cantilever surface). A tunneling tip is also acting as the mechanical tip. The quantity measured is the resonance frequency of the cantilever, which is realized by monitoring the spectrum density of the tunneling current using a spectrometer. The distance between the tip and the sample is monitored by the tunneling conductance. As shown in Fig. 15.5, a dramatic change in the resonance frequency is observed. The quantity directly measured is the *force gradient* instead of the absolute value of the force. Actually, by approximating the cantilever as a lumped-parameter system, as in Eq. (15.3), the resonance frequency is

$$f = 9.57 \sqrt{\frac{1}{M}\left(k + \frac{\partial F_z}{\partial z}\right)}, \tag{15.4}$$

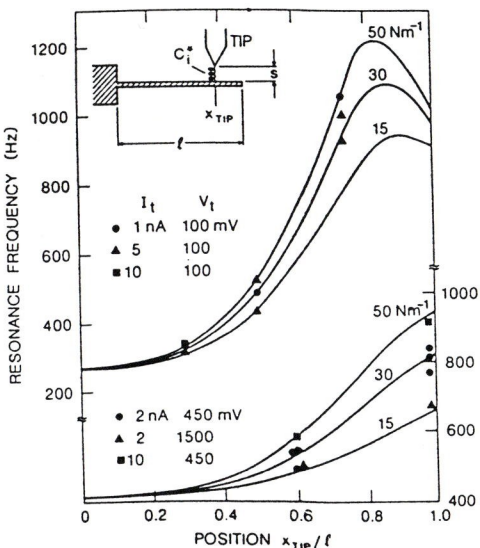

Fig. 15.5. Force-gradient detection via mechanical resonance. Experimental and cal-culated resonance frequencies of a cantilever for various positions of the tunneling tip. Upper family of curves, for the copper, and lower, for the steel beam. (Reproduced from Dürig et al., 1986, with permission.)

where k is the force constant, M is the mass of the cantilever, and F_z is the force between the tip and the sample (the cantilever surface). This method provides a direct measurement of the correlation between the tunneling conductance and the atomic force. By scanning the surface, a tunneling image and a force-gradient image can be acquired simultaneously.

15.3.3 Optical interferometry

The method of Dürig et al. (1986) uses tunneling current as the signal, which is not applicable to insulating samples. A method to detect the shift of reso-nance frequency using optical interferometry was proposed (Martin et al., 1988, Erlandsson et al., 1988). Instead of monitoring the resonance frequency excited by noise, the cantilever is *driven* by a mechanical oscillator operated at the resonance frequency of the free cantilever. The action of the force gra-dient changes the resonance frequency, and causes a *decrease* in the amplitude of the cantilever vibration. The amplitude of the cantilever is then detected by the laser interferometer (Martin et al., 1987).

Consider a cantilever with natural frequency ω_0 and quality factor Q driven by an external excitation with frequency ω. The ratio of the amplitude of the lever A relative to the maximum amplitude A_0 is

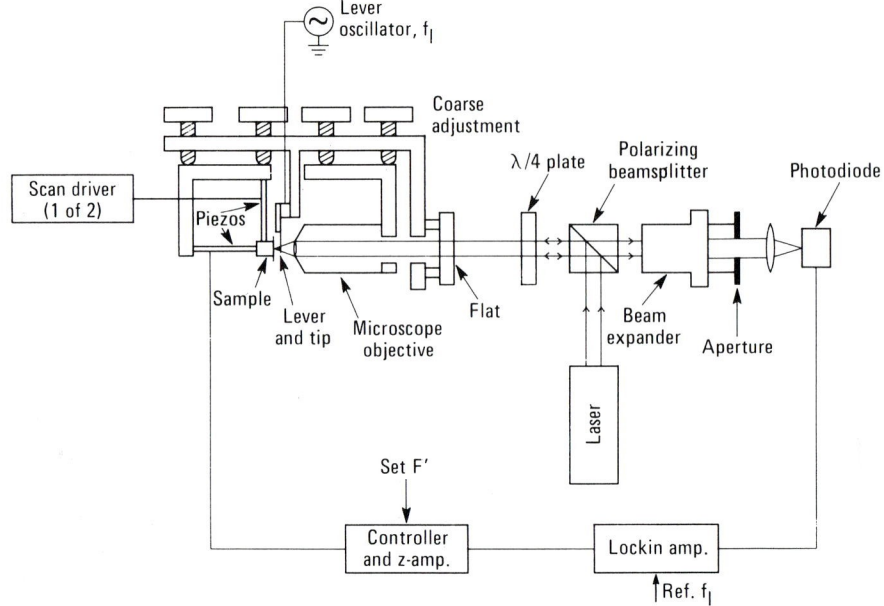

Fig. 15.6. Atomic force microscope with an optical interferometer. (Reproduced from Erlandsson, McClelland, Mate, and Chiang, 1988, with permission.)

$$\frac{1}{a} \equiv \frac{A}{A_0} = \frac{(\omega_0/\omega)}{\sqrt{1 + Q^2(\omega/\omega_0 - \omega_0/\omega)}} .\tag{15.5}$$

The dimensionless parameter a is introduced for convenience, which is almost always less than 1. The resonance frequency ω_0 is related to the force constant K through Eq. (15.3). The force gradient (in this operation mode usually negative) $-\partial F_z/\partial z$ changes the resonance frequency ω_0,

$$\omega_0 = 2.03 \sqrt{\frac{1}{M}\left(K - \frac{\partial F_z}{\partial z}\right)} .\tag{15.6}$$

The amplitude A in Eq. (15.5) changes. After a short amount of algebra, the relation between the force gradient and the amplitude reduction ratio a is

$$\frac{\partial F_z}{\partial z} = K\left(\frac{1 - 2a^2 + \sqrt{4Q^2(a^2 - 1) + 1}}{2(Q^2 - a^2)}\right).\tag{15.7}$$

Figure 15.6 is a schematic diagram of an AFM with an optical interferometer (Erlandsson et al., 1988). The lever is driven by a lever oscillator through a piezoelectric transducer. The detected force gradient F' is compared with a reference value, to drive the z piezo through a controller. In addition to the vibrating lever method, the direct detection of repulsive atomic force through the deflection of the lever is also demonstrated.

15.3.4 Optical beam deflection

The optical interferometry method is rather complicated. Meyer and Amer (1988) introduced a much simpler method by monitoring the deflection of an optical beam reflected by a small mirror on the cantilever to detect the minute deflection of the cantilever. In their original paper, they demonstrated that this method can be used to detect the force gradient in the weak-attractive-force regime using a vibrating cantilever, as shown in Fig. 15.7. The sensitivity of this method is demonstrated by the long range of the force detection. As shown, the force gradient is detectable over a distance range of more than 25 Å.

After the microfabricated cantilever was introduced (Albrecht and Quate, 1988), the optical beam deflection method was successfully applied in the repulsive-force mode, and atomic resolution on insulators as well as conductors has been achieved (Manne et al., 1990; Meyer and Amer, 1990). A schematic diagram of the optical beam deflection AFM is shown in Fig. 15.8. An

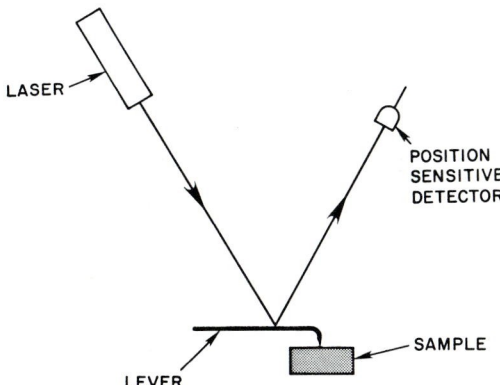

Fig. 15.7. Detection of cantilever deflection by optical beam deflection. A light beam, typically from a solid-state laser, is reflected by the top surface of the cantilever. The cantilever is vacuum deposited with gold, which reflects red light almost perfectly. The deflection of the mechanical cantilever deflects the optical beam, thus changing the proportion of light falling on the two halves of the split photodiode. The difference of the signals from the two halves of the photodiode is detected. (Reproduced from Meyer and Amer, 1988, with permission.)

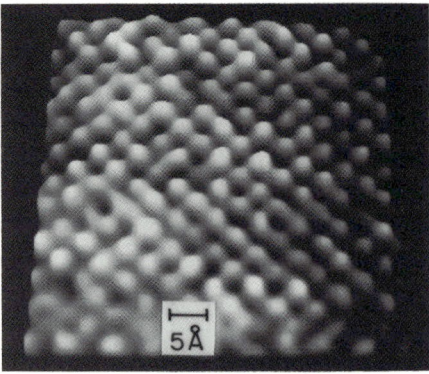

Fig. 15.8. AFM image of NaCl obtained under ultra-high vacuum condition. (Reproduced from Meyer and Amer, 1988, with permission.)

AFM image of the classical ionic crystal, NaCl, obtained under ultra-high-vacuum conditions, is shown in Fig. 15.8.

In the following, we discuss the detection limit of the optical beam deflection method. First, the relation between the variation of tip height Δz and the deflection angle θ at the end of the cantilever with length l is, according to Eq. (F.24),

$$\theta = \frac{3}{2} \frac{\Delta z}{l} . \tag{15.8}$$

The dimension of the mirror $2w$, which is the front end of the cantilever, should be a fraction of its length l. It imposes a diffraction limit on the spot size D of the laser beam at the detector, which is at a distance L from the mirror:

$$D > \frac{\lambda L}{2w} , \tag{15.9}$$

where λ is the wavelength of the light. The differential signal detected by the photodetector is

$$S = \frac{3}{2} \frac{2w}{l} \frac{P}{\lambda} \Delta z. \tag{15.10}$$

The shot noise is the major source of uncertainty. The shot noise over a bandwidth Δf is, according to Eq. (11.6),

$$I_{\text{shot}} = \sqrt{2e\bar{I}\,\Delta f} , \tag{15.11}$$

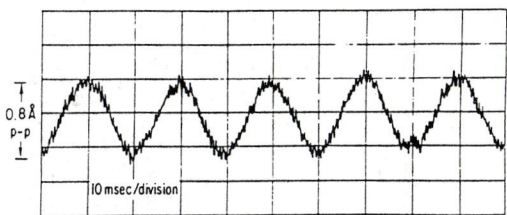

Fig. 15.9. Noise characteristics of the AFM. The optical beam deflection method is applied. (Reproduced from Meyer and Amer, 1990, with permission.)

where $\bar{I}$ is the average signal power, and e is the elementary charge. The signal-to-noise ratio is

$$\frac{S}{N} = \frac{3w}{\lambda} \sqrt{\frac{2P}{\Delta f}} \, \Delta z. \tag{15.12}$$

For a measurement bandwidth 1 kHz, using red light of $\lambda = 0.67$ μm, and $w/l \approx 0.3$, at a signal-to-noise ratio of $S/N = 10$, the detection limit is $\Delta z \approx 0.03$ Å. Such a limit has been demonstrated, as shown in Fig. 15.9.

15.4 AFM at the liquid–solid interface

An application of STM and AFM of particular interest is the in situ study of electrochemistry. Because the solid surface of interest is immersed in an electrolyte, no other microscopy method can access the liquid–solid interface except optical microscopy (which has a resolution of about 0.5 μm). The dif-

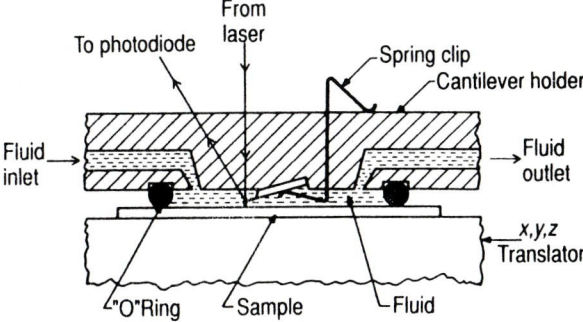

Fig. 15.10. Fluid cell for AFM study of electrochemistry. (Reproduced from Manne et al., 1991, with permission.)

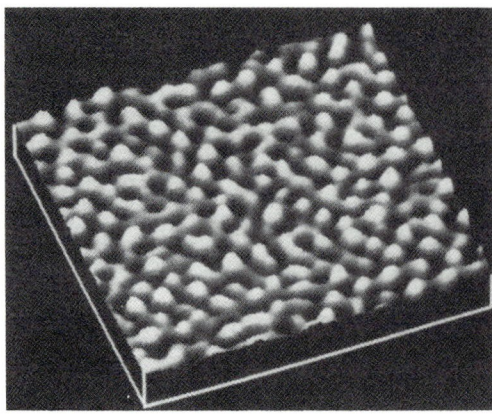

Fig. 15.11. AFM image of Au(111) under water. The reference value for the force is $\leq 10^{-7}$ N. The image has gone through a plane subtraction and a low-pass filtering. The atomic corrugation amplitude is about 1 Å. (Reproduced from Manne et al., 1990, with permission.)

ficulty of using STM is that the electrolyte is highly conducting, which makes the detection of tunneling current extremely difficult.

With the advances in AFM, especially the optical beam deflection method in the repulsive-force regime, the AFM study of solid surfaces under an electrolyte becomes practical. Atomic resolution with AFM at a liquid–solid interface has been routinely achieved (Manne et al., 1990, 1991). A typical fluid cell for the AFM study of electrochemistry is shown in Fig. 15.10. The top of the cell is made of glass to allow light to go in and out.

Figure 15.11 shows an AFM image of gold surface under water. As shown, atomic resolution is achieved.

CHAPTER 16

ILLUSTRATIVE APPLICATIONS

Over the past 10 years, the applications of STM and AFM has grown so rapidly that a thorough review would take several treatises. Here we describe some illustrative examples of the use of STM and AFM in various fields of science and technology. For more details about specific applications, there are books and conference proceedings, as well as original and review articles. The reference list at the end of this book will be helpful.

16.1 Surface structure determination

The determination of the atomic structure of surfaces is the cornerstone of surface science. Before the invention of STM, various diffraction methods are applied, such as low-energy electron diffraction (LEED) and atom beam scattering; see Chapter 4. However, those methods can only provide the Fourier-transformed information of the atomic structure averaged over a relatively large area. Often, after a surface structure is observed by diffraction methods, conflicting models were proposed by different authors. Sometimes, a consensus can be reached. In many cases, controversy remains. Besides, the diffraction method can only provide information about structures of relatively simple and perfectly periodic surfaces. Large and complex structures are out of the reach of diffraction methods. On real surfaces, aperiodic structures such as defects and local variations always exist. Before the invention of the STM, there was no way to determine those aperiodic structures.

The invention of STM completely changed the situation. As of the end of 1991, with the help of STM, combining with other techniques, a large number of surface structures have become known.

16.1.1 Periodic structures

To a good analogy, before the invention of STM, the determination of surface structure was similar to the case of speculating on the landscape of a planet from information taken through an astronomical telescope. In analogy to spacecraft, the STM sends electrons to the vicinity of the "planets" to take direct, close-up photographs.

We have already discussed the determination of the Si(111)-7 × 7 structure in Section 1.3. The example we are going to present, the Au(111)-$22 \times \sqrt{3}$ structure, is equally interesting.

The reconstruction of the Au(111) surface was discovered by LEED, and was recently studied by helium scattering (Harten et al., 1985, and references therein). All the studies show that along the $[1\bar{1}0]$ direction, the top-layer gold atoms exhibit a contraction of 4.2%. For a length of 22 Au atoms in the bulk, there are 23 Au atoms at the surface. Based on the diffraction patterns, models have been proposed for the atomic arrangement in real space. The helium scattering pattern, shown in Fig. 16.1, indicates that in the $[1\bar{1}0]$ direction, there is a 4% contraction of atomic spacing, from 2.87 to 2.75 Å. The pattern exhibits a threefold symmetry ($p3m1$). From the diffraction pattern, the reconstructed surface is determined to have a ($23 \times \sqrt{3}$) rectangular unit cell. Based on the HAS data, an improved soliton model is proposed, as shown in Fig. 16.2. Before reconstruction, the top-layer Au atoms have an fcc (face-centered cubic) stacking; that is, each one sits atop of the holes of the second-layer and the third-layer Au atoms. The reconstruction pushes the Au atoms along the $[1\bar{1}0]$ direction sideways toward an hcp (hexagonal-close-packed) position, to line up with the third-layer Au atoms.

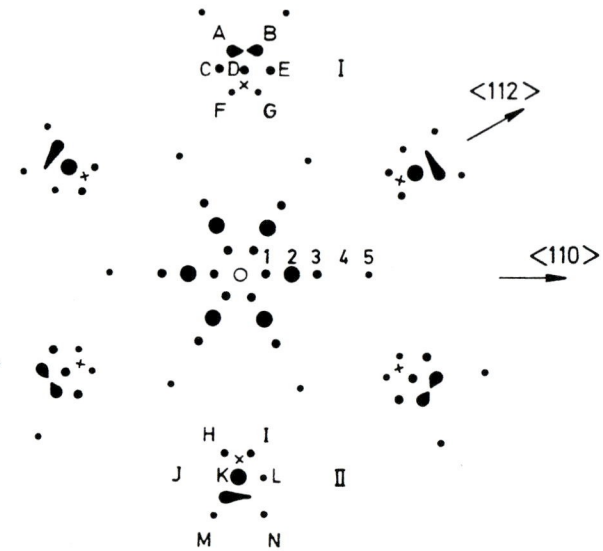

Fig. 16.1. He scattering pattern of reconstructed Au(111). Schematic representation of the full diffraction pattern. The crosses indicate the diffraction peak locations expected for the unreconstructed surface. The center spots D and K are off by ΔG=0.054 Å, which is larger than the bulk value 2.515 Å, indicating a contraction in this direction. The spot sizes [except the (112) groups] are proportional to the intensity. (Reproduced from Harten et al., 1985, with permission.)

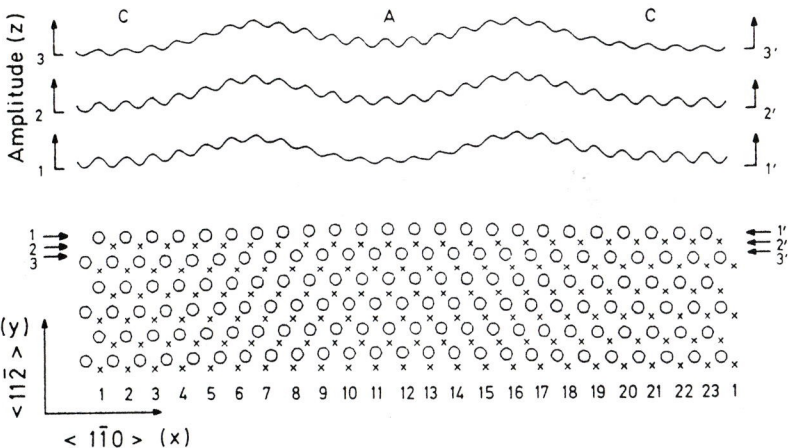

Fig. 16.2. Proposed model for the reconstructed Au(111) surface. Top: hard-wall corrugation functions at three points in the unit cell. In each period, two strips should exist as the two boundaries of fcc stacking and hcp stacking. Bottom: proposed structure. Crosses represent the second-layer and circles the surface-layer atoms. (Reproduced from Harten et al., 1985, with permission.)

On the domain boundary between the fcc stacking and the hcp stacking, the Au atoms are squeezed out from the original position to make two ridges per unit cell. The consequences of this model are in good agreement with their helium scattering data. In 1985, STM was already 3 years old. It was believed at that time the ultimate resolution of STM on metals was 6 Å, which would be insufficient to resolve the atomic structure on Au(111).

Two years later, Hallmark et al. (1987) reported atomic resolution of Au(111) by STM. The experimental evidence implied the possibility of resolving the atomic structure of the reconstructed Au(111) structure. One of the authors of the helium-scattering study of the reconstructed Au(111) surface (Ch. Wöll) participated in the STM experiment. After two more years, the report was published (Wöll et al., 1989). The major results are as follows: First, at low resolution, the pairs of strips in each unit cell are clearly observed. The corrugation height is 0.15±0.04 Å, in agreement with the helium-scattering data. Second, at high resolution, every single Au atom is resolved. The Au atoms in the row were gradually displaced sideways to reach the maximum displacement 0.7±0.3 Å, which also agrees with the surface-layer contraction model. Third, in a larger region, the strips can appear in three directions 120° apart. Fourth, the atomic corrugation, which can be as high as 0.3 Å, is larger in the regions where the atoms are raised above the surface because of the dislocation at the stacking fault edge. This

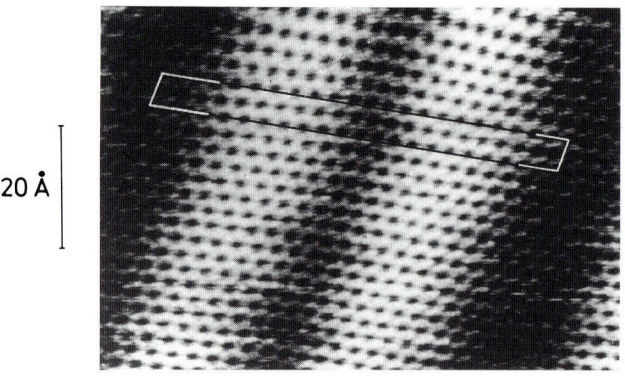

20 Å

Fig. 16.3. Atomic-resolution STM image of reconstructed Au(111). Size of the image, $80 \times 60\,\text{Å}^2$. The two lines indicate the unit cell. The lateral displacement of the individual Au atoms between fcc and hcp stacking regions is shown. (Reproduced from Barth et al., 1990, with permission.)

may indicate that the *sp* electrons are drained into the bulk, and the *d* electrons play a relatively important role in the image formation.

This work is the first demonstration that STM can determine the positions of individual atoms on metal surfaces. Therefore, the STM can be used to study lattice defects at metal surfaces with atomic resolution, a subject of considerable importance in the study of crystal growth.

The STM study of the reconstructed Au(111) surface by Wöll et al. (1989) was expanded by Barth et al. (1990), see Figures 16.3 and 16.4. The basic phenomena were completely confirmed, and more details were found.

16.1.2 Aperiodic structures

In the STM study of the reconstructed Au(111) surface, in addition to the confirmation of the model suggested from scattering experiments with unambiguous determination of the positions of atoms, aperiodic structures are also observed.

In a small (≈ 100 Å) region, the $22 \times \sqrt{3}$ reconstruction is uniaxial. The ridge runs in one direction. However, the gold surface has threefold symmetry (*p3m1*). On a large area, the reconstruction should run in three directions 120° apart with equal probability. Between regions with different reconstruction orientations, domain boundaries should be observed. While no scattering method may provide detailed information about these domain boundaries, STM may.

(a) 300 Å (b) 200 Å

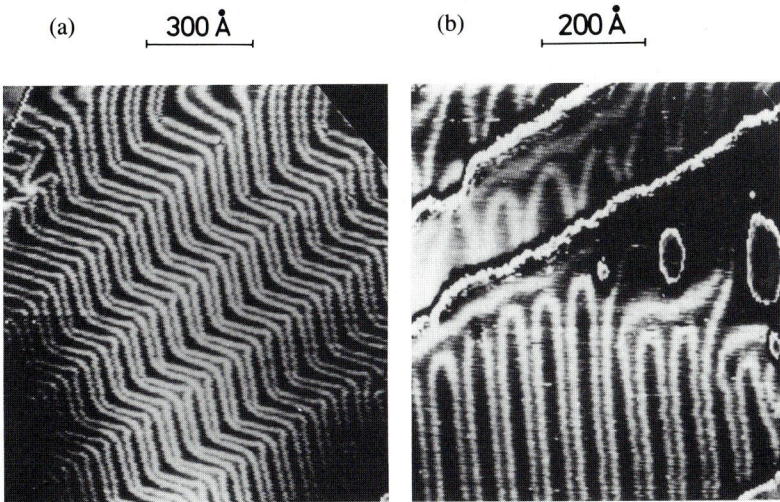

Fig. 16.4. The herringbone reconstruction and U connections on Au(111). STM images of Au(111), sputtered and annealed at 600°C. (a) The ubiquitous reconstruction forms a herringbone pattern on large terraces. The typical length of the period is about 250 Å. (b) The U-shaped connection and the effect of surface defects. A well-ordered region on the lower terrace, an alternation of fcc domains and hcp domains, is terminated by U connections between lines of neighboring line pairs. The wide (fcc) domains are terminated, and the narrow (hcp) domains are left open toward the step. The hcp region is separated from fcc stacking at the step edge by a single corrugation line along the step. Three holes on the right distort the pattern. (Reproduced from Barth et al., 1990, with permission.)

The domain boundaries between different orientations of the $(22 \times \sqrt{3})$ reconstructions appear to be extremely interesting. Over large areas, the double-ridged superstructure changes orientation about every 280 Å. A regular herringbone pattern is observed. Over even larger areas, the herringbone pattern of different orientations makes a mosaic. Near a single-atom step, the herringbone pattern tends to line up with the step edge such that the contraction is parallel to the step edge.

The herringbone structures are particularly sensitive to local defects on the surface. Near a single atomic step, the herringbone structure is often terminated by U-shaped connections.

The elbow in the herringbone pattern and the U connections are ways to relieve the surface strain created by the $22 \times \sqrt{3}$ reconstructions. Near the elbow and the U connection, the atomic arrangement is different from the uniaxial regions. Models have been proposed based on energetics considerations (Barth et al., 1990; Chambliss et al., 1991a). In the high-resolution

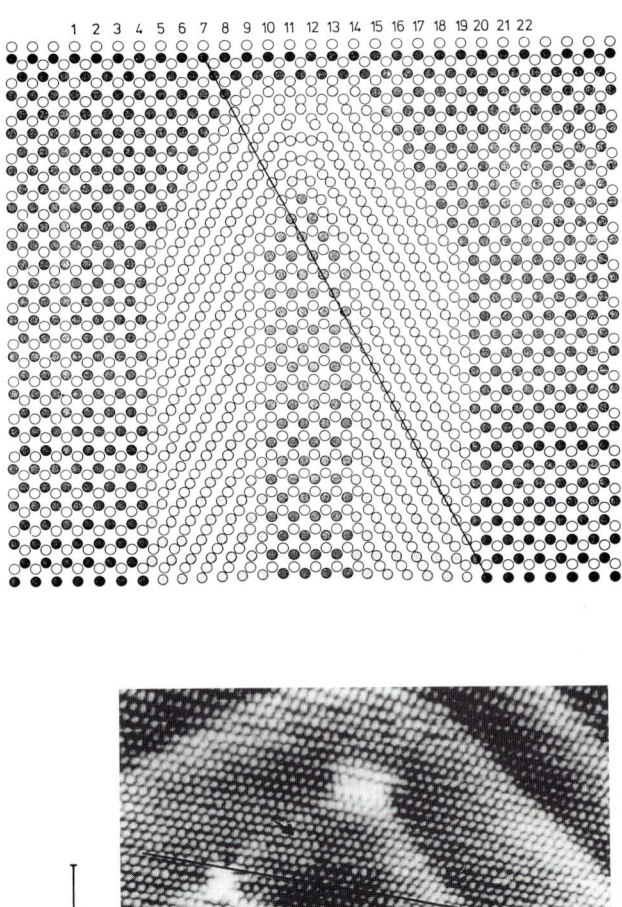

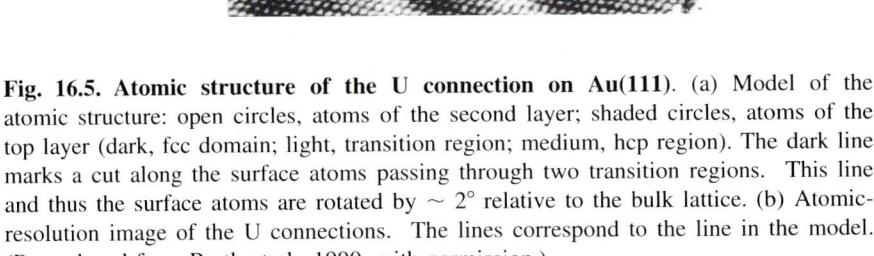

Fig. 16.5. Atomic structure of the U connection on Au(111). (a) Model of the atomic structure: open circles, atoms of the second layer; shaded circles, atoms of the top layer (dark, fcc domain; light, transition region; medium, hcp region). The dark line marks a cut along the surface atoms passing through two transition regions. This line and thus the surface atoms are rotated by $\sim 2°$ relative to the bulk lattice. (b) Atomic-resolution image of the U connections. The lines correspond to the line in the model. (Reproduced from Barth et al., 1990, with permission.)

mode, every single Au atom near an elbow or a U connection can be imaged. A complete agreement between proposed models and observed atomic-resolution images are achieved. Figure 16.5 is an example.

The STM study of Au(111) further demonstrates its imaging mechanism. For the large-scale pattern, with a periodicity of 63 Å, the STM images are almost independent of the tunneling parameters and tip conditions. The corrugation amplitude, 0.20±0.05 Å, is approximately independent of the tip–sample separation. The STM images of the superstructure are Fermi-level LDOS contours of the Au(111) surface, which are approximately parallel to each other. On the other hand, the atomic-resolution images depend dramatically on the tunneling conditions and tip states. To achieve atomic resolution, a tip treatment procedure has to precede (Wintterlin et al., 1989; see Section 13.4.1). The corrugation amplitudes decrease exponentially with tip–sample distance. The high resolution images are obtained at low tunneling resistances. An atomic corrugation as high as 1 Å was observed. The details of the image change dramatically with tip conditions under the same tunneling conditions. The detailed shape of the atomic features also depends on the tip condition. The spontaneous mode switching during a scan is often observed (Barth et al., 1990). Furthermore, in the atomic-resolution images, single atom defects are clearly seen. Near a step edge, every single atom is identified (Wöll et al., 1989).

16.2 Nucleation and crystal growth

The growth of thin films on solid surfaces is important in technology, and nucleation is one of the keys for understanding the growth mechanism. The ability of STM to image local structures down to atomic detail makes it ideal for the study of nucleation, thin film growth, and crystal growth.

An example of the nucleation phenomenon is the spontaneous formation of ordered Ni islands by vacuum deposition of Ni on Al(111) (Chambliss et al., 1991a, 1991b). The experiment starts with an Au(111) surface cleaned and annealed in ultrahigh vacuum. Then, Ni is vacuum deposited by evaporating Ni from a W basket heated with current. The gold sample is kept at room temperature during vacuum deposition. After one-tenth of a monolayer of nickel is deposited on the gold surface, the STM image of the surface is taken. A large-scale image is shown in Fig. 16.6. Instead of forming a uniform nickel layer, the Ni atoms are clustered to form ordered island arrays. The average center-to-center spacing of those Ni islands in a row along $[1\bar{2}1]$ is 140 Å between rows. Those island arrays occur in three equivalent orientations separated by 120° rotations. The typical domain size is comparable to the typical terrace size, with a width of about 2000 Å. Within a domain, some deviations from perfect ordering are seen, but the island positions remain coherent over the full domain size.

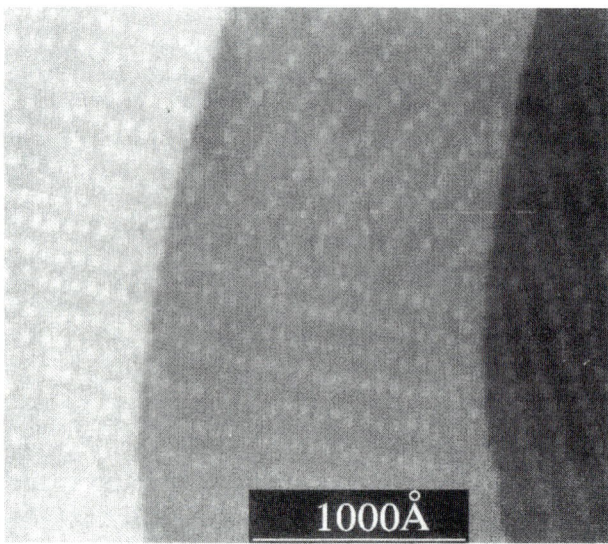

Fig. 16.6. STM image of 0.11 monolayer of Ni on Au(111). Several atomically flat Au terraces are seen, separated by steps of single-atom height. Small light dots on each terrace are monolayer Ni islands, in rows along [1$\bar{2}$1]. (Reproduced from Chambliss et al., 1991a, with permission.)

By taking STM images of a smaller area, it was found that the spontaneous formation of ordered arrays of Ni islands is determined by the herringbone reconstruction of the Au(111) surface. It is clear that the Ni islands locate at the elbows of the herringbone structure. A detailed study of the atomic-resolution STM image and the local atomic structure near the elbows indicates that at each vertex of the elbow, there is a dislocation. Energetically, the dislocation site is the most probable location for the nickel deposition to nucleate.

16.3 Local tunneling spectra of superconductors

The classical tunneling experiment of Giaever (1960) provided unambiguous proof of the BCS theory of superconductivity. The STM as a local tunneling probe is certainly suitable to probe the local properties of superconductors, such as the local structure of the Abrikosov flux lattice. The work of Hess and co-workers (1989, 1990, 1990a, 1991) is a prominent example.

Similar to the tunneling junction experiments, the choice of materials is crucial. In the tunneling junction experiment, the existence of an excellent insulator Al_2O_3 and the ease of growing it on Al surfaces is the key to its

Fig. 16.7. Abrikosov flux lattice observed by STM. Produced by a 1 T magnetic field in NbSe$_2$ at 1.8 K. the scan range is about 6000 Å. The gray scale corresponds to dI/dV ranging from approximately 1×10^{-8} mho (black) to 1.5×10^{-9} mho (white). (Reproduced from Hess et al., 1989, with permission.)

success. In the STM experiment, an easily prepared surface that can be kept clean for a long time is the key. Hess et al. (1989) owes much of the success by the use of an ideal superconducting material: 2H-NbSe$_2$. It is a layered material, with a charge-density-wave (CDW) transition temperature of 33 K and a superconducting transition temperature of 7.2 K. A clean and very inert surface can be prepared by peeling off. The samples can be prepared in air and transferred to the dewar with no need to clean again. A small STM similar to that of Smith and Binnig (1986) is used, see Fig. 12.5.

From the topographic images, a very clean and regular Abrikosov flux lattice is observed. At a magnetic field of 1 T, the spacing between adjacent vortices is a few hundred Å, as expected see Fig. 16.7. The unexpected features start with the detailed tunneling spectroscopy within a vortex. At a distance a few hundred Å from the vortex, the dI/dV curve shows a clear superconducting gap, as expected from the BCS theory. However, near and at the center of the vortex, the dI/dV curve has a pronounced peak near the zero-voltage point; see Fig. 16.7. This unexpected discovery prompted a series of theoretical studies that finally settled the enigma: The peak in the tunneling spectrum is due to a trapped state near the center of the vortex, whose LDOS is detected by the STM.

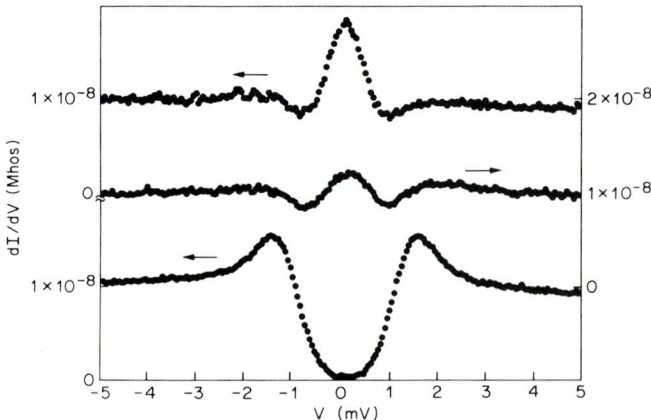

Fig. 16.8. Tunneling conductance near a vortex. Differential conductance dI/dV versus V for $2H$-NbSe$_2$ at 1.85 K and a 0.02 T magnetic field, taken at three positions: (a) on a vortex, (b) about 75 Å from a vortex, and (c) 2000 Å from a vortex. The zero of each successive curve is shifted up by one quarter of the vertical scale. (Reproduced from Hess et al., 1989, with permission.)

Subsequent studies revealed, for the first time, the anisotropic structure of the vortex, and the interactions among vortices that give rise to a well-defined, oriented lattice structure aligned with the crystallographic orientations of $2H$-NbSe$_2$ (Hess et al., 1990a). For details, see Hess et al. (1991).

16.4 Surface chemistry

A large number of processes in the microelectronics and chemical industries depend on the chemical reactions occurring on solid surfaces. The STM provides an unparalleled opportunity to study those chemical reactions *at the atomic level*. In this section, we will describe two reactions on silicon surfaces that are directly related to the understanding of the processes of manufacturing silicon devices.

The first one is the reaction of oxygen with a clean Si surface, or the initial stage of oxidation of the Si surface. On the Si(111)-7 × 7 surface, the reaction activity and the local reaction mechanism are now understood at the atom-by-atom level (Avouris and Lyo, 1990; Avouris, Lyo, and Bozso, 1991; Pelz and Koch, 1991). Two different early products of oxidation and their site selectivity are identified with STM and STS.

The topographs of the Si(111)-7 × 7 surface after oxygen exposure are studied with a bias of +2 V, that is, probing the unoccupied states. As shown in Fig. 16.9, after an exposure of ~0.2 L of oxygen to the clean Si(111)-7 × 7 surface, two new sites are observed by STM: a dark site such as those

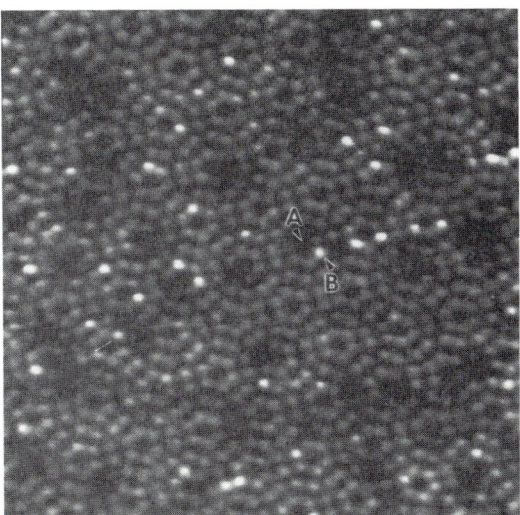

Fig. 16.9. STM topograph of a partially oxidized silicon surface. With the sample biased at +2 V relative to the tip, the unoccupied states of a Si(111)-7 × 7 surface exposed to 0.2 L of O_2 at 300 K is obtained. (Reproduced from Avouris, Lyo, and Bozso, 1991, with permission.)

labeled A and a bright site such as those labeled B. The two types of sites have different spatial distributions on the surface. The bright sites prefer corner-adatom sites over center-adatom sites by about 4:1. Moreover, there are about eight times as many bright sites in the faulted half of the 7 × 7 unit cell than in the unfaulted half. The dark sites, on the other hand, show lower selectivity. There are only ~50% more corner than center-adatom dark sites, and the faulted half is preferred by a factor of 2. Next to the dark sites, a grayish site is often observed.

The nature of those two kinds of sites is investigated with the method of scanning tunneling spectroscopy, or STS (see Chapter 14). The results are shown in Fig. 16.10. Spectra A, B, and C are found on the unreacted restatom, corner-adatom, and center-adatom, respectively. For an explanation of the nature of those sites, see Fig. 1.14. The restatom is characterized by an occupied dangling-bond surface state at -0.8 eV. Adatoms are characterized by occupied and unoccupied dangling-bond surface states at about 0.3 eV below and above E_F. The corner-adatom dangling bond has higher occupation than the center-adatom dangling bond. Adatom and restatoms on the 7 × 7 surface are coupled: This interaction leads to charge transfer from adatoms to restatoms. The STS spectrum of oxygen-induced dark sites, such as site D, shows no surface states. The grayish site F has a spectrum that is similar to

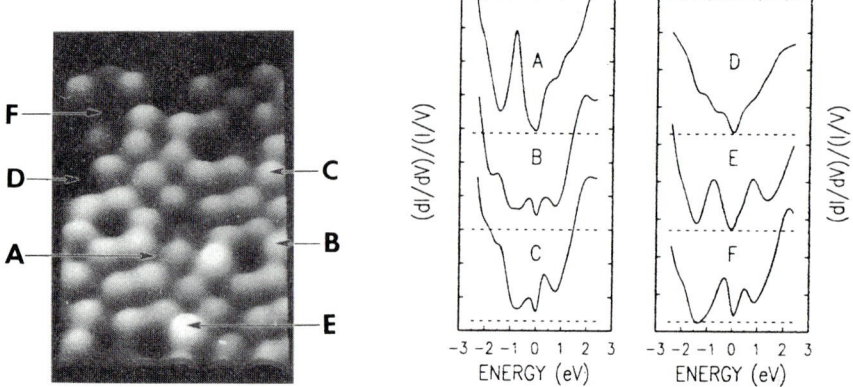

Fig. 16.10. Topograph and local spectroscopy of Si(111) with oxygen. Left, the STM image of a region of the oxygen-exposed Si(111)-7 × 7 surface. Right, the local tunneling spectra at different sites. Spectra A, B, and C are those of unreacted restatom, corner atom, and center adatom, respectively. Spectra D, E, and F are obtained over oxygen-induced dark, bright, and perturbed (gray) adatom sites, respectively. (Reproduced from Avouris, Lyo, and Bozso, 1991, with permission.)

that of an unreacted adatom but with an occupation higher than normal. Finally, bright sites such as site E in Fig. 16.10 show strong adatomlike surface states ~0.8 eV above and below the Fermi level.

To identify the chemical nature of those sites, a correlation with photoemission studies and first-principles calculations is conducted. The bright site is identified as the insertion of an oxygen atom into a Si–Si bond, and leaves the dangling bond essentially empty, which gives rise to its bright appearance in the topograph. It is difficult to identify the nature of the dark sites by STM and STS alone. The fact that at higher oxygen dosages, the dark sites increase steadily, whereas the number of the bright sites remain almost the same, and the strong tendency of conversion of the bright sites to dark sites indicate that the dark sites have an oxygen atom on the top of the Si adatom that eliminates the dangling bond (Avouris, Lyo, and Bozso, 1991a).

Another example is the reaction of hydrogen on silicon surfaces. This reaction is of scientific significance, because it has been a model system for surface scientists for a number of years (Robbins, 1991). It is also of technological significance, because it is a crucial reaction step in the epitaxial growth of silicon on silicon using disilane (Boland, 1990, 1991a, 1991b, 1991c).

Prior to the STM study, the structure of the H-saturated Si(100) surface has been in dispute for many years (Boland, 1990, and references therein). The STM study of clean Si(100) has been discussed previously; see Fig. 1.15. After hydrogen exposure, there are several possible structures, as shown in

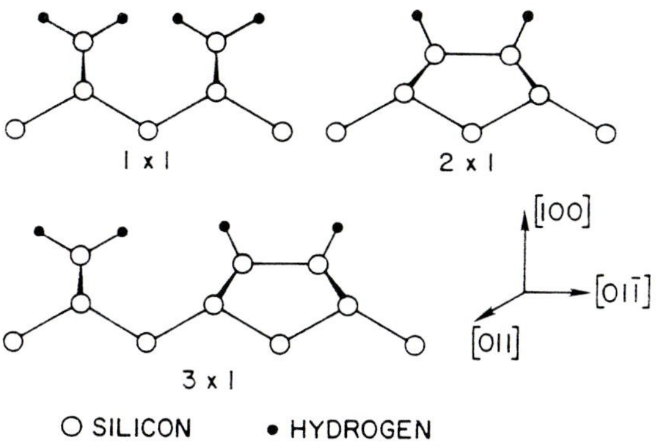

O SILICON   • HYDROGEN

Fig. 16.11. Various possible structures of hydrogenated Si(100) surfaces. (Reproduced from Boland, 1990, with permission.)

Fig. 16.11. The two structures in the top row were the subject of speculation by the previous authors, which are easy to understand. However, various studies showed that under certain circumstances, neither of these two structures can explain the observed phenomena. The STM experiment of Boland (1990) unambiguously identified a third structure, as shown in the bottom of

Fig. 16.12. Topograph of the Si(100) surface exposed to hydrogen. The image of the unoccupied states with bias +2 V on the sample. The area shown is $100 \times 107 \, Å^2$. An antiphase boundary, marked AB, is observed. (Reproduced from Boland, 1990, with permission.)

Fig. 16.11 (see also Fig. 16.12). On the STM images, the alternation of monohydride rows and dihydride rows made a very regular 3×1 pattern. The 3×1 structure only occurs in a limited temperature range (380 ± 20 K). Prolonged exposure at room temperature generates a 1×1 structure on this surface.

16.5 Organic molecules

The STM has successfully resolved a number of organic molecules adsorbed on various conducting substrates (Sleator and Tycko, 1988; Ohtani et al., 1988; Foster and Frommer, 1988). A commonly used substrate is graphite, which is easy to prepare, defect free on large areas (several micrometers), and inert. Under favorable conditions, nearly atomic or atomic resolution can be achieved. In many cases, the organic molecules adsorbed on crystalline substrates form regular patterns, which are of scientific interest in and of themselves.

As an example, we describe an STM study of certain liquid-crystal molecules by Smith et al. (1989), n-alkylcyanobiphenyl (nCB), where $n = 8, 10, 12$. The molecules are sublimated onto a freshly cleaved graphite surface. The STM images are found to be independent of the film thickness, indicating that the STM tip pushes away the top-layer molecules to image only the layer in direct contact with the graphite surface. The molecules are found to form regular patterns which can be understood in term of the structures of them and graphite surface, as shown in Fig. 16.13.

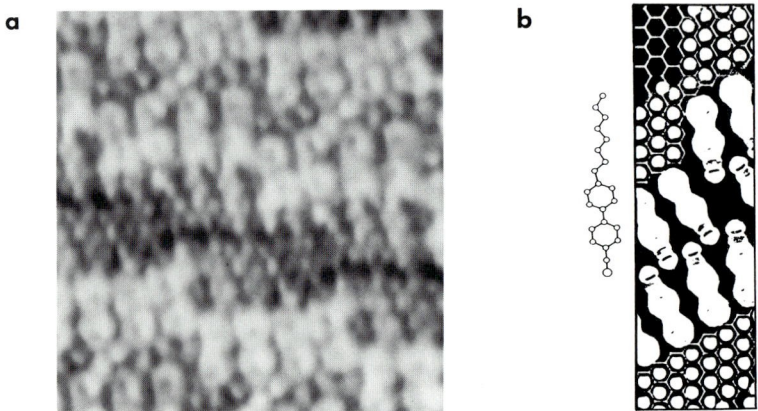

Fig. 16.13. Imaging organic molecules with STM. (a) STM image of 8CB on graphite (about 60×60 Å^2). The holes of the benzene rings and the alkyl groups in the tails are clearly resolved. (b) Model showing the packing of the 8CB molecules. The hexagonal pattern is the lattice of graphite. (After Smith et al., 1989. Original image by courtesy of the authors.)

16.6 Electrochemistry

The chemical process at the interface of an electrolyte and a solid surface has been a fertile field of science and technology for centuries, which includes applications to electrolytic plating, electroless plating, corrosion and corrosion prevention, batteries, and many more. For a long time, it was known that the electrochemical process depends dramatically on the atomic details of the electrode surfaces. For example, the plating rate on different crystallographic orientations can differ by two orders of magnitude. The results of voltammetry depend dramatically on the atomic arrangement of the electrodes, as perfect surfaces usually generate sharp and reproducible features in the voltammograms. The STM and the AFM, which work at the liquid–solid interface as well, provide natural tools for the study of the electrochemical process at an atomic level.

The demonstration of STM to image solid surfaces immersed in electrolytes (Sonnenfeld and Hansma, 1986; Liu et al., 1986) opened a way of studying electrochemisty. For a review, see Sonnenfeld, Schneir, and Hansma (1990). The major difficulty of implementing the STM in an electrolyte is that the tunneling current is mixed with Faradaic current, which makes the signal-to-noise ratio worse. In a well-defined electrochemical cell, there are at least three electrodes: the reference electrode, the counterelectrode, and the working electrode. By setting the potential of the tip equal to that of the reference electrode, and covering the tip except for a tiny end, the influence of Faradaic current to the measured total current can be minimized. However, the bias between the tip and the working electrode — which is usually the STM sample, becomes out of control. Nevertheless, atomic resolution was achieved, and some new information on the atomic arrangement on the electrodes has been obtained.

The underpotential deposition of a single layer of copper on single-crystal copper is a prototype system for a number of electrochemical studies. The atomic structure on the electrode surface is the key to understanding the electrochemical process. Using STM, Magnussen et al. (1990) revealed in real space the atomic structure of the Cu monolayer on Au(111) and Au(100) surfaces in copper sulfate ($CuSO_4$) solution. The clean Au surfaces, without Cu overlayer, are imaged by STM for reference. The relative registration of the Cu atoms on Au surfaces are determined from the subsequent STM images.

Figure 16.14 shows two atomic-resolution images of Cu overlayer on Au and the structure models. On Au(111), a $(\sqrt{3} \times \sqrt{3})R30°$ structure is observed. On Au(100), the structure is incommensurate, but appears quasihexagonal. For both, the nearest-neighbor distance is 4.9 ± 0.2 Å. According to the authors (Magnussen et al., 1990), this is most likely caused by repulsive interactions within the adlayer due to the presence of coadsorbed anions.

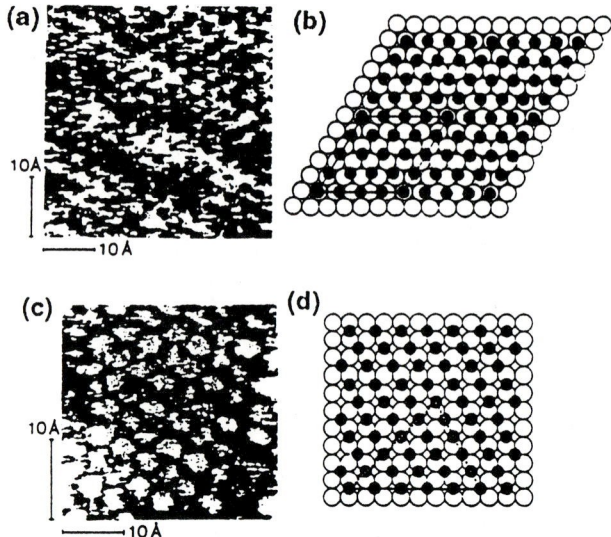

Fig. 16.14. Ordered structures of Cu layers on Au electrodes. (a) STM image of Cu adlayer on Au(111). A ($\sqrt{3} \times \sqrt{3}$)$R30°$ structure is observed. The Cu coverage is 0.33 monolayer. (b) Model of the adlayer structure. (c) STM image of Cu adlayer on Au(100). A quasihexagonal structure is observed. (d) Model of the adlayer structure (○ − Au atoms in the topmost layer. • − Cu adatoms). (Reproduced from Magnussen et al., 1990, with permission.)

With advances in AFM, especially optical beam deflection in the repulsive-force regime, AFM studies at the solid surfaces under an electrolyte became practical. Atomic resolution with AFM at the liquid–solid interface has been routinely achieved (Manne et al., 1990, 1991). A typical fluid cell for the AFM study of electrochemistry is shown in Fig. 15.9. The top of the cell is made of glass to allow light to go in and out. With such an AFM, the study of electrochemistry with atomic resolution is greatly simplified. There is no more tunneling current. The AFM tip, which is an insulator, has much less interference to the electrochemical process than an STM tip.

Using the electrochemical AFM, Manne et al. (1991a) studied the problem of the underpotential deposition of Cu on Au in more detail. Some of their results are shown in Fig. 16.15. Several unexpected phenomena were discovered. First, the structure of the Cu adlayer on Au(111) depends dramatically on the nature of the electrolyte. With $CuSO_4$ solution, the observation of Magnussen et al. (1990) was completely confirmed. However, in $Cu(ClO_4)_2$ solution, the Cu adlayer becomes close packed with a 30±10° rotation relative to the top-layer Au structure. Second, in some cases, a monolayer step of the Cu adlayer with a well-defined border is observed. At

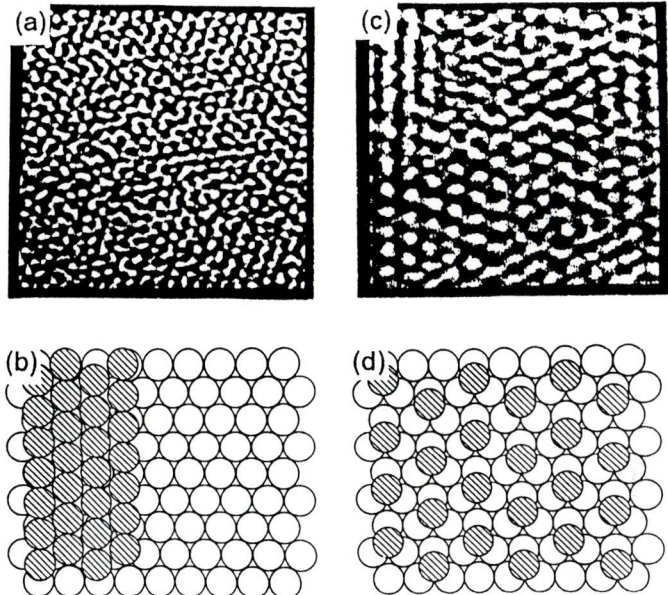

Fig. 16.15. AFM images of Cu adlayer on Au(111) in different electrolytes. (a) AFM image of Cu adlayer on Au(111) in copper perchlorate solution, showing close-packed structure with a rotation to the Au(111) substrate. (b) Schematic of the incommensurate structure of the Cu adlayer. (c) $(\sqrt{3} \times \sqrt{3})R30°$ structure of the Cu adlayer on Au(111), in copper sulfate solution. (d) Schematic diagram of that structure. The open circle represents Au atom at the topmost layer, the hatched circle represents the Cu adatom. (Reproduced from Manne et al., 1991a, with permission.)

the time this book was is being written, the nature of those phenomena was still a puzzle to be resolved.

16.7 Biological applications

Although many authors reported observations of atomic structures of large biological molecules with STM and AFM, for example DNA (Beebe et al. 1988), the validity of those results is still under investigation (Clemmer and Beebe, 1991). Here we present a biological application of AFM that is unambiguous and provides a significant piece of new information not obtained previously: an observation of the entire process of a single living cell infected by a virus, using the AFM, with a resolution of 10 nm (Häberle et al., 1992).

Although electron microscopy can resolve details in a single cell, the cell must be dried and stained before the observation. With STM and AFM, it is

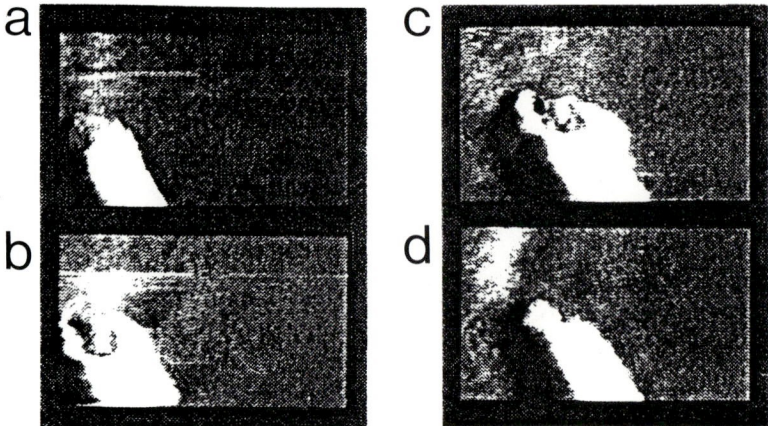

Fig. 16.16. AFM observation of a living cell infected by a virus. (a) A cell is infected by a virus. (b) The infected site becomes a protrusion. (c) The protrusion eventually erupts. (d) A scar is left on the cell. (Reproduced from Häberle et al., 1992, with permission.)

possible to observe biological objects alive, that is, in real time under physiological conditions. Therefore, with STM and AFM, the processes in living cells can be imaged with nanometer resolution.

The experiment of Häberle et al. was with monkey kidney cells cultured in standard growth solution. An AFM was used for imaging. The sequence of the AFM images is shown in Fig. 16.16. Initially, the AFM images of the cell are very similar to those obtained by SEM on dried, uncoated cells. By adding a drop of liquid containing orthopox viruses, the cell starts to become infected. A few minutes after the invasion of the virus, the call membrane is significantly softened. After 2–3 hours, a large protrusion is observed on the cell. More than 6 hours later, the protrusion is inflated dramatically. Finally, the protrusion abruptly disappears, leaving a scar on the cell. Evidently, the viruses exited the cell (see Fig. 16.16).

APPENDIX A

REAL WAVEFUNCTIONS

Wigner (1930) has shown that if time is reversible in a quantum-mechanical system, then all wavefunctions can be made real. This theorem enables us to use real wavefunctions whenever possible, which are often more convenient than complex ones. Here we present a simplified proof of Wigner's theorem, with some examples of its applications.

In the absence of a magnetic field, the Schrödinger equation of a particle moving in an external potential $U(\mathbf{r})$ is

$$-\frac{\hbar^2}{2m_e} \nabla^2 \psi + U(\mathbf{r})\psi = E\psi. \tag{A.1}$$

This equation is explicitly time-reversal invariant, because it only contains the square of the momentum $\mathbf{p} = -i\hbar\nabla$. By taking the complex conjugate of Eq. (A.1), we have

$$-\frac{\hbar^2}{2m_e} \nabla^2 \psi^* + U(\mathbf{r})\psi^* = E\psi^*. \tag{A.2}$$

Obviously, if ψ is a solution of Eq. (A.1), then ψ^* is also a solution of Eq. (A.1) *with the same energy eigenvalue E.* Consequently, the linear combinations of ψ and ψ^* are also solutions of Eq. (A.1) with the same energy eigenvalue. If ψ and ψ^* are linearly independent (that is, they do not differ by a constant multiplier), then the following two real wavefunctions,

$$\psi_1 = \frac{1}{2}(\psi + \psi^*), \tag{A.3}$$

$$\psi_2 = \frac{1}{2i}(\psi - \psi^*), \tag{A.4}$$

represent two states with the same energy eigenvalue. In other words, that energy level is degenerate.

We discuss a few examples in the following. A plane wave

$$\psi = e^{ikx} \tag{A.5}$$

represents a particle running in the positive x direction. In the absence of magnetic fields, its complex conjugate

$$\psi^* = e^{-ikx}, \tag{A.6}$$

which represents a particle running in the negative x direction, should be a solution of the Schrödinger equation with the same energy eigenvalue. The wavefunctions can be written in real form:

$$\psi_1 = \cos kx, \tag{A.7}$$

$$\psi_2 = \sin kx. \tag{A.8}$$

Those wavefunctions represent a pair of standing-wave states with a 90° phase difference in space.

In a central field of force, the wavefunctions can always be written in terms of spherical harmonics,

$$\psi_{nlm} = u(r)P_l^m(\cos \theta) e^{im\phi}, \tag{A.9}$$

where $P_l^m(x)$ is an associated Legendre function (Arfken, 1968), and $u(r)$ is a radial function. A solution with $m > 0$ represents a counterclockwise circular motion of the particle around the z axis. A solution with $m < 0$ represents a clockwise circular motion of the particle around the z axis. In the presence of an external magnetic field, those states might have different energy levels. In the absence of an external magnetic field, that is, when the system exhibits time reversal symmetry, two states with equal and opposite quantum number m are at the same energy level. Therefore, all those wavefunctions can be written in forms of $\sin m\phi$ or $\cos m\phi$, which represent standing waves with respect to the azimuth ϕ.

The spherical harmonics in real form have explicit nodal lines on the unit sphere. Morse and Feshbach (1953) have given a detailed description of those real spherical harmonics, and gave them special names. Here we list those real spherical harmonics in normalized form. In other words, we require

$$\int |Y_l^m(\theta, \phi)|^2 \cos \theta \, d\theta \, d\phi = 1. \tag{A.10}$$

The first of the real spherical harmonics is a constant, which does not have any nodal line:

$$Y_0^0(\theta, \phi) = \frac{1}{\sqrt{4\pi}} \, . \tag{A.11}$$

The ones with $l \neq 0$ and $m = 0$ have nodal lines dividing the sphere into horizontal zones, which are called *zonal harmonics*. The first two are

$$Y_1^0(\theta, \phi) = \sqrt{\frac{3}{4\pi}} \, \cos\theta, \tag{A.12}$$

$$Y_2^0(\theta, \phi) = \sqrt{\frac{5}{16\pi}} \left(3\cos^2\theta - 1\right). \tag{A.13}$$

The ones with $m = l$ divide the sphere into sections with vertical nodal lines, which are called *sectoral harmonics*. Those are:

$$Y_1^1(\theta, \phi) = \sqrt{\frac{3}{4\pi}} \, \sin\theta \cos\phi, \tag{A.14}$$

$$Y_1^{-1}(\theta, \phi) = \sqrt{\frac{3}{4\pi}} \, \sin\theta \sin\phi, \tag{A.15}$$

$$Y_2^2(\theta, \phi) = \sqrt{\frac{15}{16\pi}} \, \sin^2\theta \cos 2\phi, \tag{A.16}$$

$$Y_2^{-2}(\theta, \phi) = \sqrt{\frac{15}{16\pi}} \, \sin^2\theta \sin 2\phi. \tag{A.17}$$

The rest are called *tesseral harmonics,* which have both vertical and horizontal nodal lines on the unit sphere:

$$Y_2^1(\theta, \phi) = \sqrt{\frac{15}{16\pi}} \, \sin 2\theta \cos\phi, \tag{A.18}$$

$$Y_2^{-1}(\theta, \phi) = \sqrt{\frac{15}{16\pi}} \, \sin 2\theta \sin\phi. \tag{A.19}$$

Those real spherical harmonics are graphically shown in Fig. A.1.

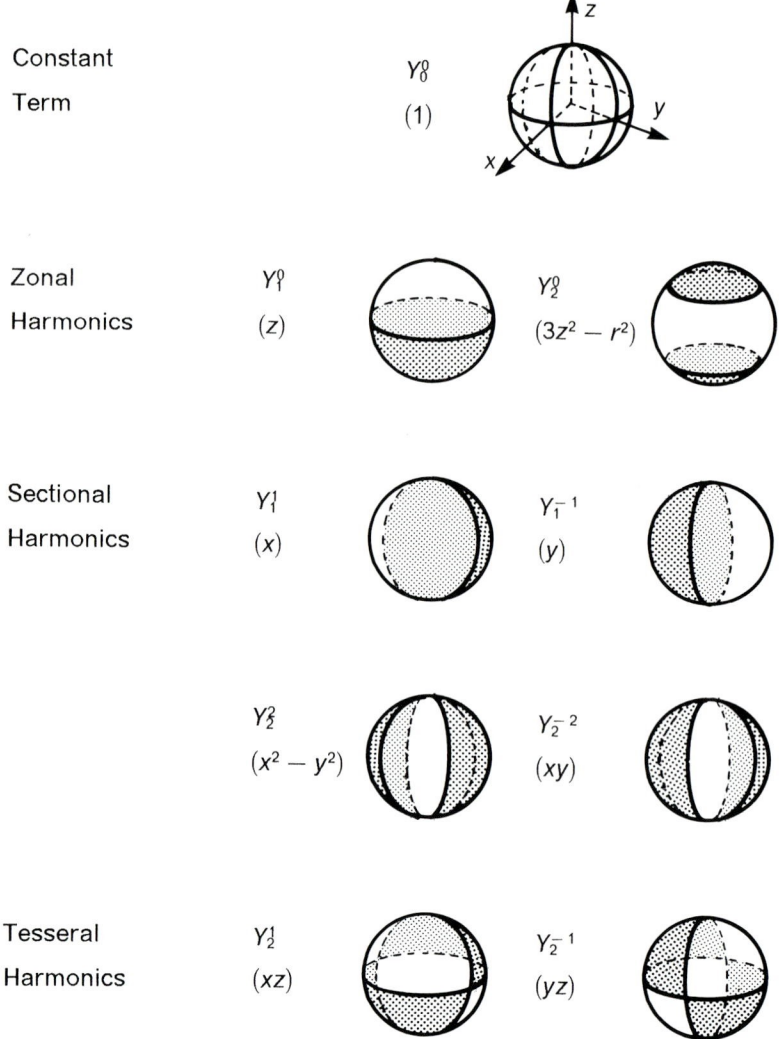

Fig. A.1. Real spherical harmonics. The first one, Y_0^0, is a constant. The coordinate system attached to the unit sphere is shown. The two zonal harmonics, Y_1^0 and Y_2^0, section the unit sphere into vertical zones. The unshaded area indicates a positive value for the harmonics, and the shaded area indicates a negative value. The four sectoral harmonics are sectioned horizontally. The two tesseral harmonics have both vertical and horizontal nodal lines on the unit sphere. The corresponding "chemists notations," such as $(3z^2 - r^2)$, are also marked.

APPENDIX B
GREEN'S FUNCTIONS

The concept of Green's functions can be illustrated by the example of the electrostatic problem. The potential $U(\mathbf{r})$ for a given charge distribution $\rho(\mathbf{r})$ is determined by the Poisson equation

$$\nabla^2 U(\mathbf{r}) = -4\pi\rho(\mathbf{r}). \tag{B.1}$$

It is the inhomogeneous counterpart of the Laplace equation, $\nabla^2 U(\mathbf{r}) = 0$, which describes the potential in the vacuum. A convenient method to solve Eq. (B.1) is to introduce a *Green's function*, denoted as $G(\mathbf{r}, \mathbf{r}_0)$, which is the potential of a unit point charge located at $\mathbf{r}_0$:

$$\nabla^2 G(\mathbf{r}, \mathbf{r}_0) = -4\pi\,\delta(\mathbf{r} - \mathbf{r}_0). \tag{B.2}$$

The solution of the Poisson equation, Eq. (B.1), is then

$$U(\mathbf{r}) = \int G(\mathbf{r}, \mathbf{r}_0)\rho(\mathbf{r}_0)\,d^3\mathbf{r}_0. \tag{B.3}$$

By directly solving Eq. (B.2) in spherical coordinates, which leads to an ordinary differential equation, we find the Green's function is simply the Coulomb potential of a unit charge,

$$G(\mathbf{r}, \mathbf{r}_0) = \frac{1}{|\mathbf{r} - \mathbf{r}_0|}. \tag{B.4}$$

Therefore, the general solution of the Poisson equation is

$$U(\mathbf{r}) = \int \frac{\rho(r_0)}{|\mathbf{r} - \mathbf{r}_0|}\,d^3\mathbf{r}_0. \tag{B.5}$$

Eq. (B.4), the *Green's function* of the Poisson equation, has an obvious physical meaning. By labeling the point charge at $\mathbf{r}_0$ as the "cause," the potential it creates at a point $\mathbf{r}$ is its "effect," and the influence of the point charge at $\mathbf{r}_0$ to the potential at $\mathbf{r}$ is described by the Green's function $G(\mathbf{r}, \mathbf{r}_0)$, which is the Coulomb potential of a point charge.

The Schrödinger equation in vacuum with a negative energy eigenvalue E is

$$\nabla^2 \psi(\mathbf{r}) - \kappa^2 \psi(\mathbf{r}) = 0, \qquad (B.6)$$

where $\kappa = \sqrt{2m_e |E|} / \hbar$. It is a homogeneous differential equation that is often called the *modified Helmholtz equation*.

The Green's function of the modified Helmholtz equation is defined in analogy to Eq. (B.2),

$$\nabla^2 G(\mathbf{r}, \mathbf{r}_0) - \kappa^2 G(\mathbf{r}, \mathbf{r}_0) = -4\pi \, \delta(\mathbf{r} - \mathbf{r}_0). \qquad (B.7)$$

Similarly, the Green's function should depend only on r; that is,

$$\frac{1}{r} \frac{d^2}{dr^2} [rG(r)] - \kappa^2 G(r) = -4\pi \, \delta(\mathbf{r} - \mathbf{r}_0). \qquad (B.8)$$

The general solution of Eq. (B.8), except for the point $\mathbf{r} = \mathbf{r}_0$, can be obtained by direct integration. It is

$$G(\mathbf{r}, \mathbf{r}_0) = \frac{A}{r} e^{\kappa r} + \frac{B}{r} e^{-\kappa r}. \qquad (B.9)$$

For large r, $G(\mathbf{r}, \mathbf{r}_0)$ must vanish, which requires that $A = 0$. For small distances, where $\kappa r << 1$, it should be identical to the Coulomb potential, which requires that $B = 1$. Finally, we find that the Green's function of the Schrödinger equation in vacuum is the *Yukawa potential*,

$$G(\mathbf{r}, \mathbf{r}_0) = \frac{1}{|\mathbf{r} - \mathbf{r}_0|} e^{-\kappa |\mathbf{r} - \mathbf{r}_0|}. \qquad (B.10)$$

The physical meaning of this Green's function is as follows. If in the entire space, the potential equals the potential of the vacuum, then the solution of Eq. (B.6) is identically zero in the entire space. In order to have a meaningful physical situation, some place in the space must have a certain potential well and nonvanishing wavefunction. The Green's function Eq. (B.10) then describes the influence of the potential well and the wavefunction in the non-vacuum region on the wavefunction in the vacuum region.

APPENDIX C
SPHERICAL MODIFIED BESSEL FUNCTIONS

The solution of the Schrödinger equation in vacuum with a negative energy eigenvalue, in terms of spherical coordinates, is in the form of spherical modified Bessel functions. Therefore, in treating tunneling problems in three-dimensional space, these functions are of fundamental importance. Although considered as a kind of special function, the spherical modified Bessel functions are actually *elementary functions,* that is, simple combinations of exponential functions and power functions. Here is a list of important formulas of those functions following the definition and notation of Arfken (1968).

The radial part of the Schrödinger equation in spherical coordinates is

$$\frac{d}{dz}\left(z^2\,\frac{d f(z)}{dz}\right) - \left[z^2 + n(n+1)\right] f(z) = 0, \qquad (C.1)$$

where $z = \kappa r$. By making a substitution

$$f(z) = \frac{1}{\sqrt{z}}\, g(z), \qquad (C.2)$$

the differential equation for $g(z)$ is

$$r^2\,\frac{d^2 g}{dr^2} + r\,\frac{dg}{dr} - \left[r^2 + \left(n+\frac{1}{2}\right)^2\right] g = 0. \qquad (C.3)$$

This is the modified Bessel equation of order $\nu = n + \frac{1}{2}$. The solutions of Eq. (C.3) are modified Bessel functions of the first kind, which is defined through the Bessel function $J_\nu(x)$ as

$$I_\nu(z) \equiv e^{-2\nu\pi i} J_\nu(iz), \qquad (C.4)$$

and the modified Bessel function of the second kind, which is defined through the Hankel function $H_\nu^{(1)}(x)$ as

$$K_\nu(z) \equiv \frac{\pi}{2} e^{2(\nu + 1)\pi i} H_\nu^{(1)}(iz). \tag{C.5}$$

The solutions of Eq. (C.1) are defined through the modified Bessel functions. Those are spherical modified Bessel functions of the first kind

$$i_n(z) = \sqrt{\frac{\pi}{2z}} \, I_{n + \frac{1}{2}}(z), \tag{C.6}$$

and of the second kind,

$$k_n(z) = \sqrt{\frac{2}{\pi z}} \, K_{n + \frac{1}{2}}(z), \tag{C.7}$$

see Fig. C.1. These two functions are linearly independent. The function $i_n(z)$ is the only solution of Eq. (C.1) that is regular at $z = 0$, and the function $k_n(z)$ is the only solution which is regular at $z = \infty$. These so-called special functions are actually *elementary functions,* with the following general expression

$$i_n(z) = z^n \left(\frac{d}{z \, dz} \right)^n \frac{\sinh z}{z}, \tag{C.8}$$

$$k_n(z) = (-1)^n z^n \left(\frac{d}{z \, dz} \right)^n \frac{\exp(-z)}{z}. \tag{C.9}$$

The first three pairs of these functions are

$$i_0(z) = \frac{\sinh z}{z} \tag{C.10}$$

$$i_1(z) = -\frac{\sinh z}{z^2} + \frac{\cosh z}{z} \tag{C.11}$$

$$i_2(z) = \left(\frac{3}{z^3} + \frac{1}{z} \right) \sinh z - \frac{3}{z^2} \cosh z \tag{C.12}$$

$$k_0 = \frac{1}{z} e^{-z} \tag{C.13}$$

$$k_1 = \left(\frac{1}{z} + \frac{1}{z^2} \right) e^{-z} \tag{C.14}$$

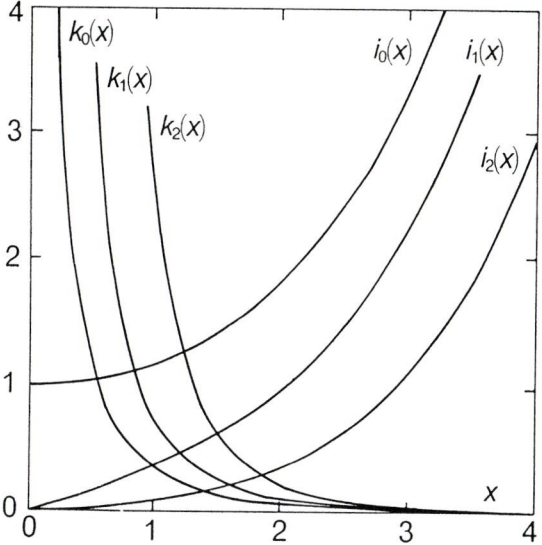

Fig. C.1. Spherical modified Bessel functions

$$k_2 = \left(\frac{1}{z} + \frac{3}{z^2} + \frac{3}{z^3} \right) e^{-z} \qquad (C.15)$$

Actually, by starting with Equations (C.8) and (C.9) as the definitions, all the properties of the spherical modified Bessel functions can be obtained, without tracing back to the formal definition, Equations (C.6) and (C.7).

Following the standard series expansion of Bessel functions, the power-series expansion of the function $i_n(z)$ near $z = 0$ has the following form:

$$i_n(z) = z^n \sum_{k=0}^{\infty} \frac{1}{k! \, (2n + 2k + 1)!!} \left(\frac{z^2}{2} \right)^k. \qquad (C.16)$$

The first term is proportional to z^n:

$$i_n(z) \approx \frac{z^n}{(2n + 1)!!} . \qquad (C.17)$$

An $i_n(z)$ of even order only has even powers of z, and an $i_n(z)$ of odd order only has odd powers of z. These properties are essential in a unified derivation of the tunneling matrix elements.

On the other hand, following the standard asymptotic expansion of Bessel functions, it is easy to prove that the functions $k_n(z)$ have the following exact general expression,

$$k_n(z) = \frac{e^{-z}}{z} \sum_{k=0}^{n} \frac{(n+k)!}{k!\,(n-k)!} \frac{1}{(2z)^k}.$$ $$(C.18)$$

Also, following the standard recursion relations of Bessel functions, the recursion relations for both i_n and k_n are:

$$(2n+1)\,f_n(z) = z\,f_{n-1}(z) - z\,f_{n+1}(z),$$ $$(C.19)$$

$$\frac{d}{dz}\,f_n(z) = f_{n+1}(z) + \frac{n}{z}\,f_n(z).$$ $$(C.20)$$

Finally, the Wronskian of the pair is

$$W\{i_n(z),\ k_n(z)\} \equiv i_n(z)\,k'_n(z) - i'_n(z)\,k_n(z) = -\frac{1}{z^2}.$$ $$(C.21)$$

APPENDIX D

TWO-DIMENSIONAL FOURIER SERIES

For a solid surface with two-dimensional periodicity, such as a defect-free crystalline surface, all the measurable quantities have the same two-dimensional periodicity, for example, the surface charge distribution, the force between a crystalline surface and an inert-gas atom (Steele, 1974; Goodman and Wachman, 1976; Sakai, Cardino, and Hamann, 1986), tunneling current distribution, and STM topographic images (Chen, 1991). These quantities can be expanded into two-dimensional Fourier series. Usually, only the few lowest Fourier components are enough for describing the physical phenomenon, which requires a set of Fourier coefficients. If the surface exhibits an additional symmetry, then the number of independent Fourier coefficients can be further reduced.

The two-dimensional periodicity can always be described by two primitive vectors $\mathbf{a}_1$ and $\mathbf{a}_2$, as shown in Fig. D.1. A periodic function F has the property

$$F(\mathbf{r} + m\mathbf{a}_1 + n\mathbf{a}_2) = F(\mathbf{r}), \tag{D.1}$$

where m and n are arbitrary integers. It is always convenient to define a pair of primitive vectors in reciprocal space,

$$\mathbf{b}_1 = 2\pi \frac{\mathbf{a}_2 \times \mathbf{e}_z}{|\mathbf{a}_1 \times \mathbf{a}_2|}, \tag{D.2}$$

and

$$\mathbf{b}_2 = 2\pi \frac{\mathbf{e}_z \times \mathbf{a}_1}{|\mathbf{a}_1 \times \mathbf{a}_2|}. \tag{D.3}$$

The vector $\mathbf{e}_z$ is the unit vector in the z direction. These two pairs of vectors satisfy the following relations:

$$\mathbf{a}_1 \bullet \mathbf{b}_1 = \mathbf{a}_1 \bullet \mathbf{b}_2 = 2\pi, \tag{D.4}$$

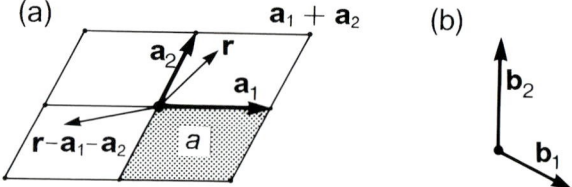

Fig. D.1. A surface with two-dimensional periodicity. (a) in real space, the function has two-dimensional periodicity, which is indicated by two primitive vectors, a_1 and a_2. (b) In reciprocal space, two primitive vectors b_1 and b_2 are introduced.

$$\mathbf{a}_1 \bullet \mathbf{b}_2 = \mathbf{a}_2 \bullet \mathbf{b}_1 = 0. \qquad (D.5)$$

The periodic function $F(\mathbf{r})$ can always be expanded into a two-dimensional Fourier series,

$$F(\mathbf{r}) = \sum_{\mathbf{g}} \tilde{F}_{\mathbf{g}}(z)\, e^{i\mathbf{g} \bullet \mathbf{x}}, \qquad (D.6)$$

where $\mathbf{g} = g_1\mathbf{b}_1 + g_2\mathbf{b}_2$, and g_1, g_2 are integers. The sum runs over all possible integers. The Fourier coefficients are

$$\tilde{F}_{\mathbf{g}}(z) = \frac{1}{a} \int_a d^2\mathbf{x}\, F(\mathbf{r})\, e^{-i\mathbf{g} \bullet \mathbf{x}}, \qquad (D.7)$$

where $a = |\mathbf{a}_1 \times \mathbf{a}_2|$ is the area of the unit cell. On the other hand, a periodic function $F(\mathbf{r})$ can always by expressed as a sum of a nonperiodic functions, $f(\mathbf{r})$,

$$F(\mathbf{r}) = \sum_{m,n = -\infty}^{\infty} f(\mathbf{r} - m\,\mathbf{a}_1 - n\,\mathbf{a}_2). \qquad (D.8)$$

Such a decomposition has physical meanings in surface science. For example, the potential between an atom and a crystalline surface can be calculated as the sum of pairwise potentials; the tunneling current from a tip state to a crystalline surface can be calculated as the sum of the tunneling currents to the individual orbitals on surface atoms, etc. The Fourier coefficients $\tilde{F}_g(z)$ can be calculated directly by integrating the individual interaction function $f(\mathbf{r})$ over the entire surface. Actually, substituting Eq. (D.8) into Eq. (D.7), noting that the individual interaction observed at the present cell originates from a center at another cell is equal to the individual interaction originated from the present cell but observed from another cell, the integral over the unit cell can be extended to an integral over the entire surface area A,

$$
\begin{aligned}
\tilde{F}_g(z) &= \frac{1}{a} \int_a \sum_{m,n=-\infty}^{\infty} f(\mathbf{r} - m\,\mathbf{a}_1 - n\,\mathbf{a}_2)\, e^{-i\mathbf{g}\cdot\mathbf{x}}\, d^2\mathbf{x} \\
&= \frac{1}{a} \int_A f(\mathbf{r})\, e^{-i\mathbf{g}\cdot\mathbf{x}}\, d^2\mathbf{x}.
\end{aligned}
\tag{D.9}
$$

For many practically important interaction functions, the Fourier coefficients in Eq. (D.9) have finite analytic forms, for example, the Lennard–Jones potential, the Yukawa potential, the Morse potential, and functions that can be derived from those functions. For a power-law interaction

$$
f(\mathbf{r}) = \frac{C}{|\mathbf{r}|^n},
\tag{D.10}
$$

the integral can be evaluated by transforming the Cartesian coordinates (x, y) into polar coordinates (ρ, ϕ). Denoting $g = |\mathbf{g}|$ and the polar angle of $\mathbf{g}$ as ξ,

$$
\begin{aligned}
\tilde{F}_g(z) &= \frac{C}{a} \int_0^\infty \rho\, d\rho\, \frac{1}{(\rho^2 + z^2)^{n/2}} \int_0^{2\pi} e^{i g\rho \cos(\phi - \xi)}\, d\phi \\
&= \frac{2\pi C}{a} \int_0^\infty \frac{\rho\, d\rho}{(\rho^2 + z^2)^{n/2}}\, J_0(g\rho) \\
&= \frac{2\pi C}{(n-1)!a} \left(\frac{g}{2z} \right)^{n/2} K_{n-1}(gz),
\end{aligned}
\tag{D.11}
$$

where $J_0(x)$ is a Bessel function, and $K_n(x)$ is a modified Bessel function. For a Yukawa potential,

$$
f(\mathbf{r}) = \frac{C}{r}\, e^{-\kappa r},
\tag{D.12}
$$

the integral is

$$\tilde{F}_{\mathbf{g}}(z) = \frac{2\pi C}{a} \int_0^\infty J_0(g\rho) \, e^{-\kappa\sqrt{\rho^2 + z^2}} \, d\rho$$

$$= \frac{2\pi C}{a\sqrt{g^2 + \kappa^2}} \, e^{-z\sqrt{g^2 + \kappa^2}}. \tag{D.13}$$

An alternative proof of Eq. (D.13) is to integrate it in Cartesian coordinates. By denoting $\mathbf{g} = (g_x, \, g_y)$, we obtain

$$\tilde{F}_{\mathbf{g}}(z) = C \int_{-\infty}^\infty dy \int_{-\infty}^\infty dx \, \frac{e^{i(g_x x + g_y y)}}{\sqrt{x^2 + y^2 + z^2}} \, \exp\left(-\kappa\sqrt{x^2 + y^2 + z^2}\right)$$

$$= 2C \int_{-\infty}^\infty dy \, e^{ig_y y} K_0\left(\sqrt{(y^2 + z^2)(\kappa^2 + g_x^2)}\right) \tag{D.14}$$

$$= \frac{2\pi C \exp\left(-z\sqrt{\kappa^2 + g^2}\right)}{\sqrt{\kappa^2 + g^2}},$$

using two identities listed in Gradsteyn and Ryzhik (1980), pp. 498 and 736.

APPENDIX E

PLANE GROUPS AND INVARIANT FUNCTIONS

For most of the surfaces of interest, in addition to the two-dimensional translational symmetry, there are additional symmetry operations that leave the lattice invariant. If the tip has axial symmetry, then the STM images and the AFM images should exhibit the same symmetry as that of the surface. The existence of those symmetry elements may greatly reduce the number of independent parameters required to describe the images.

The collection of all symmetry operations that leave a crystalline lattice invariant forms a *space group*. Each type of crystal lattice has its specific space group. The problem of enumerating and describing all possible space groups, both two dimensional three dimensional, is a pure mathematical problem. It was completely resolved in the mid-nineteenth century. A contemporary tabulation of the properties of all space groups can be found in Hahn (1987). Burns and Glazer (1990) wrote an introductory book to that colossal table.

Compared with the problem in three-dimensional space, which has 14 Bravais lattices and 230 space groups, the problem of surface symmetry is truly a dwarf: It only has 5 Bravais lattices and 17 different groups. The five Bravais lattices are listed in Table E.1.

Table E.1. Two-dimensional Bravais lattices

Monoclinic Primitive	Orthorhombic Primitive	Orthorhombic Centered	Tetragonal Primitive	Hexagonal Primitive
mp	*op*	*oc*	*tp*	*hp*

As shown in Table E.1, there is only one centered lattice, *oc*. It is easy to show that for monoclinic, orthorhombic, and hexagonal cases, the centered lattice reduces to primitive lattices with halved unit cells.

E.1 A brief summary of plane groups

The symbols for plane groups, the Hermann–Mauguin symbol, have been the standard in crystallography. The first place indicates the type of lattice, p indicates primitive, and c indicates centered. The second place indicates the axial symmetry, which has only 5 possible vales, 1-, 2-, 3-, 4-, and 6-fold. For the rest, the letter m indicates a symmetry under a mirror reflection, and the letter g indicates a symmetry with respect to a glide line, that is, one-half of the unit vector translation followed by a mirror reflection. For example, the plane group $p4mm$ means that the surface has fourfold symmetry as well as mirror reflection symmetries through both x and y axes.

The 17 plane groups are not mutually unrelated. Some of them are subgroups of other plane groups, as shown in Fig. E.1. The order of the factor group, that is, the number of different symmetry operations other than translational symmetry, is also shown for each plane group.

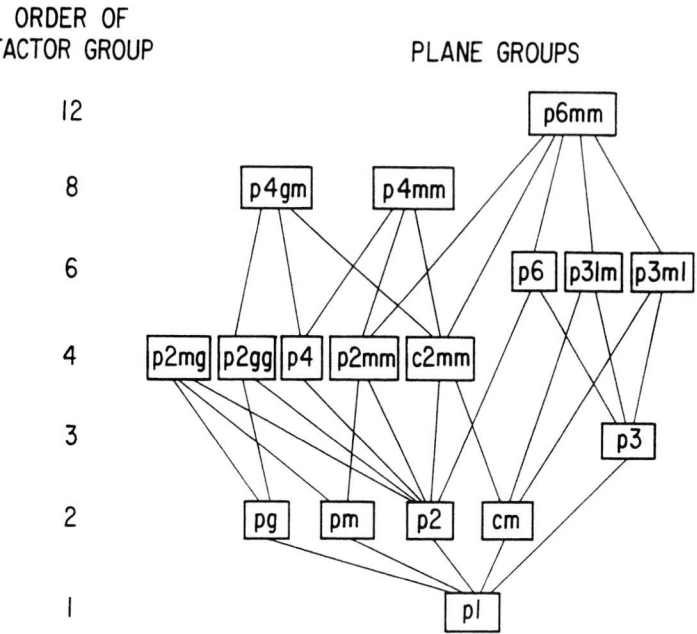

Fig. E.1. Relations among plane groups. In this figure, the plane groups are shown in their degrees of symmetry, as indicated by the order of the factor groups. A plane group with high symmetry always has one or several subgroup(s). The chart shows such relations within the same lattice.

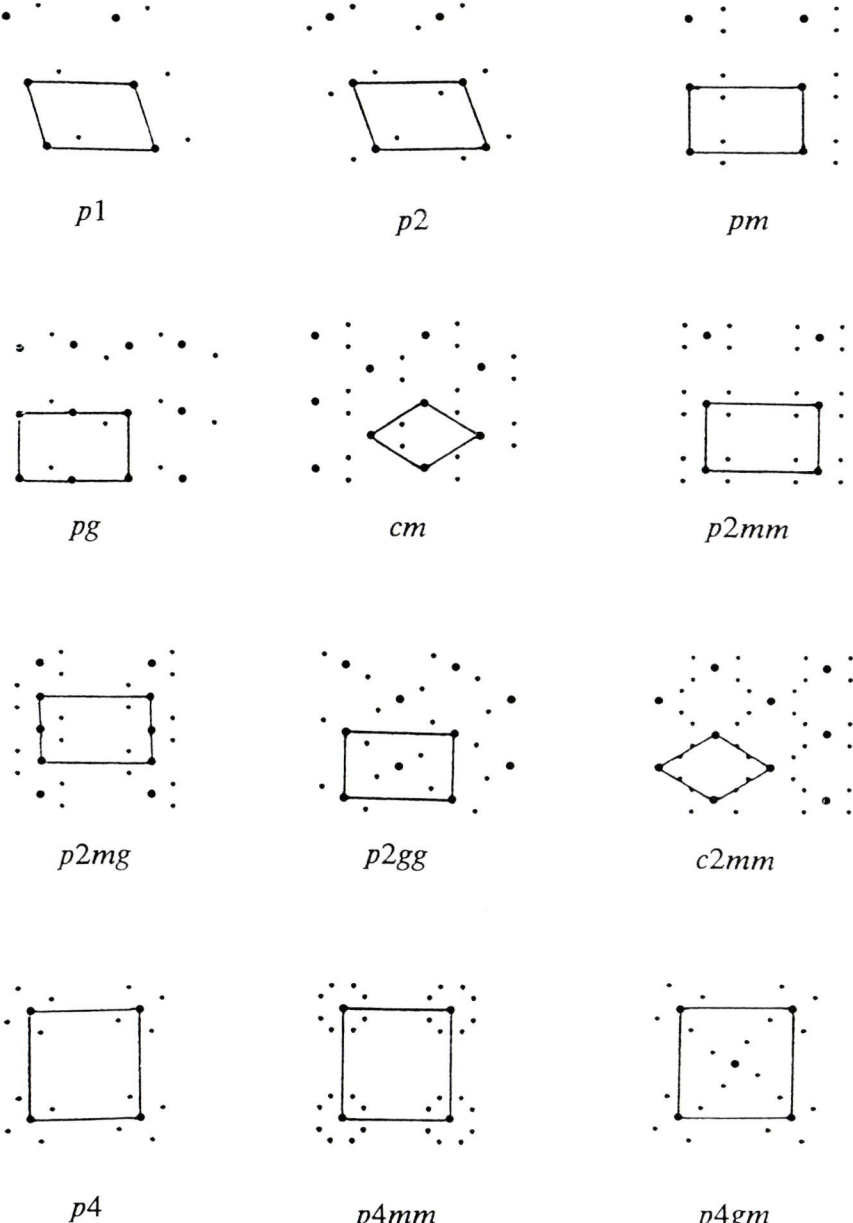

Fig. E.2. The plane groups I:. Plane groups in the monoclinic, orthorhombic, and tetragonal families.

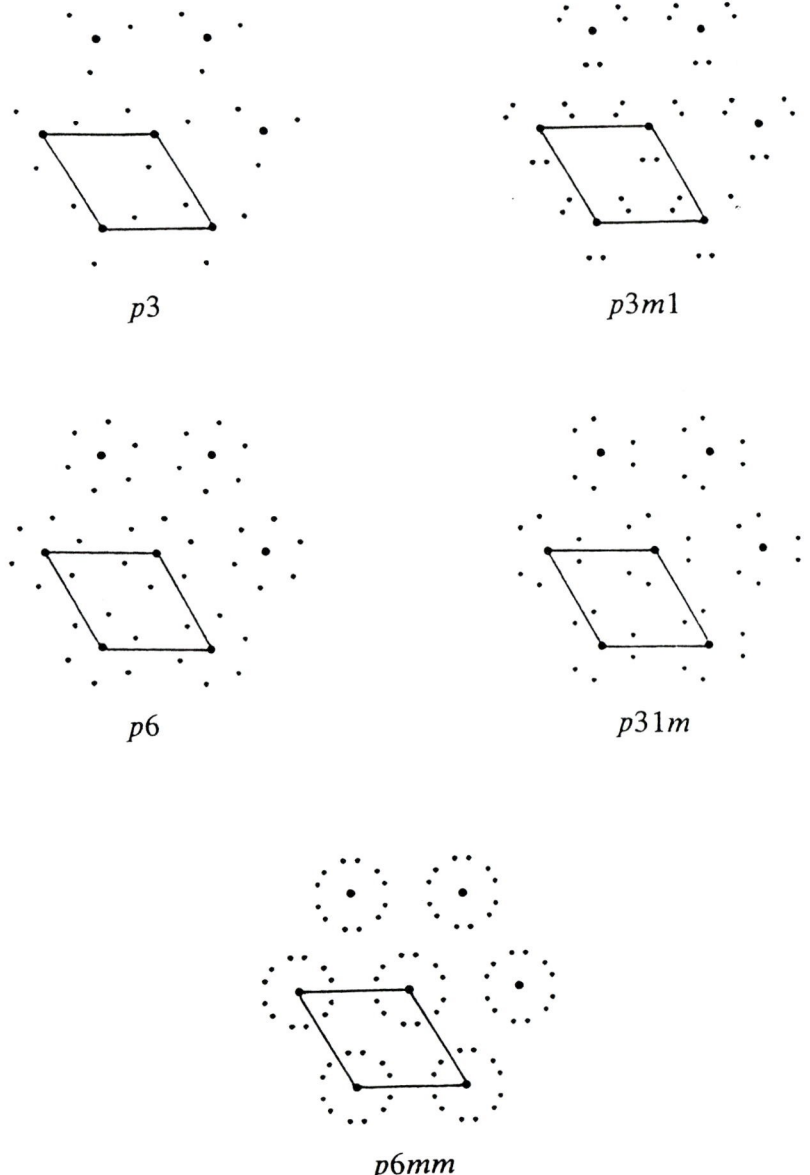

Fig. E.3. The plane groups II. Plane groups in the hexagonal family.

E.2 Invariant functions

As we have discussed previously, any function with two-dimensional periodicity can be expanded into two-dimensional Fourier series. If a function has additional symmetry other than translational, then some of the terms in the Fourier expansion vanish, and some nonvanishing Fourier coefficients equal each other.   The number of independent parameters is then reduced.   In general, the form of a quantity periodic in x and y would be

$$Q(\mathbf{r}) = Q_0(z) + Q_1(z)f_1(x, y) + Q_2(z)f_2(x,y) + \ldots, \qquad (E.1)$$

where $f_1(x,y)$, $f_2(x,y)$, are symmetrized sums of sinusoidal functions.   At a given distance z, the first term $Q_0(z)$ is a constant.   We call the function $f_1(x,y)$ etc. an *invariant function* of that group, which is normalized to be

$$\max[f_1(x,y)] - \min[f_1(x,y)] = 1. \qquad (E.2)$$

The quantity $Q_1(z)$ is the *corrugation amplitude* of the quantity $Q(\mathbf{r})$ with $f_1(x,y)$ describing the way it varies with x and y.   In the following, we will derive and list explicitly the invariant functions for several important plane groups.

Plane group pm

An example is the (110) plane of III−V semiconductors, such as GaAs(110). The only nontrivial symmetry operation is a mirror reflection through a line connecting two Ga (or As) nuclei in the $[001]$ direction, which we labeled as the x axis.   The Bravais lattice is orthorhombic primitive (op).   In terms of real Fourier components, the possible corrugation functions are

$$f_1(x,y) = \frac{1}{2} \cos \frac{2n\pi x}{a} \cos \frac{2m\pi y}{b}, \qquad (E.3)$$

$$f_2(x,y) = \frac{1}{2} \sin \frac{2n\pi x}{a} \cos \frac{2m\pi y}{b}, \qquad (E.4)$$

$$f_3(x,y) = \frac{1}{2} \cos \frac{2n\pi x}{a} \sin \frac{2m\pi y}{b}, \qquad (E.5)$$

$$f_4(x,y) = \frac{1}{2} \sin \frac{2n\pi x}{a} \sin \frac{2m\pi y}{b}. \qquad (E.6)$$

The mirror-image symmetry with respect to the x axis eliminates f_3 and f_4. Also, the functions f_1 and f_2 should have the same z dependence. Thus, we only need one type of invariant function:

$$f_{nm}(x,y) = \frac{1}{2} \cos\left(\frac{2n\pi x}{a} + \phi_{nm}\right) \cos \frac{2m\pi y}{b} , \qquad (E.7)$$

where ϕ_{nm} are phase constants.

Plane group p2gm

By replacing Ga and As atoms with the same species, such as Si or Ge, the symmetry becomes higher. In Fig. E.4 the Si(110) plane is shown as an example. The additional gliding symmetry operation means that by letting $y \rightarrow y + b/2$ and $x \rightarrow -x$ simultaneously, the function should not change. The only Fourier components satisfying this condition are

$$f_1(x,y) = \frac{1}{2} \cos \frac{2n\pi x}{a} \cos \frac{4m\pi y}{b} , \qquad (E.8)$$

and

$$f_2(x,y) = \frac{1}{2} \sin \frac{2n\pi x}{a} \cos \frac{(4m+2)\pi y}{b} . \qquad (E.9)$$

The general form of a corrugation function that contains the first four terms (in order of the absolute value of $\mathbf{g}$) is:

$$\begin{aligned} Q(\mathbf{r}) = Q_0(z) &+ \frac{Q_1(z)}{2} \cos \frac{2\pi x}{a} + \frac{Q_2(z)}{2} \cos \frac{4\pi x}{a} \\ &+ \frac{Q_3(z)}{2} \sin \frac{2\pi x}{a} \cos \frac{2\pi y}{b} . \end{aligned} \qquad (E.10)$$

The term Q_0 is a constant. The next two terms describe the corrugation in the $[001]$ direction. The last term describes the corrugation in the $[1\bar{1}0]$ direction, which should be the smallest among all the terms in Eq. (E.10).

Plane group p2mm.

A number of surfaces of interest exhibit *p2mm* symmetry, for example, Si(111)2×1, Si(100)2×1, etc. The symmetric function is similar to that of plane group *p2gm*. We leave the derivation of it as an exercise.

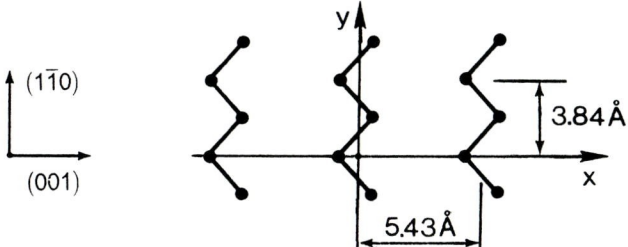

Fig. E.4. The Si(110) surface. An example of plane group p2gm.

Plane group p4mm.

By setting the origin of the coordinate system at the intersection of the two mirror reflection lines, it is easy to see that only Eq. (E.3) of the four corrugation functions is invariant under the mirror reflection operation. The fourfold rotational symmetry further requires $n = m$, and $a = b$. To the lowest nontrivial corrugation component, the general form of the corrugation function is

$$Q(\mathbf{r}) = Q_0(z) + \frac{1}{2} Q_1(z) \cos \frac{2\pi x}{a} \cos \frac{2\pi y}{a} . \tag{E.11}$$

This function describes a simple tetragonal surface with atoms located at (na, ma), where n and m are arbitrary integers.

Plane group p6mm.

The general form is listed in Chapter 5, Equations (5.41) and (5.42):

$$Q(\mathbf{r}) = Q_0(z) + Q_1(z) \phi^{(6)}(k\mathbf{x}), \tag{E.12}$$

where $k = 4\pi/\sqrt{3}\, a$ is the length of a primitive reciprocal lattice vector, and

$$\phi^{(6)}(\mathbf{X}) \equiv \frac{1}{3} + \frac{2}{9} \sum_{n=0}^{2} \cos \omega_n \cdot \mathbf{X} \tag{E.13}$$

is the hexagonal cosine function, where $\omega_0 = (0, 1)$, $\omega_1 = (-\frac{1}{2}\sqrt{3}, -\frac{1}{2})$, and $\omega_2 = (\frac{1}{2}\sqrt{3}, -\frac{1}{2})$, respectively, see Fig. 5.5.

APPENDIX F
ELEMENTARY ELASTICITY THEORY

Excellent treatises and textbooks on elasticity are abundant, for example, Landau and Lifshitz (1986), Timoshenko and Goodier (1970), and Timoshenko and Young (1962). It takes a lot of time to read and find useful information. This appendix contains an elementary treatment of elasticity regarding STM and AFM.

F.1 Normal stress and normal strain

Imagine that a small bar along the z direction is isolated from a solid body, as shown in Fig. F.1. The force in the x direction per unit area on the x facet is denoted as σ_x, a *normal stress:*

$$\sigma_x \equiv \frac{F_x}{S_x} . \qquad (F.1)$$

A body under stress deforms. In other words, a strain is generated. With a normal stress σ_x, the bar elongates in the x direction. The standard notation to describe strain is by introducing *displacements, u, v,* and *w,* in the x, y, and z directions, respectively, as shown in Fig. F.1. The dimensionless quantity

$$\epsilon_x \equiv \frac{\partial u}{\partial x} \qquad (F.2)$$

is called the *unit elongation,* which is a component of the strain tensor. Hooke's law says that the unit elongation is proportional to the normal stress in the same direction:

$$\epsilon_x = \frac{\sigma_x}{E} , \qquad (F.3)$$

where the quantity E is the *Young's modulus.*

Under the same stress σ_x, the y and z dimensions of the bar contract:

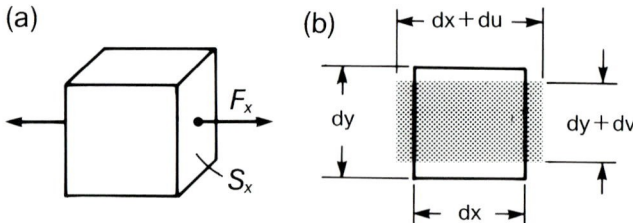

Fig. F.1. Normal stress and normal strain. (a) The normal force per unit area in the x direction is the x component of the stress tensor. (b) The normal stress causes an elongation in the x direction and a contraction in the y, and z directions.

$$\epsilon_y \equiv \frac{\partial v}{\partial y} = -\nu \frac{\sigma_x}{E}, \tag{F.4}$$

$$\epsilon_z \equiv \frac{\partial w}{\partial z} = -\nu \frac{\sigma_x}{E}. \tag{F.5}$$

This effect was discovered by Poisson, and the dimensionless constant ν is called *Poisson's ratio*. For most materials, $\nu \approx 0.25$.

F.2 Shear stress and shear strain

The x component of the force per unit area on a facet in the y direction is denoted as τ_{xy}, and is called a component of the *shear stress:*

$$\tau_{xy} \equiv \frac{F_x}{S_y}. \tag{F.6}$$

The condition of equilibrium, that is, the absence of a net torque on a small volume element, requires that $\tau_{xy} = \tau_{yx}$. A shear stress causes a shear strain, defined as (see Fig. F.2)

$$\gamma_{xy} \equiv \frac{\partial v}{\partial x} + \frac{\partial u}{\partial y}. \tag{F.7}$$

The relation between shear stress and shear strain can be established based on the relation between normal stress and normal strain, Equations (F.3) and (F.4). Actually, by rotating the coordinate system $45°$, it becomes a problem of normal stress and normal strain. Using geometrical arguments, it can be shown that (see, for example, Timishenko and Goodier, 1970):

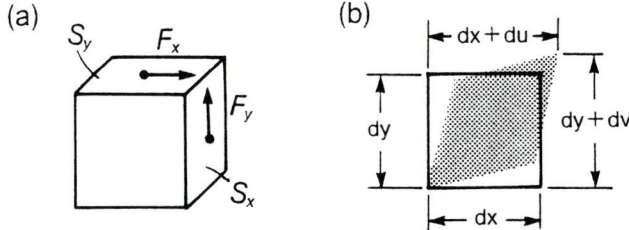

Fig. F.2. Shear stress and shear strain. (a) The shear force per unit area is a component of the stress tensor. (b) The shear stress causes a shear strain.

$$\tau_{xy} = \frac{E}{2(1 + v)}\, \gamma_{xy} \equiv G\gamma_{xy}. \tag{F.8}$$

The quantity $G \equiv E/2(1 + v)$ is called the *modulus of elasticity in shear.*

F.3 Small deflection of beams

The deflection of a beam under a vertical distribution of force, as shown in Fig. F.3, is a very basic problem in the theory of elasticity. Under the condition that the deflection is small and there is no force parallel to the main axis, the deflection u as a function of distance z satisfies a simple differential equation, as we will show in this section.

To start with, we show two simple relations in statics; see Fig. F.3. First, consider the equilibrium of the net force in the vertical direction. At each cross section of the beam, there is a vertical force $V(z)$, which should compensate the external force $F(z)$:

$$dV(z) \equiv V(z + dz) - V(z) = -F(z)\, dz. \tag{F.9}$$

Second, at each cross section of the beam, the normal stress forms a torque $M(z)$. The equilibrium condition of a small section of the beam dz with respect to rotation requires (see Fig. F.3)

$$dM(z) \equiv M(z + dz) - M(z) = V(z)\, dz. \tag{F.10}$$

Combining Equations (F.9) and (F.10), we have

$$\frac{d^2M(z)}{dz^2} = -F(z). \tag{F.11}$$

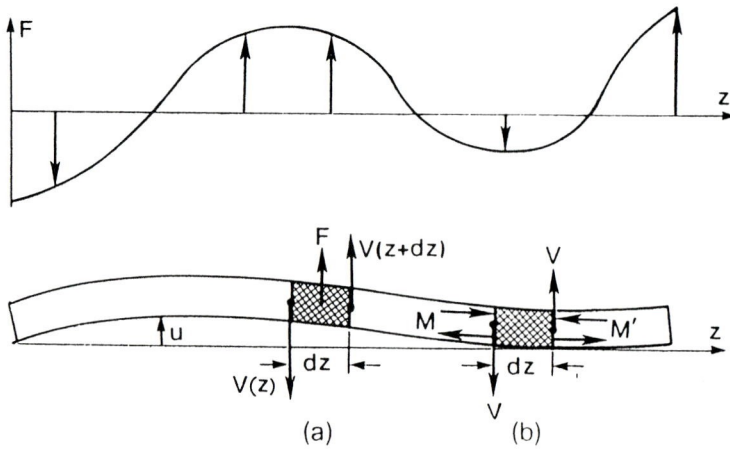

Fig. F.3. Bending of a beam. Under the action of a vertical force F distributed along the axis of the beam, the beam bends. The deflection u as a function of position z is determined by the force $F(z)$. (a). the equilibrium of a small section dz of the beam respect to the force in the vertical direction. (b). the equilibrium of a small section dz of the beam with respect to the torque.

Under the influence of a torque, the beam deforms. In other words, the slope $\tan \theta$ changes with z, as shown in Eq. (F.4). For small deflections, $\tan \theta \approx \theta$. In other words,

$$\frac{du(z)}{dz} = \theta. \qquad (F.12)$$

Consider a small section Δz of the beam. The total force horizontal must be zero. For symmetrical cross sections, such as rectangles, circular bars, and tubes, symmetry conditions require that the neutral line of force must be in the median plane, denoted as $x = 0$ (see Fig. F.4). The distribution of normal strain is then

$$\epsilon_z \equiv \frac{\Delta w}{\Delta z} = \frac{x\, d\theta}{dz} = x\, \frac{d^2 u}{dz^2}. \qquad (F.13)$$

Using Eq. (F.3) and neglecting the contraction of the width $A(x)$, the total torque acting on a cross section is

$$M = \frac{d^2 u}{dz^2}\, E \int A(x)\, x^2\, dx \equiv \frac{d^2 u}{dz^2}\, EI. \qquad (F.14)$$

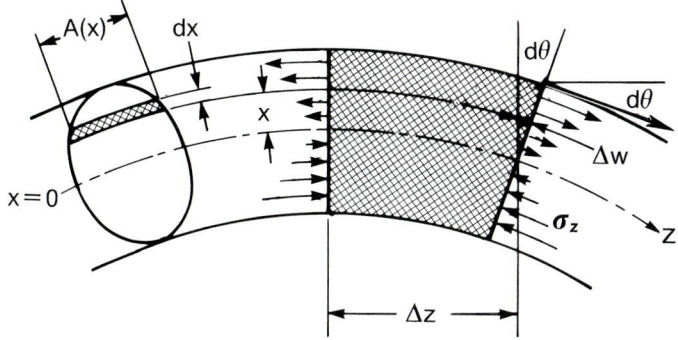

Fig. F.4. Deformation of a segment of a beam. Under the influence of a torque acting on a cross section, a beam bends. For small deformations, the slope is $\theta = du(z)/dz$. The change of slope with distance $d\theta/dz$ is connected with a strain distribution in the beam, $\Delta w/\Delta z$. The strain is connected with a distribution of normal stress σ_z in the beam. The total torque is obtained by integration over the cross section of the beam.

The quantity I is the *moment of inertia* of the cross section. For simple shapes, the integral can be evaluated easily, which is left as an exercise. For a rectangular bar of width b and height h,

$$I = \frac{1}{12} bh^3. \tag{F.15}$$

For a rod of diameter D, it is

$$I = \frac{\pi}{64} D^4. \tag{F.16}$$

For a tube with outer radius D and inner radius d,

$$I = \frac{\pi}{64} (D^4 - d^4). \tag{F.17}$$

Combining Equations (F.10) and (F.14), we obtain a differential equation for the deflection $u(z)$,

$$EI \frac{d^4u(z)}{dz^4} = - F(z). \tag{F.18}$$

We will give an example of a concentrated force W acting at a point on the beam, $z = L$, which is clamped at the origin, $z = 0$. Thus, $F(z) = W\,\delta(z - L)$. Integrating Eq. (F.18) once, we find

$$EI \frac{d^3u(z)}{dz^3} = \begin{cases} W, & 0 < z < L, \\ 0, & z > L \text{ or } z < 0. \end{cases} \qquad (F.19)$$

Integrating it again and using the condition that at $z = L$, the torque is zero, $d^2u(z)/dz^2 = 0$, we get

$$EI \frac{d^2u(z)}{dz^2} = W(z - L). \qquad (F.20)$$

Integrating twice and using the condition that at $z = 0$, $u = 0$ and $du/dz = 0$, we obtain

$$EI \frac{du(z)}{dz} = \frac{1}{2} Wz^2 - WLz, \qquad (F.21)$$

$$EIu(z) = \frac{1}{6} Wz^3 - \frac{1}{2} WLz^2. \qquad (F.22)$$

The deflection at $z = L$ is

$$u(L) = - \frac{WL^3}{3EI}. \qquad (F.23)$$

The angle at $z = L$ is

$$\theta = \left(\frac{du(z)}{dz} \right)_{z = L} = \frac{3}{2} \frac{u(L)}{L}. \qquad (F.24)$$

F.4 Vibration of beams

If a deflection exists in a beam in the absence of external force, the elastic force will cause an acceleration in the vertical direction. From Eq. (F.18), using Newton's second law, the equation of motion of the beam is

$$EI \frac{d^4u(z,t)}{dz^4} = - \rho S \frac{d^2u(z,t)}{dt^2}, \qquad (F.25)$$

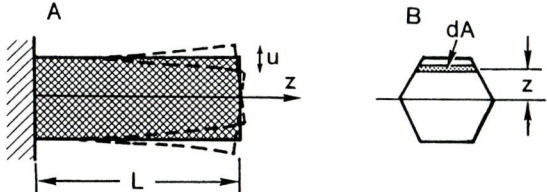

Fig. F.5. Vibration of a beam. One end of the beam is clamped; the displacement and the slope are zero. Another end of the beam is free; the force and the torque are zero. The lowest resonance frequency, which corresponds to a vibration without a node, is determined by Eq. (F.32).

where ρ is the density of the material, and A is the cross-sectional area of the beam. For a sinusoidal vibration, $u(z,t) = u(z) \cos(\omega t + \alpha)$, Eq. (F.25) becomes

$$EI \frac{d^4 u(z)}{dz^4} = \rho S \omega^2 u(z). \qquad (F.26)$$

Denoting

$$\kappa^4 = \frac{\omega^2 \rho S}{EI}, \qquad (F.27)$$

the general solution of Eq. (F.26) is

$$u(z) = A \cos \kappa z + B \sin \kappa z + C \cosh \kappa z + D \sinh \kappa z, \qquad (F.28)$$

which can be verified by direct substitution. The most important case in STM and AFM is the vibration of a beam with one end clamped and another end free; see Fig. F.5. At the clamped end, $z = 0$, the boundary conditions are

$$u = 0; \quad \frac{du}{dz} = 0. \qquad (F.29)$$

At the free end, $z = L$, the vertical force V and the torque M vanish. The boundary conditions are

$$\frac{d^2 u}{dz^2} = 0; \quad \frac{d^3 u}{dz^3} = 0. \qquad (F.30)$$

By going through simple algebra (we leave it as an exercise for the reader), the vibration frequency is found to be determined by the equation

$$\cos \kappa L \cosh \kappa L + 1 = 0. \tag{F.31}$$

The lowest resonance frequency is determined by the lowest solution of Eq. (F.31), which is $\kappa L \approx 1.875$. Substituting into Eq. (F.27), the frequency is (see Landau and Lifshitz, 1986, p. 102):

$$f = \frac{0.56}{L^2} \sqrt{\frac{EI}{\rho S}}. \tag{F.32}$$

F.5 Torsion

Another important case in the theory of elasticity is the shearing stress and the angles of twist of circular bars and tubes subjected to twisting torque T, as shown in Fig. F.6. Consider a small section of a circular bar with length L and radius r. If the twist angle θ per unit length is not too large, every cross section perpendicular to the axis remains unchanged. A small, initially rectangular volume element at radius ρ will have a shear strain

$$\gamma = \frac{\theta \rho}{L}. \tag{F.33}$$

The shear stress is, according to Eq. (F.8),

$$\tau = G \frac{\theta \rho}{L}. \tag{F.34}$$

The torque acting on a ring of radius ρ and width $d\rho$ is

$$dT = \frac{2\pi G\theta}{L} \rho^3 dr. \tag{F.35}$$

By integrating over the entire area, the total torque is

$$T = \frac{2\pi G\theta}{L} \int_0^r \rho^3 d\rho = \frac{\pi G\theta}{2L} r^4. \tag{F.36}$$

In terms of diameter $d = 2r$, the torque is

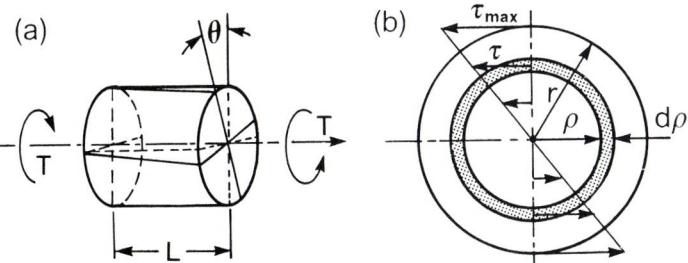

Fig. F.6. Torsion of a circular bar. (a) Under the action of a torque T, a bar of length L is twisted through an angle θ. (b) The twist results in a distribution of shear strain γ in a cross section, which generates a distribution of shear stress τ.

$$T = \frac{\pi G \theta}{32 L} d^4. \tag{F.37}$$

F.6 Helical springs

Helical springs are common elements of almost every mechanical device. The important parameters of a helical spring are: wire diameter r, coil diameter (from an axis of the wire to another axis) D, number of coils n, and the modulus of elasticity of shear of the material G. To provide an understanding of the helical springs, we give a simple derivation of the total stretch f and the maximum shear stress τ_{max} as a function of the axial load P acting on the spring. The stretch f of a spring is the increase of the pitch h times the number of coils n. To simplify the derivation, we imagine that every coil in the spring is a flat ring at rest; that is, the pitch at rest is zero. Because all the quantities are linear in the pitch, an incremental value of pitch results in an incremental value of axial load and maximum shear stress. When the pitch is increased by $h = f/n$, the cross section of the wire is twisted by an angle $\phi = f/2nD$. For every one half of a coil, the length of wire is

$$L = \frac{\pi D}{2}, \tag{F.38}$$

which has a total twist angle (see Fig. F.7)

$$\theta = 2\phi = \frac{f}{nD}. \tag{F.39}$$

On the other hand, to generate a torque T in a wire, the axial load at the center of the spring should be

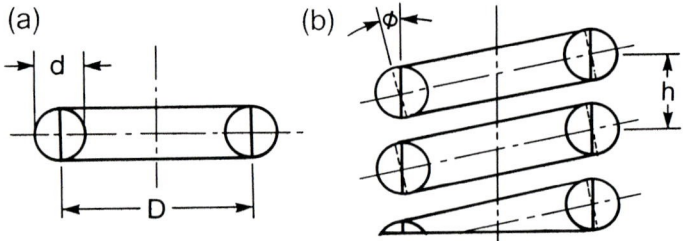

Fig. F.7. Stiffness of a helical spring. (a) A fictitious position with a zero pitch. (b) When the pitch h of a spring increases, the twisting angle of the wire increases. The torque also increases. Using the formula for torsion, the stiffness of a helical spring is obtained.

$$P = \frac{2T}{D}. \qquad (F.40)$$

Substituting these values into Eq. (F.37), we obtain

$$P = \frac{Gd^4}{8nD^3} f. \qquad (F.41)$$

The quantity $(GD^4/8nD^3)$ is often called the *stiffness* of a spring.
  The shear stress at the periphery of the wire, after Eq. (F.34), is

$$\tau_{max} = \frac{G\theta r}{L} = \frac{Gd}{n\pi D^2} f$$
$$= \frac{8D}{\pi d^3} P. \qquad (F.42)$$

  In designing a spring, the maximum stress must not exceed the *allowable stress* or *working stress* of the material (Roark and Young, 1975). These formulas are useful in the design of vibration isolation systems.

F.7 Contact stress: The Hertz formulas

By pressing a sphere upon a planar surface, a deformation occurs near the original point of contact, see Fig. F.8. This problem was first solved by Hertz in 1881 (see Landau and Lifshitz, 1986; Timoshenko and Goodier, 1970). The derivations are complicated. We state the results without proof.

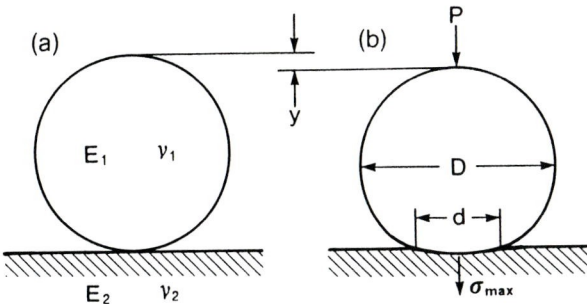

Fig. F.8. Contact stress and contact deformation. When a sphere (a) is pressed upon a flat surface with a vertical load P, deformation occurs (b). The yield y, the maximum stress σ_{max}, and the diameter of the contact area d were analyzed by Hertz in 1881.

Consider a sphere of diameter D of material 1 in contact with a planar surface of material 2. The reaction of the load P is determined by an effective Young's modulus E^*, which is defined as

$$\frac{1}{E^*} = \frac{1 - v_1^2}{E_1} + \frac{1 - v_2^2}{E_2},$$ (F.43)

in terms of the Young's modulus and Poisson ratios of material 1 and material 2, respectively.

The diameter of contact is

$$d = \left(\frac{3PD}{E^*} \right)^{1/3}.$$ (F.44)

The total yield, or total relative motion, is

$$y = \frac{1}{2} \left(\frac{9P^2}{DE^{*2}} \right)^{1/3}.$$ (F.45)

The average normal stress in the contact area is

$$\sigma = \frac{4P}{\pi d^2} = \frac{4}{\pi} \left(\frac{PE^{*2}}{9D^2} \right)^{1/3}.$$ (F.46)

The maximum normal stress at the point of contact, according to Hertz, is 1.5 times the average normal stress:

$$\sigma_{max} = \frac{2}{\pi} \left(\frac{3PE^{*2}}{D^2} \right)^{1/3}. \tag{F.47}$$

These formulas are important in the discussion of tip and sample deformation (see, for example, Pethica and Oliver, 1987), as well as some mechanical design problems of STM and AFM.

APPENDIX G

A SHORT TABLE OF LAPLACE TRANSFORMS

The following is a short table of Laplace transforms for quick reference.

Time Function		Laplace Transform
Unit Impulse	$\delta(t)$	1
Unit Step	$u(t)$	$1/s$
Unit Ramp	t	$1/s^2$
Polynomial	t^n	$n!/s^{n+1}$
Exponential	e^{-at}	$\dfrac{1}{s+a}$
Sine Wave	$\sin \omega t$	$\dfrac{\omega}{s^2 + \omega^2}$
Cosine Wave	$\cos \omega t$	$\dfrac{s}{s^2 + \omega^2}$
Damped Sine Wave	$e^{-at} \sin \omega t$	$\dfrac{\omega}{(s+a)^2 + \omega^2}$
Damped Cosine Wave	$e^{-at} \cos \omega t$	$\dfrac{s+a}{(s+a)^2 + \omega^2}$

APPENDIX H
OPERATIONAL AMPLIFIERS

This appendix is written for readers who are not familiar with electronic circuits. The book of Horowitz and Hill (1982) is readable and resourceful.

The operational amplifier or in short, op-amp, is used so extensively in modern electronic circuits that it is called a panacea. Op-amps are always used with negative feedback so that the circuits are essentially determined by the feedback networks only. Within certain limits, the characteristics of the op-amps can often be neglected (Fig. H.2).

A typical op-amp has:

1. Very high input impedance, with a typical input current of a few pA.

2. Very low output impedance, typically a few ohms.

3. An inverting and an noninverting input, with

$$V_{OUT} = A(V_+ - V_-). \qquad (H.1)$$

4. For low-frequency signals, the open-loop gain A is very large, typically $10^4 \sim 10^6$.

5. The open-loop gain decreases with increasing frequency.

Figure H.1 shows two simple applications of the op-amp. In the inverting amplifier, the output voltage tends to make the voltage difference between the inputs zero while keeping the input current zero: Therefore,

$$\text{voltage gain} = \frac{V_{OUT}}{V_{IN}} = -\frac{R_2}{R_1}. \qquad (H.2)$$

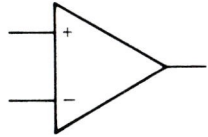

Fig. H.2. The operational amplifier (op-amp). It has two inputs with very high impedance, an output with very low impedance, and very high gain.

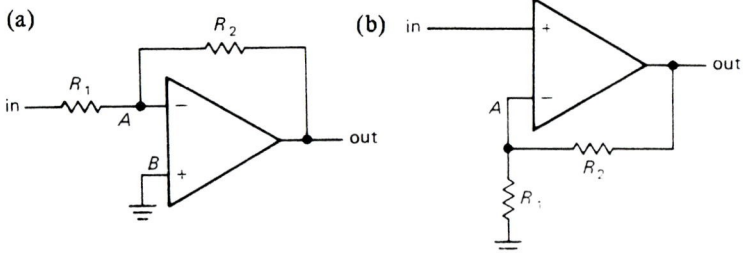

Fig. H.1. Two simple applications of the op-amp. (a) An inverting amplifier. (b) A noninverting amplifier.

In the noninverting amplifier, the condition of zero difference between the inputs requires $V_A = V_{IN}$. Thus,

$$\text{voltage gain} = \frac{V_{OUT}}{V_{IN}} = 1 + \frac{R_2}{R_1}. \tag{H.3}$$

Tables H-1 and H-2 provide information on some useful op-amps in STM and AFM. Table H-1 lists op-amps for tunneling current amplifiers. The requirements are, low bias current I_B, low input noise level, i_n and e_n. The typical power supply voltage is $\pm 5 - \pm 18$ V.

Table H.1. Op-amps recommended for tunneling current amplifiers

Manufacturer (Product)	I_B (fA)	V_ϕ (μV)	i_n, $\dfrac{fA}{\sqrt{Hz}}$	e_n, $\dfrac{nV}{\sqrt{Hz}}$	f_T (MHz)
Burr Brown (OPA128LM)	75	500	0.12	92	0.5
Analog Devices (AD549L)	60	500	0.11	90	0.7

Table H-2 lists high-voltage op-amps for driving the piezos. The requirements are: a high supply voltage range and a high slew rate (SR). The usable current I_Q is also an important parameter for high-voltage op-amps. In STM, it is not critical. Most of the useful high-voltage op-amps are manufactured by Apex.

Table H.2. High-voltage op-amps

Product	Voltage range (V)	f_T (MHz)	SR $V/\mu s$	I_Q (mA)
PA84	15 – 150	75	200	7.5
PA85	15 – 225	110	1000	25
PA88	15 – 225	1	30	2

REFERENCES

Akamine, S., Barrett, R. C., and Quate, C. F. (1990). Improved atomic force microscope images using microcantilevers with sharp tips. *Appl. Phys. Lett.* **57,** 316–318.

Albrecht, T. R., and Quate, C. F. (1988). Atomic resolution with the atomic force microscope on conductors and nonconductors. *J. Vac. Sci. Technol. A* **6,** 271–274.

Albrecht, T. R., Akamine, S., Carver, T. E., and Quate, C. F. (1990). Microfabrication of cantilever styli for the atomic force microscope *J. Vac. Sci. Technol. A* **8,** 3386–3396.

Albrektsen, O., Arent, D. J., Meier, H. P., and Salemink, H. W. M. (1990). Tunneling microscopy and spectroscopy of molecular beam epitaxy grown GaAs–AlGaAs interfaces. *Appl. Phys. Lett.* **57,** 31–33.

Appelbaum, J. A., and Hamann, D. R. (1972). Variational calculation of the image potential near a metal surface. *Phys. Rev. B* **6,** 1122–1130.

Appelbaum, J. A., and Hamann, D. R. (1972a). Self-consistent electronic structure of solid surfaces. *Phys. Rev. B* **6,** 2166–2177.

Appelbaum, J. A., and Hamann, D. R. (1973). Surface states and surface bonds of Si(111). *Phys. Rev. Lett.* **31,** 106–109.

Appelbaum, J. A., and Hamann, D. R. (1973a). Surface potential, charge density, and ionization potential of Si(111) − a self-consistent calculation. *Phys. Rev. Lett.* **32,** 225–228.

Appelbaum, J. A., and Hamann, D. R. (1976). The electronic structure of solid surfaces. *Rev. Mod. Phys.* **48,** 479–496.

Arfken, G. (1968). *Mathematical Methods for Physicists.* Second edition, Academic Press, New York.

Arlinghaus, F. J., Gay, J. G., and Smith, J. R. (1980). Chemisorption on d-band metals. in *Theory of Chemisorption,* edited by J. R. Smith, Springer-Verlag, 71–113.

Ashcroft, N. W., and Mermin, N. D. (1985). *Solid State Physics,* Sanders College, Philadelphia.

Atlan, D., Gardet, G., Binh, V. T., Garcia, N., and Sáenz, J. J. (1992). 3D calculations at atomic scale of the electrostatic potential and field created by a teton tip. *Ultramicroscopy* **42–44,** 154–162.

Avouris, Ph., and Wolkow, R. (1989). Atom resolved surface chemistry studied by scanning tunneling microscopy and spectroscopy. *Phys. Rev. B* **39,** 5091–5100.

Avouris, Ph. (1990). Atom-resolved surface chemistry using the scanning tunneling microscope. *J. Phys. Chem.* **94,** 2246–2256.

Avouris, Ph., and Lyo, I. W. (1990). Studying surface chemistry atom-by-atom using the scanning tunneling microscope. *Chemistry and Physics of Solid Surfaces,* edited by R. Vauselow, Springer, Berlin.

Avouris, Ph., Lyo, I. W., Bozso, F., and Kaxiras, E. (1990). Adsorption of boron on Si(111): Physics, chemistry, and atomic-scale electronic devices. *J. Vac. Sci. Technol. A* **8,** 3405–3411.

Avouris, Ph., and Lyo, I. W. (1991). Probing and inducing surface chemistry with the STM: The reactions of Si(111)-7×7 with $H_2 O$ and O_2. *Surf. Sci.* **242,** 1–11.

Avouris, Ph., Lyo, I. W., and Bozso, F. (1991a). Atom-resolved surface chemistry: The early steps of Si(111)-7 × 7 oxidation. *J. Vac. Sci. Technol. B* **9,** 424–430.

Avouris, Ph., and Lyo, I. W. (1991b). Probing and inducing surface chemistry on the atomic scale using the STM. in *Scanned Probe Microscopies,* edited by H. K. Wickramasinghe, American Institute of Physics, 1991.

Avouris, Ph., and Cahill, D. (1992). STM studies of Si(100)-2×1 oxidation: defect chemistry and Si ejection. *Ultramicroscopy* **42–44,** 838-844.

Avouris, Ph., and Lyo, I.-W. (1992). Probing the chemistry and manipulating surfaces at the atomic scale with the STM. *Appl. Surf. Sci.* **60/61,** 426–436.

Bardeen, J. (1936). Theory of the work function. *Phys. Rev.* **49,** 653–663.

Bardeen, J. (1960). Tunneling from a many-body point of view. *Phys. Rev. Lett.* **6,** 57–59.

Baratoff, A. (1984). Theory of scanning tunneling microscopy – Methods and approximations. *Physica (Amsterdam)* **127B,** 143–150.

Barrett, R. C. Nogami, J., and Quate, C. F. (1990). Charge density waves of 1T TaS_2 imaged by atomic force microscopy. *Appl. Phys. Lett.* **57,** 992–994.

Barth, J. V., Burne, H., Ertl, G., and Behm, R. J. (1990). Scanning tunneling microscopy observations on the reconstructed Au(111) surface: Atomic structure, long-range superstructure, rotational domains, and surface defects. *Phys. Rev. B* **42,** 9307–9318.

Bates, D. R., Ledsman, K., and Stewart, A. L. (1953). Wave functions of the hydrogen molecule ion. *Phil. Trans. Roy. Soc. London* **246,** 215–240.

Batra, I. P., and Ciraci, S. (1988). Theoretical scanning tunneling microscopy and atomic force microscopy study of graphite including tip–surface interaction. *J. Vac. Sci. Technol. A* **6,** 313–318.

Becker, R. S., Golovchenko, J. A., and Swartzentruber, B. S. (1985). Electron interferometry at crystal surfaces. *Phys. Rev. Lett.* **55,** 987–990.

Becker, R. S., Golovchenko, J. A., McRae, E. G., and Swartzentruber, B. S. (1985a). Tunneling images of atomic steps on the Si(111) 7 × 7 surface. *Phys. Rev. Lett.* **55,** 2028–2031.

Becker, R. S., Golovchenko, J. A., Hamann, D. R., and Swartzentruber, B. S. (1985b). Real-space observation of surface states on Si(111)7 × 7 with the scanning tunneling microscope. *Phys. Rev. Lett.* **55,** 2032–2035.

Becker, R. S., Golovchenko, J. A., and Swartzentruber, B. S. (1985b). Tunneling images of germanium surface reconstruction and phase boundaries. *Phys. Rev. Lett.* **54,** 2678–2680.

Beebe, T. P., Jr., Troy, E. W., Ogletree, D. F., Katz, J. E., Balhorn, R., Salmeron, M. B., and Siekhaus, W. J. (1988). Direct observation of native DNA structures with the scanning tunneling microscope. *Science* **243,** 370–372.

Behm, R. J. (1990). Scanning tunneling microscopy: Metal surfaces, adsorption and surface reactions. In Behm, R. J., Garcia, N., and Rohrer, H., *Scanning Tunneling Microscopy and Related Methods,* Kluwer, Dordrecht, 173–210.

Behm, R. J., Garcia, N., and Rohrer, H. (1990). *Scanning Tunneling Microscopy and Related Methods,* Kluwer, Dordrecht.

Berndt, R., Gimzewski, J. K., and Schlittler, R. R. (1992). Tunneling characteristics at atomic resolution on close-packed metal surfaces. *Ultramicroscopy,* **42–44,** 528–532.

Besocke, K. (1987). An easily operable scanning tunneling microscope. *Surf. Sci.* **181,** 145–155.

Besocke, K. H., Teske, M., and Frohn, J. (1988). Investigation of silicon in air with a fast scanning tunneling microscope. *J. Vac. Sci. Technol. A* **6,** 408–411.

Biegelson, D. K., Ponce, F. A., and Tramontana, J. C. (1987). Ion milled tips for scanning tunneling microscopy. *Appl. Phys. Lett.* **50,** 696–698.

Biegelson, D. K., Ponce, F. A., and Tramontana, J. C. (1989). Simple ion milling preparation of (111) tungsten tips. *Appl. Phys. Lett.* **54,** 1223–1225.

Biegelson, D. K., Bringans, R. D., Northrup, J. E., and Swartz, L. E. (1990). Reconstructions of GaAs(111) surfaces observed by scanning tunneling microscopy. *Phys. Rev. Lett.* **65,** 452–455.

Binh, V. T. (1988). Characterization of microtips for scanning tunneling microscopy. *Surf. Sci.* **202,** L539–L549.

Binh, V. T. (1988a). In situ fabrication and regeneration of microtips for scanning tunneling microscopy. *J. Microscopy,* **152,** Pt 2, 355–361.

Binh, V. T., and Garcia, N. (1991). Atomic metallic ion emission, field surface melting, and scanning tunneling microscopy tips. *J. Phys. I.* **1,** 605–612.

Binh, V. T., and Garcia, N. (1992). On the electron and metallic ion emission from nanotips fabricated by field-surface-melting technique: experiments on W and Au tips. *Ultramicroscopy,* **42–44,** 80–90.

Binh, V. T., Purcell, S. T., Garcia, N., and Doglini, J. (1992). Field emission spectroscopy of single-atom tips. *Phys. Rev. Lett.* **69,** 2527–2530.

Binnig G., and Rohrer, H. (1982). Scanning tunneling microscopy. *Helv. Phys. Acta.* **55,** 726–735.

Binnig G., and Rohrer, H. (1983). Scanning tunneling microscopy. *Surface Science.* **126,** 236–244.

Binnig, G., Rohrer, H., Gerber, Ch., and Weibel, E. (1982a). Tunneling through a controllable vacuum gap. *Appl. Phys. Lett.* **40,** 178–180.

Binnig, G., Rohrer, H., Gerber, Ch., and Weibel, E. (1982b). Surface studies by scanning tunneling microscopy. *Phys. Rev. Lett.* **49,** 57–61.

Binnig, G., Rohrer, H., Gerber, Ch., and Weibel, E. (1983). 7×7 reconstruction on Si(111) resolved in real space. *Phys. Rev. Lett.* **50,** 120–123.

Binnig, G., Fuchs, H., Gerber, Ch., Rohrer, H., Stoll, E., and Tosatti, E. (1983a). Energy-dependent state-density corrugation of a graphite surface as seen by scanning tunneling microscopy. Europhys. Lett. **1,** 31–36.

Binnig, G., Rohrer, H., Gerber, Ch., and Weibel, E. (1983b). (111) facets as the origin of reconstructed Au(110) surfaces. *Surf. Sci.* **131,** L379–L384.

Binnig, G., Garcia, N., Rohrer, H., Soler, J. M., and Flores, F. (1984). Electron-metal-surface interaction potential with vacuum tunneling: Observation of the image force. *Phys. Rev. B* **30,** 4816–4818.

Binnig, G., Rohrer, H., Gerber, Ch., and Stoll, E. (1984a). Real-space observation of the reconstruction of Au(110). *Surface Sci.* **144,** 120–123.

Binnig, G., and Rohrer, H. (1985). The scanning tunneling microscope. *Scientific American,* August 1985, 3–8.

Binnig, G., Frank, K. H., Fuchs, H., Garcia, N., Reihl, B., Rohrer, H., Salvan, F., and Williams, A. R. (1985a). Tunneling spectroscopy and inverse photoemission: Image and field states. *Phys. Rev. Lett.* **55**, 991–994.

Binnig, G., and Rohrer, H. (1986). Scanning tunneling microscopy. *IBM J. Res. Develop.* **30**, 355–369.

Binnig, G., Fuchs, H., Gerber, Ch., Rohrer, H., Stoll, E., and Tosetti, E. (1986a). Energy-dependent state-density corrugation of a graphite surface as seen by scanning tunneling microscopy. *Europhys. Lett.* **1**, 31–36.

Binnig, G., and Rohrer, H. (1987), Scanning tunneling microscopy – from birth to adolescence. *Rev. Mod. Phys.* **56**, 615–625.

Binnig, G. K. (1987a). Atomic-force microscopy. *Physica Scripta,* **T19**, 53–54.

Binnig. G., Quate, C. F., and Gerber, Ch. (1986). Atomic force microscope. *Phys. Rev. Lett.* **56**, 930–933.

Binnig, G., and Smith, D. P. E. (1986). Single-tube three-dimensional scanner for scanning tunneling microscopy. *Rev. Sci. Instrum.* **57**, 1688–1689.

Boland, J. J. (1990). Structure of the H-saturated Si(111) surface. *Phys. Rev. Lett.* **65**, 3325–3328.

Boland, J. J and Villarrubia, J. S. (1990). Identification of the products from the reaction of chlorine with the silicon(111)-7 × 7 surface. *Science* **248**, 838–840.

Boland, J. J. (1991). The importance of structure and bonding in semiconductor surface chemistry: hydrogen on the Si(111)-7 × 7 surface. *Surf. Sci.* **244**, 1–14.

Boland, J. J. (1991a). Role of hydrogen desorption in the chemical vapor deposition of Si(100) epitaxial films using disilane. *Phys. Rev. B* **44**, 1383–1386.

Boland. J. J. (1991b). Evidence of pairing and its role in the recombinative desorption of hydrogen from the Si(100)-2 × 1 structure. *Phys. Rev. Lett.* **67**, 1539–1542.

Boland, J. J. (1991c). Erratum to "Evidence of pairing and its role in the recombinative desorption of hydrogen from the Si(100)-2 × 1 surface." *Phys. Rev. Lett.* **67**, 2201.

Boland, J. J. (1992). Hydrogen as a probe of semiconductor surface structure: The Ge(111)-$c(2 \times 8)$ structure. *Science,* **255**, 186-188.

Boland, J. J. (1992a). Role of bond-strain in the chemistry of hydrogen on the Si(100) surface. *Surf. Sci.* **261**, 17–28.

Brodie, I. (1978). The visibility of atomic objects in the field electron microscope. *Surf. Sci.* **70**, 186–196.

Brunner H., and van der Houwen, P. J. (1986). *The Numerical Solution of Volterra Equations,* North Holland, Amsterdam.

Burns, G., and Glazer, A. M. (1990). *Space Group for Solid State Scientists.* Academic Press, Boston.

Burrau, Ø. (1927). Berechnung des Energiewertes des Wasserstoffmolekel-ions (H_2^+) im Normalzustand. *Danske Videnskabernes Selskab,* **7**, No. 14, 1–18.

Burstein, E., and Lundquist, S. (1969). *Tunneling Phenomena in Solids,* Plenum, New York.

Cady, W. G. (1946). *Piezoelectricity.* McGraw-Hill, New York.

Calori, C., Combescot, R., Nozieres, P., and Saint-James, D. (1972). A direct calculation of the tunneling current: IV. Electron–phonon interaction effects, *Solid State Physics* **5**, 21–42.

Cabrera, N., and Goodman, F. O. (1972). Summation of pairwise potentials in gas–surface interaction calculations. *J. Chem. Phys.* **56**, 4899–4902.

Cardona, M., and Ley, L. (1978). *Photoemission in Solids I: General Principles,* Springer-Verlag, Berlin.

Carr, R. G. (1988). Finite element analysis of PZT tube scanner for scanning tunneling microscopy. *J. Microscopy* **152,** Part 2, 379-385.

Chambliss, D. D., and Wilson, R. J. (1991). Relaxed diffusion limited aggregation of Ag on Au(111) observed by scanning tunneling microscopy. *J. Vac. Sci. Technol. B* **9,** 928-932.

Chambliss, D. D., Wilson, R. J., and Chiang, S. (1991a). Ordered nucleation of Ni and Au islands on Au(111) studied by scanning tunneling microscopy. *J. Vac. Sci. Technol. B* **9,** 933-937.

Chambliss, D. D., Wilson, R. J., and Chiang, S. (1991b). Nucleation of ordered island arrays on Au(111) by surface-lattice dislocations. *Phys. Rev. Lett.* **66,** 1721-1724.

Chen, C. H., and Gewirth, A. (1992). AFM study of the structure of underpotentially deposited Ag and Hg on Au(111). *Ultramicroscopy* **42-44,** 437-444.

Chen, C. J. (1988). Theory of scanning tunneling spectroscopy. *J. Vac. Sci. Technol. A* **6,** 319-322.

Chen, C. J. (1990). Origin of atomic resolution on metals in scanning tunneling microscopy. *Phys. Rev. Lett.* **65,** 448-451.

Chen, C. J. (1990a). Tunneling matrix elements in three-dimensional space: The derivative rule and the sum rule. *Phys. Rev. B* **42,** 8841-8857.

Chen, C. J. (1991). Microscopic view of scanning tunneling microscopy. *J. Vac. Sci. Technol. A* **9,** 44-50.

Chen, C. J., and Hamers, R. J. (1991a). Role of atomic force in tunneling barrier measurements. *J. Vac. Sci. Technol. A* **9,** 230-233.

Chen, C. J. (1991b). Perturbation approach for quantum transmission. *Mod. Phys. Lett. B* **5,** 107-115.

Chen, C. J. (1991c). Attractive atomic force as a tunneling phenomenon. *J. Phys. Cond. Matter* **3,** 1227-1245.

Chen, C. J. (1992). Electromechanical deflections of piezoelectric tubes with quartered electrodes. *Appl. Phys. Lett.* **60,** 132-134.

Chen, C. J. (1992a). In situ testing and calibration of tube piezoelectric scanners. *Ultramicroscopy,* **42-44,** 1653-1658.

Chen, C. J. (1992b). In situ characterization of tip electronic structure in scanning tunneling microscopy. *Ultramicroscopy,* **42-44** 147-153.

Chen, C. J. (1992c). Effect of m≠0 tip states in scanning tunneling microscopy: the explanations of corrugation reversal. *Phys. Rev. Lett.* **69,** 1656-1659.

Cherry, W. L. Jr., and Adler, R. (1947). Piezoelectric effect in polycrystalline barium titanate. *Phys. Rev.* **72,** 981-982.

Chiang, S., Wilson, R. J., Gerber, Ch., and Hallmark, V. M. (1988). An ultrahigh vacuum scanning tunneling microscope with interchangeable sample and tips. *J. Vac. Sci. Technol. A* **6,** 386-389.

Chiang, S., Wilson, R. J., Mate, C. M., and Ohtani, H. (1988). Real space imaging of co-adsorbed CO and benzene molecules on Rh(111). *J. Microscopy,* **152** Pt. 2, 567-571.

Chung, M. S., Feuchtwang, T. E., and Cutler, P. H. (1987). Spherical tip model in the theory of the scanning tunneling microscope. *Surf. Sci.* **187,** 559-568.

Ciraci, S., Baratoff, A., and Batra, I. P. (1990). Tip-sample interaction effects in scanning-tunneling and atomic-force microscopy. *Phys. Rev. B* **41,** 2763-2775.

Ciraci, S., Baratoff, A., and Batra, I. P. (1990a). Site-dependent electronic effects, forces, and deformations in scanning tunneling microscopy of flat metal surfaces. *Phys. Rev. B* **42**, 7618–7621.

Cizek, J., Damburg, R., Graffi, S., Grecchi, V., Harrell, E. M., Harris, J., Nakai, S., Paldus, J. Propin, R. K., and Silverstone, H. J. (1986). 1/R expansion for H_2^+: Calculation of exponentially small terms and asymptotics. *Phys. Rev. A* **33**, 12–54.

Clementi, E., and Raimondi, D. L. (1963). Atomic screening constants from SCF functions. *J. Chem. Phys.* **38**, 2686–2689.

Clementi, E., Raimondi, D. L., and Reinhardt, W. P. (1967). Atomic screening constants from SCF functions. II. Atoms with 37 to 86 electrons. *J. Chem. Phys.* **47**, 1300–1307.

Clementi, E., and Roetti, C. (1974). Roothaan-Hartree-Fock atomic wavefunctions: basis functions and their coefficients for ground and certain excited states of neutral and ionized atoms, $Z \leq 54$. *Atomic Data and Nuclear Data Tables* **14**, 177–320.

Clemmer, C. R., and Beebe, T. P. (1991). Graphite: a mimic for DNA and other biomolecules in scanning tunneling microscopy studies. *Science* **251**, 640–642.

Coleman, R. V., Drake, B., Hansma, P. K., and Slough, G. (1985). Charge-density waves observed with a tunneling microscope. *Phys. Rev. Lett.* **55**, 394–397.

Coleman, R. V., Giambattista, B., Hansma, P. K., Johnson, A., McNairy, W. W., and Slough, C. G. (1988). Scanning tunneling microscopy of charge-density waves in transition metal chalcogenides. *Adv. Phys.* **37**, 559–644.

Coombs, J. H., and Pethica, J. B. (1986). Properties of vacuum tunneling currents: anomalous barrier heights. *IBM J. Res. Develop.* **30**, 455–459.

Coombs, J. H., Welland, M. E., and Pethica, J. B. (1988) Surface Science **198**, L353 (1988).

Colton, R. J., Baker, S. M., Driscoll, R. J., Youngquist, M. G., Baldeschwieler, J. D., and Kaiser, W. J. (1988). Imaging graphite in air by scanning tunneling microscopy: Role of the tip. *J. Vac. Sci. Technol. A* **6**, 349–353.

Crewe, A. V., Wall, J., and Langmore, J. (1970). Visibility of single atoms. *Science,* **168**, 1338–1340.

Curie, J., and Curie, P. (1880). Sur l'électricité polaire dans les cristaux hèmièdres à faces inclinées. *Comptes Rendus* **91**, 383–386.

Curie, J., and Curie P. (1882). Déformations électriques du quartz. *Comptes Rendus,* **95,** 914–917.

Curie, J. (1889). Quartz piézo-électrique. *Annales de Chimie et de Physique,* **17,** 392–401.

Curie, J. (1889a). Dilatation électrique du quartz. *Journal de Physique,* 2e série, **8,** 149–171.

Dalton, J. (1808). *A New System of Chemical Philosophy.* Reprinted by Philosophical Library, New York, 1964.

Damburg, R. J., and Propin, R. K. (1968). On asymptotic expansions of electronic terms of the molecular ion H_2^+. *J. Phys. B, Ser. 2,* **1,** 681–691.

Demuth, J. E., Hamers, R. J., Tromp, R. M., and Welland, M. E. (1986). A scanning tunneling microscope for surface science studies. *IBM J. Res. Develop.* **30**, 396–402.

Demuth, J. E., Hamers, R. J., Tromp, R. M., and Welland, M. E. (1986a). A simplified scanning tunneling microscope for surface science studies. *J. Vac. Sci. Technol. A* **4,** 1320–1323.

Demuth, J. E., Koehler, U., and Hamers, R. J. (1988). The STM learning curve and where it may take us. *J. Microscopy.* **151,** 289–302.

Demuth, J. E., van Loenen, E. J., Tromp, R. M., and Hamers, R. J. (1988a). Local electronic structure and surface geometry of Ag on Si(111). *J. Vac. Sci. Technol.* **B6,** 18–25.

Dionne, N. J., and Rhodin, T.N. (1974). *d*-band contributions to the energy distribution of field-emitted electrons from platinum-group metals. *Phys. Rev. Lett.* **32,** 1311–1314.

DiStefano, J. J., Stubberud, A. R., and Willaims, I. J. (1967). *Feedback and Control Systems,* McGraw–Hill, New York.

Doyen, G., Kötter, E., Vigneron, J. P., and Scheffler, M. (1990). Theory of scanning tunneling microscopy. *Appl. Phys. A* **51,** 281–288.

Drake, B., Sonnenfeld, R., Schneir, J., and Hansma, P. K. (1986). Tunneling microscope for operating in air and fluids. *Rev. Sci. Instrum.* **57,** 441–445.

Drakova, D., Doyen. G., and Trentini, F. V. (1985). Self-consistent calculations of rare-gas–transition-metal interaction potentials. *Phys. Rev. B* **32,** 6399–6423.

Drake, B., Sonnenfeld, R., Schneir, J., and Hansma, P. K. (1987). Scanning tunneling microscopy of processes at liquid–solid interfaces. *Surf. Sci.* **181,** 92–97.

Drechsler, M. (1978). Erwin Müller and the early development of field emission microscopy. *Surf. Sci.* **70,** 1–18.

Drube, W., Straub, D., Himpsel, F. J., Soukiassian, P., Fu, C. L., and Freeman, A. J. (1986). Unoccupied surface states on W(001) and Mo(001) by inverse photoemission. *Phys. Rev.* **B34,** 8989–8992.

Duke, C. B. (1969) *Tunneling in Solids,* Academic Press, New York.

Dürig, U., Gimzewski, J. K., and Pohl, D. W. (1986), Experimental observation of forces acting during scanning tunneling microscopy. *Phys. Rev. Lett.* **57,** 2403–2406.

Dürig, U., Züger, O., and Pohl, D. W. (1988). Force sensing in scanning tunneling microscopy: Observation of adhesion forces on clean metal surfaces. *J. Microscopy.* **152,** Part 1, 259–267.

Dürig, U., Züger, O., and Pohl, D. W. (1990). Observation of metal adhesion using scanning tunneling microscopy. *Phys. Rev. Lett.* **65,** 349–352.

Eigler, D. M., Weiss, P. S., Schweizer, E. K., and Lang, N. D. (1991). Imaging Xe with a low-temperature scanning tunneling microscope. *Phys. Rev. Lett.* **66,** 1189–1192.

Erlandsson, R., McClelland, G. M., Mate, C. M., and Chiang, S. (1988). Atomic force microscopy using optical interferometry. *J. Vac. Sci. Technol. A* **6,** 266–270.

Esbjerg, N., and Nørskov, J. K. (1980). Dependence of the He-scattering potential at surfaces on the surface-electron-density profile. *Phys. Rev. Lett.* **45,** 807–810.

Escapa, L., and Garcia, N. (1990). On the quantized conductance of small contacts. In Behm, R. J., Garcia, N., and Rohrer, H., *Scanning Tunneling Microscopy and Related Methods,* Kluwer, Dordrecht, 143–156.

Estermann, I., and Stern, O. (1930). Beugung von Molekülarstrahlen. *Z. Physik* **61,** 95–125.

Feenstra, R. M., and Fein, A. P. (1985). Surface morphology of GaAs(110) by scanning tunneling microscopy. *Phys. Rev. B* **32,** 1394–1396.

Feenstra, R. M., Thompson, W. A., and Fein, A. P. (1986). Real-space observation of π-bonded chains and surface disorder on Si(100)2 × 1. *Phys. Rev. Lett.* **56**, 608–611.

Feenstra, R. M., Thompson, W. A., and Fein, A. P. (1986a). Scanning tunneling microscopy studies of Si(100)-2 × 1 surfaces. *J. Vac. Sci. Technol.* A **4**, 1315–1319.

Feenstra, R. M., and Stroscio, J. A. (1987). Real-space determination of surface structure by scanning tunneling microscopy. *Physica Scripta.* **T19**, 55–60.

Feenstra, R. M., Stroscio, J. A., and Fein, A. P. (1987a). Tunneling spectroscopy of the Si(111)2 × 1 surface. *Surface Sci.* **181**, 295–306.

Feenstra, R. M., Stroscio, J. A., Tersoff, J., and Fein, A. P. (1987b). Atom-selective imaging of the GaAs(110) surface. *Phys. Rev. Lett.* **58**, 1192–1195.

Feenstra, R. M., and Martensson, P. (1988). Fermi level pinning at the Sb/GaAs(110) surface studied by scanning tunneling spectroscopy. *Phys. Rev. Lett.* **61**, 447–450.

Feenstra, R. M. (1991). Band gap pf the Ge(111)2 × 1, and Si(111)2 × 1 surfaces by scanning tunneling microscopy. *Phys. Rev. B* **44**, 13791–13794.

Feenstra, R. M., Slavin, A. J., Held, G. A., and Lutz, M. A. (1991). Surface diffusion and phase transition on the Ge(111) surface studied by scanning tunneling microscopy. *Phys. Rev. Lett.* **66**, 3257–3260.

Feibelman, P. J. (1990). Theory of adsorbate interactions. *Ann. Rev. Phys. Chem.* **40**, 261–290.

Feuchtwang, T. E., and Cutler, P. (1987). Tunneling and scanning tunneling microscopy: a critical review. *Phys. Scr.* **35**, 132–140.

Feuerbacher, B., Fitton, B. (1972). Photoemission from surface states on tungsten. *Phys. Rev. Lett.* **29**, 786–789.

Feuerbacher, B., Fitton, B., and Willis, R. F. (1978). *Photoemission and the electronic properties of surfaces,* John Wiley and Sons, Chichester.

Feuchtwang, T. E. (1979). Tunneling theory without the transfer Hamiltonian formalism: V. A theory of inelastic electron tunneling spectroscopy. *Phys. Rev. B* **20**, 430–455, and references therein.

Feynman, R. P., Leighton, R. B., and Sands, M. (1965). *The Feynman Lectures on Physics III, Quantum Mechanics,* (Addison–Wesley, Reading, MA).

Fink, H. W. (1986). Mono-atomic tips for scanning tunneling microscopy. *IBM J. Res. Develop.* **30**, 460–465.

Flores, F., Martin-Rodeno, A., Goldberg, E. C., and Duran, J. C. (1988). Molecular orbital theory and tunneling currents. *Nuovo Cimento* **10 D**, 303–311.

Fowler, R. H., and Nordheim, L. (1928). Electron emission in intense electric fields. *Proc. R. Soc. London* **A119**, 173–181.

Frolov, K. V., and Furman, F. A. (1990). *Applied Theory of Vibration Isolation Systems,* Hemisphere, New York.

Fu, C. L., and Ho, K. M. (1989). External-charge-induced surface reconstruction on Ag(110). *Phys. Rev. Lett.* **63**, 1617–1620.

Furuya, N., and Koide, S. (1989). Hydrogen adsorption on platinum single-crystal surfaces. *Surf. Sci.* **220**, 18–28.

Furuya, N., and Koide, S. (1990). Hydrogen adsorption on iridium single-crystal surfaces. *Surf. Sci.* **226**, 221–225.

Gadzuk, J. W., and Plummer, E. W. (1973). Field emission energy distribution. *Rev. Mod. Phys.* **45**, 487–548.

Garcia, N., Ocal, C., and Floers, F. (1983). Model theory for scanning tunneling microscopy: Application to Au(110)-2 × 1. *Phys. Rev. Lett.* **50,** 2002–2005.

Garcia, N. (1986). Theory of scanning tunneling microscopy and spectroscopy: Resolution, image and field states, and thin oxide layers. *IBM J. Res. Develop.* **30,** 533–542.

Gay, J. G., Smith, J. R., and Arlinghaus, F. J. (1977). Self-consistent calculation of work function, charge densities, and local densities of states for Cu(100). *Phys. Rev. Lett.* **38,** 561–564.

Gay, J. G., Smith, J. R., and Arlinghaus, F. J. (1979). Self-consistent calculation of work function, charge densities, and local densities of states for Cu(100). *Phys. Rev. B* **42,** 332–335.

Gaylord, R. H., Jeong, K. H., and Kevan, S. D. (1989). Experimental Fermi surfaces of clean and hydrogen-covered W(110). *Phys. Rev. Lett.* **62,** 2036–2039.

Gerber, Ch., Binnig, G., Fuchs, H., Marti, O., and Rohrer, H. (1986). Scanning tunneling microscope combined with a scanning electron microscope. *Rev. Sci. Instrum.* **57,** 221–224.

Germano, C. P. (1959). A study of a two-channel cylindrical PZT ceramic transducer for use in stereo phonograph cartridges. *IRE Transactions on Radio,* July–August 1959, 96–100.

Giaever, I. (1960). Energy gap in superconductors measured by electron tunneling. *Phys. Rev. Lett.* **5,** 147–148.

Giaever, I. (1960a). Electron tunneling between two superconductors. *Phys. Rev. Lett.* **5,** 464–466.

Giaever, I., and Megerle, K. (1961). Study of superconductors by electron tunneling. *Phys. Rev.* **122,** 1101–1111.

Giambattista, B., Slough, C. G., McNairy, W. W., and Coleman, R. V. (1990). Scanning tunneling microscopy of atoms and charge-density waves in $1T$-TaS$_2$, $1T$-TaSe$_2$, and $1T$-VSe$_2$. *Phys. Rev. B* **41,** 10082–10103.

Gimzewski, J. K., Möller, R., Pohl, D. W., and Schlittler, R. R. (1987). Transition from tunneling to point contact investigated by scanning tunneling microscopy and spectroscopy. *Surface Sci.* **189/190,** 15–23.

Glazman, L. I., Lesovik, G. B., Khmelnitskii, D. E., and Shekhter, R. I. (1988). Reflectionless quantum transport and fundamental ballistic-resistance steps in microscopic constrictions. *JETP Lett.* **48,** 238–241.

Golovchenko, J. A. (1986). The tunneling microscope: A new look at the atomic world. *Science* **232,** 48–53.

Good, R. H., and Müller, E. W. (1956). Field emission, in *Handbuch der Physik.* Springer-Verlag, Berlin, **21,** 181–250.

Goodman, F. O. (1976). Summation of the Morse pairwise potential in gas–surface interaction calculations. *J. Chem. Phys.* **65,** 1561–1564.

Goodman, F. O., and Wachman, H. Y. (1976). *Dynamics of Gas-Surface Scattering,* Academic Press, New York.

Goodwin, E. T. (1939). Electronic states at the surfaces of crystals. *Proc. Camb. Phil. Sco.* **35,** 205–231.

Gradsteyn, I. S., and Ryzhik, I. M. (1980). *Table of Integrals, Series and Products.* Academic Press, New York.

Güntherodt, H.-J., and Wiesendanger, R. (1992). *Scanning Tunneling Microscopy,* Vols. I through III, Springer Verlag.

Häberle, W., Hörber, J. K. H., Ohnesorge, F., Smith, D. P. E., and Binnig, G. (1992). In situ investigations of single living sells infected by viruses. *Ultramicroscopy*, **42–44**, 1161–1167.

Haefke, H., Meyer, E., Howald, L., Schwarz, U., Gerth, G., and Krohn, M. (1992). Atomic surface and lattice structures of AgBr thin films. *Ultramicroscopy* **42–44**, 290-297.

Hahn, T. (1987). *International Tables for Crystallography,* Volume A, Reidel, Dordrecht.

Hallmark, V. M., Chiang, S., Rabolt, J. F., Swalen, J. D., and Wilson, R. J. (1987). Observations of atomic corrugations on Au(111) by scanning tunneling microscopy. *Phys. Rev. Lett.* **59**, 2879–2882.

Hamann, D. R. (1981). Surface charge densities and atom diffraction. *Phys. Rev. Lett.* **46**, 1227–1230.

Hamers, R. J. (1989). Effects of coverage on the geometry and electronic structure of Al overlayers on Si(111). *Phys. Rev. B* **40**, 1657–1671.

Hamers, R. J. (1989a). Atomic-resolution surface spectroscopy with the scanning tunneling microscope. *Ann. Rev. Phys. Chem.* **40**, 531–559.

Hamers, R. J., Tromp, R. M., and Demuth, J. E. (1986). Surface electronic structure of Si(111)7 × 7 resolved in real space. *Phys. Rev. Lett.* **56**, 1972–1975.

Hamers, R. J., Tromp, R. M., and Demuth, J. E. (1987). Electronic and geometric structure of Si(111)-7×7 and Si(001) surfaces. *Surf. Sci.* **181**, 346-355.

Hamers, R. J., Avouris, Ph., and Bozso, F. (1988). A scanning tunneling microscope study of the reaction of Si(001)2 × 1 with NH_3. *J. Vac. Sci. Technol.* **A6**, 508–511.

Hamers, R. J., and Demuth, J. E. (1988b). Electronic structure of localized Si dangling bond defects by tunneling spectroscopy. *Phys. Rev. Lett.* **60**, 2527–2530.

Hamers, R. J., and Demuth, J. E. (1988a). Atomic structure and bonding of Si(111)-($\sqrt{3} \times \sqrt{3}$) Al. *J. Vac. Sci. Technol.* **A6**, 512–516.

Hamers, R. J., Koehler, U. K., and Demuth, R. J. (1990). Epitaxial growth of silicon on Si(001) by scanning tunneling microscopy, *J. Vac. Sci. Technol.* **A8**, 195–200.

Haneman, D., and Haycock, R. (1982). Estimation of surface charge densities for low-energy atom diffraction. *J. Vac. Sci. Technol.* **21**, 330–332.

Hansma, P. (1982). *Tunneling Spectroscopy: Capabilities, Applications, and New Techniques.* Plenum, New York.

Hansma, P. K. (1986). Squeezable tunneling junctions. *IBM J. Res. Develop.* **30**, 370–373.

Hansma, P. K., Elings, V. B., Marti, O, and Bracker, C. E. (1988). Scanning tunneling microscopy and atomic force microscopy: Application to biology and technology. *Science* **242**, 209–242.

Hansma, P. K., and Tersoff, J. (1987). Scanning tunneling microscopy. *J. Appl. Phys.* **61**, R1–R23.

Harris, J., and Liebsch, A. (1982). Interaction of helium with a metal surface: Determination of corrugation profile of Cu(110). *Phys. Rev. Lett.* **49**, 341–344.

Harris, J., and Liebsch, A. (1982a). Interaction of helium with a metal surface. *J. Phys. C: Solid State Phys.* **15**, 2275–2291.

Harten, U., Lahee, A. M., Toennies, J. P., and Wöll, Ch. (1985). Observation of a soliton reconstruction of Au(111) by high-resolution helium-atom diffraction. *Phys. Rev. Lett.* **54**, 2619–2622.

Hasegawa, Y., and Avouris, Ph. (1992). Manipulation of the Au(111) surface with the STM. *Science* **258,** 1763–1765.

Heine, V. (1963). On the general theory of surface states and scattering of electrons in solids. *Proc. Phys. Soc.* **81,** 300–310.

Heine, V. (1980). Electronic structure from the point of view of the local atomic environment. *Solid State Physics* **35,** 1–127.

Heisenberg, W. (1926). Mehrkörperproblem und Resonanz in der Quantunmechanik. *Z. Phys.* **38,** 411–426.

Heizelmann, H., Meyer, E., Grütter, P., Hidber, H.-R., Rosenthaler, L., and Güntherodt, H.-J. (1988). Atomic force microscopy: General aspects and application to insulators. *J. Vac. Sci. Technol. A* **6,** 275–278.

Herbert, J. M. (1982). *Ferroelectric Transducers and Sensors.* Gorden and Beach, New York.

Herman, F., and Skillman, S. (1963). *Atomic Structure Calculations.* Prentice–Hall, Englewood Cliffs, New Jersey.

Herring, C. (1962). Critique of the Heitler–London method of calculating spin couplings at large distances. *Rev. Mod. Phys.* **34,** 631–645.

Herring, C. (1992). Recollections from the early years of solid-state physics. *Physics Today,* April 1992, 26–33.

Hess, H. F., Robinson, R. B., Dynes, R. C., Valles, J. M. Jr., and Waszczak, J. V. (1989). Scanning tunneling microscope observation of the Abrikosov flux lattice and the density of states near and inside a fluxoid. *Phys. Rev. Lett.* **62,** 214–216.

Hess, H. F., Robinson, R. B., Dynes, R. C., Valles, J. M., Jr., and Waszczak, J. V. (1990). Spectroscopic and spatial characterization of superconducting vortex core states with a scanning tunneling microscope. *J. Vac. Sci. Technol. A* **8,** 450–454.

Hess, H. F., Robinson, R. B., and Waszczak, J. V. (1990a). Vortex-core structure observed with a scanning tunneling microscope *Phys. Rev. Lett.* **64,** 2711–2714.

Hess, H. F., Robinson, R. B., and Waszczak, J. V. (1991). STM spectroscopy of vortex cores and the flux lattice. *Physica B* **169,** 422–431.

Higashi, G. S., Becker, R. S., Chabal, Y. J., and Becker, A. J. (1991). Comparison of Si(111) surfaces prepared using aqueous solutions of NH_4 F versus HF. *Appl. Phys. Lett.* **58,** 1656–1658.

Himpsel, F. J. (1983). Angle-resolved measurements of the photoemission of electrons in the study of solids. *Adv. Phys.* **32,** 1–51.

Hoffmann, R. (1988). A chemical and theoretical way to look at bonding on surfaces. *Rev. Mod. Phys.* **60,** 601–628.

Hohenberg, P., and Kohn, W. (1964). Inhomogeneous electron gas. *Phys. Rev.* **136,** B864–871.

Holstein, T. (1952). Mobilities of positive ions in their parent gases. *J. Phys. Chem.* **56,** 832–836.

Holstein, T. (1955). Charge-exchange interaction between ions and parent atoms, Westinghouse Research Report 60–94698–3–R9. Copies of this research report can be obtained from Westinghouse Science and Technology Center, Pittsburgh, Pennsylvania.

Horowitz, P., and Hill, W. (1982). *The Art of Electronics.* Cambridge University Press, Cambridge.

Ibe, J. P., Bey, P. P. Jr., Brandow, S. L., Brizzolara, R. A., Burnham, N. A., DiLella, D. P., Lee, K. P., Marrian, C. R. K., and Colton, R. J. (1990). On the electrochemical etching of tips for scanning tunneling microscopy. *J. Vac. Sci. Tachnol. A* **8**, 3570–3575.

Ihm, J., Cohen, M. L., and Chadi, D. J. (1980). The electronic structure of Si(100) 2 × 1. *Phys. Rev. B* **21**, 4592–4602.

Inglefield, J. E. (1982). Surface electronic structure. *Rep. Prog. Phys.* **45**, 223–284.

Ivanchenko, Yu. M., and Riseborough, P. S. (1991). Quantum uncertainty effect in tunneling. *Phys. Rev. Lett.* **67**, 338–341.

Iwasaki, F., Tomitori, M., and Nishikawa, O. (1992). *Ultramicroscopy* **42–44** 902–909.

Jaffe, B., Cook, W. R., and Jaffe, H. (1971). *Piezoelectric Ceramics,* Academic Press, London.

Jaffe, B., Roth, R. S., and Marzullo, S. (1954). Piezoelectric properties of lead zirconate–lead titanate solid-solution ceramics. *J. Appl. Phys.* **25**, 809–810.

Jaklevic, R. C., and Lambe, J. (1966). Molecular vibration spectra by electron tunneling. *Phys. Rev. Lett.* **17**, 1139–1140.

Jaklevic, R. C., and Elie, L. (1988). Scanning tunneling microscopy observation of surface diffusion on an atomic scale: Au on Au(111). *Phys. Rev. Lett.* **60**, 120–123.

Kais, S., Morgan, J. D., and Herschbach, D. R. (1991). Electronic tunneling and exchange energy in the D-dimensional hydrogen-molecule ion. *J. Chem. Phys.* **95**, 9028–9041.

Keithley, J. F., Yeager, J. R., and Erdman, R. J. (1984). *Low Level Measurements,* Keithley Instruments, Inc., Cleveland, Ohio.

Kerker, G. P., Ho, K. M., and Cohen, M. (1978). Mo(001) surface: A self-consistent calculation of the electronic structure. *Phys. Rev. Lett.* **24**, 1593–1596.

Kihlborg, L. (1979). *Nobel Symposium 47: Direct Imaging of Atoms in Crystals and Molecules.* The Royal Swedish Academy of Sciences, Stockholm.

Kirk, M. D., Albrecht, T. R., and Quate, C. F. (1988). Low temperature atomic force microscopy. *Rev. Sci. Instrum.* **59**, 833–835.

Kirtley J. (1982). The interaction of tunneling electrons with molecular vibrations, in Hansma. P. K., *Tunneling Spectroscopy: Capabilities, Applications, and New Techniques,* Plenum, New York.

Kittel, C. (1963). *Quantum Theory of Solids,* John Wiley and Sons, New York.

Kittel, C. (1986). *Introduction to Solid State Physics,* sixth edition. John Wiley and Sons, New York.

Klitsner, T., Becker, R. S., and Vickers, J. S. (1990). Observation of the effect of tip electronic states on tunneling spectra acquired with the scanning tunneling microscope. *Phys. Rev.* **B41**, 3837–3840.

Koehler, U. K., Demuth, J. E., and Hamers, R. J. (1988). Surface reconstruction and the nucleation of palladium silicide on Si(111). *Phys. Rev. Lett.* **60**, 2499–2502.

Kohn, W., and Sham, L. J. (1965). Self-consistent equations including exchange and correlation effects. *Phys. Rev.* **140**, A1133–1138.

Kohn, W., and Vashishta, P. (1983). General density functional theory. In *Theory of the Inhomogeneous Electron Gas,* edited by S. Lundqvist and N. H. March, Plenum, New York, pp. 79–147.

Kortan, A. R., Becker, R. S., Theil, F. A., and Chen, H. S. (1990). Real-space atomic structure of a two-dimensional decagonal quasicrystal. *Phys. Rev. Lett.* **64**, 200–203.

Kuk, Y., and Silverman, P. J. (1986). Role of tip structure in scanning tunneling microscopy. *Appl. Phys. Lett.* **48**, 1597–1599.

Kuk, Y., Silverman, P. J., and Nguyen, H. Q. (1988). Study of metal surfaces by scanning tunneling microscopy with field ion microscopy. *J. Vac. Sci. Technol. A* **6**, 524–527.

Kuk, Y., and Silverman, P. J. (1989). Scanning tunneling microscope instrumentation. *Rev. Sci. Instrum.* **60**, 165–180.

Lambe, J., and Jaklevic, R. C. (1968). Molecular vibration spectra by inelastic electron tunneling. *Phys. Rev.* **165**, 821–832.

Landau, L. D., and Lifshitz, L. M. (1977). *Quantum Mechanics,* third edition, Pergamon Press, Oxford. The treatment of the hydrogen molecular ion is presented as a problem on page 312.

Landau, L. D., and Lifshitz, E. M. (1986). *Theory of Elasticity.* Pergamon Press, Oxford.

Lang, N. D., and Kohn, W. (1970). Theory of metal surfaces: charge density and surface energy. *Phys. Rev. B* **1**, 4555–4568.

Lang, N. D. (1973). The density-functional formalism and the electronic structure of metal surfaces. In *Solid State Physics,* edited by H. Ehrenreich, F. Seitz, and D. Turnbull, Vol. 28, Academic, New York.

Lang, N. D. (1985). Vacuum tunneling current from an adsorbed atom. *Phys. Rev. Lett.* **55**, 230–233.

Lang, N. D. (1986). Theory of single atom imaging in scanning tunneling microscopy. *Phys. Rev. Lett.* **56**, 1164–1167.

Lang, N. D. (1987). Apparent size of an atom in the scanning tunneling microscope as a function of bias. *Phys. Rev. Lett.* **58**, 45–48.

Lang, N. D. (1988). Apparent barrier height in scanning tunneling microscopy. *Phys. Rev. B* **37**, 10395–10398.

Lapujoulade, J., Salanon, B., and Gorse, D. (1984). Surface structure analysis by atomic beam diffraction. In *The Structure of Surfaces,* edited by Van Hove, M. A., and Tong, S. Y., Springer-Verlag, Berlin.

Lawunmi, D., and Payne, M. C. (1990). Theoretical investigation of the scanning tunneling microscope image of graphite. *J. Phys.: Condens. Matter.* **2**, 3811–3821.

Lemke, H., Göddenhenrich, T., Bochem, H. P., Hartman, U., and Heiden, C. (1990). Improved microtips for scanning tunneling microscopy. *Rev. Sci. Instrum.* **61**, 2538–2541.

Ley, L., and Cardona, M. (1979). *Photoemission in Solids II: Case Studies.* Springer-Verlag, Berlin.

Liebsch, A., Harris, J., and Weinert, M. (1984). Interaction of helium with a graphite surface. *Surf. Sci.* **145**, 207–222.

Lieske, N. P. (1984). The electronic structure of semiconductor surfaces. *J. Phys. Chem. Solids* **45**, 821–870.

Lifshitz, E. M., and L. P. Pitaevskii, L. P. (1980). *Statistical Physics II, Theory of the Condensed State,* Pergamon Press, Oxford, pp. 331–347.

Lipari, N. O. (1987). STM applications for semiconductor materials and devices. *Surf. Sci.* **181**, 285–294.

Lippel, P. H., Wilson, R. J., Miller, M. D., Wöll, Ch., and Chiang, S. (1989). High-resolution imaging of copper-phthalocyanine by scanning tunneling microscopy. *Phys. Rev. Lett.* **62**, 171–174.

Lippmann, G. (1881). Principe de la conservation de l'électricité, ou second principe de la théorie des phénomènes électriques. *J. de Phys.* **10**, 381–394.

Liu, H.-Y., Fan, F.-R., Lin. C. W., and Bard, A. J. (1986). Scanning electrochemical and tunneling ultramicroelectrode microscope for high-resolution examination of electrode surfaces in solution. *J. Am. Chem. Soc.* **108**, 3838–3839.

Loucks, T. L. (1965). Relativistic energy bands for tungsten. *Phys. Rev. Lett.* **14**, 693–694.

Lyding, J. W., Skala, S., Hubacek, J. S., Brockenbrough, R., and Gammie, G. (1988). Variable-temperature scanning tunneling microscope. *Rev. Sci. Instrum.* **59**, 1897–1902.

Lyo, I. W., and Avouris, Ph. (1989). Negative differential resistance on the atomic scale: Implications for atomic scale devices. *Science* **245**, 1369–1371.

Lyo, I. W., and Avouris, Ph. (1990). Atomic scale processes induced by the scanning tunneling microscope. *J. Chem. Phys.* **93**, 4479–4480.

Lyo, I. W., and Avouris, Ph. (1991). Field induced nanometer to atomic scale manipulation of silicon surfaces with the STM. *Science* **253**, 173–176.

Magnussen, O. M., Hotlos, J., Nichlos, R. J., Kolb, D. M., and Behm, R. J. (1990). Atomic structure of Cu adlayers on Au(100) and Au(111) electrodes observed by *in situ* scanning tunneling microscopy. *Phys. Rev. Lett.* **64**, 2929–2932.

Mamin, H. J., Ganz, E., Abraham, D. W., Thompson, R. E., and Clarke, J. (1986). Contamination-mediated deformation of graphite by the scanning tunneling microscope. *Phys. Rev. B* **34**, 9015–9018.

Mamin, H. J., Guenter, P. H., and Ruger, D. (1990). Atomic emission from a gold scanning tunneling microscope tip. *Phys. Rev. Lett.* **65**, 2418–2421.

Manne, S. Butt, H. J. Gould, S. A. C. and Hansma, P. K. (1990). Imaging metal atoms in air and water using the atomic force microscope. *Appl. Phys. Lett.* **56**, 1758–1759.

Manne, S., Massie, J., Elings, V. B., Hansma, P. K., and Gewirth, A. A. (1991). Electrochemistry on a gold surface observed with the atomic force microscope. *J. Vac. Sci. Technol. B* **9**, 950–954.

Manne, S., Hansma, P. K., Massie, J., Elings, V. B., and Gewirth, A. A. (1991a). Atomic-resolution electrochemistry with the atomic force microscope – Copper deposited on gold. *Science,* **251**, 183–186.

Mårtensson, P., and Feenstra, R. M. (1989). Geometric and electronic structure of antimony on the GaAs(110) surface studied by scanning tunneling microscopy. *Phys. Rev. B* **39**, 7744–7753.

Martin, Y., Williams, C. C., and Wickramasinghe, H. K. (1988). Scanning tip microscopies. *Scanning Microscopy* **2**, 3–12.

Mate, C. M., Erlandsson, R., McClelland, G. M., and Chiang, S. (1989). Direct measurement of forces during scanning tunneling microscope imaging of graphite. *Surface Science* **208**, 473–486.

Mattheiss, L. F., and Hamann, D. (1984). Electronic structure of the tungsten (001) surface. *Phys. Rev. B* **29**, 5372–5381.

McGonigal, G. C., Bernhardt, R. H., and Thomson, D. J. (1990). Imaging alkane layers at the liquid/graphite interface with the scanning tunneling microscope. *Appl. Phys. Lett.* **57**, 28–30.

McLean, A. D., and McLean, R. S. (1981). Roothaan-Hartree-Fock atomic wavefunctions. Slater basis set expansions for A=55–92. *Atomic Data and Nuclear Data Tables* **26**, 197–401.

Mednick, K., and Kleinman, L. (1980). Self-consistent Al(111) film calculations. *Phys. Rev. B* **22**, 5768–5773.

Melmed, A. J. (1991). The art and science and other aspects of making sharp tips. *J. Vac. Sci. Technol. B* **9**, 601–608.

Meyer, E., and Frommer, J. (1991). Forcing surface issues. *Physcis World,* April 1991, 46–49.

Meyer, E., Heinzelmann, H., Brodbeck, D., Overney, G., Overney, R., Howald, L., Hug, H., Jung, T., Hidber, H. R., and Guntherodt, H. J. (1991). Atomic resolution on the surface of LiF(100) by atomic force microscopy. *J. Vac. Sci. Technol. B* **9**, 1329–1332.

Meyer, G., and Amer, N. M. (1988). Novel optical approach to atomic force microscopy. *Appl. Phys. Lett.* **53**, 1045–1047.

Meyer, G., and Amer, N. M. (1990). Optical-beam-deflection atomic force microscopy: The NaCl(001) surface. *Appl. Phys. Lett.* **56**, 2100–2101.

Mo, Y. W., Kleiner, J., Webb, M. B., and Lagally, M. G. (1991). Activation energy for surface diffusion of Si on Si(001): A scanning tunneling microscopy study. *Phys. Rev. Lett.* **66**, 1998–2001.

Mo, Y. W. (1992). Precursor states in the adsorption of Sb_4 on Si(001). *Phys. Rev. Lett.* **69**, 3643-3646.

Modinos, A. (1984). *Field, Thermionic, and Secondary Electron Emission Spectroscopy.* Plenum, New York.

Modinos, A., and Nicolaou, N. (1976). Surface density of states and field emission. *Phys. Rev. B* **13**, 1536–1547.

Morse, P. M., and Feshbach, H. (1953). *Methods of Theoretical Physics.* McGraw–Hill, New York.

Moruzzi, V. L., Janak, J. F., and Williams, A. R. (1978), *Calculated Electronic Properties of Metals,* Pergamon, New York.

Müller, E. W. (1964). The effect of polarization, field stress, and gas impact on the topography of field evaporated surfaces. *Sur. Sci.* **2**, 484–494.

Müller, E. W., and Tsong, T. T. (1969). *Field-Ion Microscopy.* American Elsevier, New York.

Muralt, P., Meier, H., Pohl, D. W., and Salemink, W. M. (1987). Scanning tunneling microscopy and potentiometry on a semiconductor heterojunction. *Appl. Phys. Lett.* **50**, 1352–1355.

Muralt, P., Pohl, D. W., and Denk, W. (1986). Wide-range, low-operating-voltage, bimorph STM: Application as potentiometer. *IBM J. Res. Develop.* **30**, 443–450.

Nagahara, L. A., Thundat, T., and Lindsay, S. M. (1989). Preparation and characterization of STM tips for electrochemical studies. *Rev. Sci. Instrum.* **60**, 3128–3130.

Nagaya, K. (1984). On a magnetic damper consisting of a circular magnetic flux and a conductor of arbitrary shape. Part II: Applications and numerical results. *Transactions of the ASME: J. Dynam. Syst. Meas. Control.* **106**, 52–55.

Nagaya, K., and Kojima, H. (1984). On a magnetic damper consisting of a circular magnetic flux and a conductor of arbitrary shape. Part I: Derivation of the damping coefficients. *Transactions of the ASME: J. Dynam. Syst. Meas. Control.* **106**, 46–51.

Neddermeyer, H., and Drechsler, M. (1988). Electric field-induced changes of W(110) and W(111) tips. *J. Microscopy* **151–2**, 459–466.

Neddermeyer, H., and Tosch, S. (1988). Scanning tunneling spectroscopy on Si. *Ultramicroscopy* **25**, 135–148.

Nishikawa, O., Koyama, H., Tomitori, M., and Iwawaki, F. (1991). Tunneling characteristics of silicon covered molybdenum tip apex. *J. Vac. Sci. Technol. B* **9**, 789–793.

Nordheim, L. W. (1928). Effect of the image force on the emission and reflection of electrons by metals. *Proc. Roy. Soc. Lond., Ser. A* **121**, 626–639.

Northrop, J. (1986). Origin of surface states on Si(111)-7 × 7. *Phys. Rev. Lett.* **57**, 154–157.

Ohnishi, S., and Tsukada, M. (1989). Molecular orbital theory for scanning tunneling microscopy. *Solid State Commun.* **71**, 391–394.

Ohtani, H., Wilson, R. J., Chiang, S., and Mate, C. M. (1988). Scanning tunneling microscopy observations of benzene molecules on the Rh(111)-(3 × 3)(C_6 H_6+2CO) surface. *Phys. Rev. Lett.* **60**, 2398–2401.

Okano, M., Kajimura, K., Wakiyama, S., Sakai, F., Mizutani, W., and Ono, M. (1987). Vibration isolation for scanning tunneling microscopy. *J. Vac. Sci. Technol. A* **5**, 3313–3320.

Oppenheimer, J. R. (1928). Three notes on the quantum theory of aperiodic effects. *Phys. Rev.* **13**, 66–81.

Ott, H. W. (1976). *Noise Reduction Techniques in Electronic Systems.* John Wiley and Sons, New York.

Pandey, K. C. (1981). New π-bonded chain model for Si(111)-(2X1) surface. *Phys. Rev. Lett.* **47**, 1913–1917.

Pandey, K. C. (1982). Reconstruction of semiconductor surfaces: Buckling, ionicity, and π-bonded chains. *Phys. Rev. Lett.* **49**, 223–226.

Papaconstantopoulos, D. A. (1986). *Handbook of the Band Structures of Elemental Solids,* Plenum, New York.

Park, S. I., and Quate, C. F. (1986). Tunneling microscopy of graphite in air. *Appl. Phys. Lett.* **48**, 112–114.

Park, S. I., and Quate, C. F. (1987). Theories of the feedback and vibration isolation systems for the scanning tunneling microscope. *Rev. Sci. Instrum.* **58**, 2004–2009.

Pauli, W., Jr. (1922). Über das Modell des Wasserstoffmolekülions. *Ann. d. Phys.* **68**, 177–240.

Pauling, L. (1977). *The Nature of the Chemical Bond,* third edition. Cornell University Press, Ithaca.

Pauling, L., and Wilson, E. B. (1935). *Introduction to Quantum Mechanics,* McGraw-Hill, New York.

Payne, M. C., and Inkson, J. C. (1985). Measurement of work functions by tunneling and the effect of the image potential. *Surface Science* **159**, 485–495.

Pelz, J. P. (1991). Tip-related electronic artifacts in scanning tunneling spectroscopy. *Phys. Rev.* **B43**, 6746–6749.

Pelz, J. P., and Koch, R. H. (1991). Successive oxidation stages and annealing behavior of the Si(111)-7 × 7 surface observed with scanning tunneling microscopy and spectroscopy. *J. Vac. Sci. Technol. B* **7**, 775–778.

Penn, D. R., and Plummer, E. W. (1974). Field emission as a probe of the surface density of states. *Phys. Rev. B* **9**, 1216–1222.

Pethica, J. B. (1986). Comment on "Interatomic forces in scanning tunneling microscopy: Giant corrugations of the graphite surface." *Phys. Rev. Lett.* **57**, 3235.

Pethica, J. B., and Oliver, W. C. (1987). Tip-surface interactions in STM and AFM. *Physica Scripta* **T19**, 61–66.

Pethica, J. B., and Sutton, A. P. (1988). On the stability of a tip and flat at very small separations. *J. Vac. Sci. Technol. A* **6**, 2490–2494.

Piner, R., and Reifenberger, R. (1989). Computer control of the tunnel barrier width for the scanning tunneling microscope. *Rev. Sci. Instrum.* **60**, 3123–3127.

Plummer, E. W. (1975). Photoemission and field emission spectroscopy. *Interactions on Metal Surfaces*, edited by R. Gomer, Springer-Verlag, New York, 143–223.

Plummer, E. W., and Gadzuk, J. W. (1970). Surface states on tungsten. *Phys. Rev. Lett.* **25**, 1493–1495.

Plummer, E. W., Gadzuk, J. W., and Penn, D. R. (1975a). Vacuum-tunneling spectroscopy. *Physics Today,* April 1975, 63–71.

Pohl, D. W. (1986). Some design criteria in scanning tunneling microscopy. *IBM J. Res. Develop.* **30**, 417–427.

Pohl, D. W. (1986a). Dynamic piezoelectric translation devices. *Rev. Sci. Instrum.* **58**, 54–57.

Pohl, D. W. (1987). Sawtooth nanometer slider: A versatile low voltage piezoelectric translation device. *Surf. Sci.* **181**, 174–175.

Poirier, G. E., and White, J. M. (1989). A new ultra-high vacuum scanning tunneling microscope design for surface science studies. *Rev. Sec. Instrum.* **60**, 3113–3118.

Posternak, M., Krakauer, H., Freeman, A. J., and Koelling, D. D. (1980). Self-consistent electronic structure of surfaces: Surface states and surface resonances on W(001). *Phys. Rev. B* **21**, 5601–5612.

Promisel, N. E. (1964). *Tungsten, Tantalum, Molybdenum, Niobium and their alloys.* Pergamon Press, New York, page 293.

Quate, C. F. (1986). Vacuum tunneling: A new technique for microscopy. *Physics Today* August 1986, 26–33.

Rayleigh, J. W. S., Baron (1895). *Theory of Sound.* A paperback of this classic treatise was published by Dover Publications, New York, 1945.

Rieder, K. H., and Engel, T. (1979). Structural investigation of an adsorbate-covered diffraction: Ni(110) -(1 × 2) H. *Phys. Rev. Lett.* **43**, 373–376.

Roark, R. J., and Young, W. C. (1975). *Formulas for Stress and Strain,* fifth edition. McGraw–Hill, New York.

Robinson, H. J. (1984). A theorist's philosophy of science. *Physics Today,* March, 24–32.

Roetti, C., and Clementi, E. (1974). Simple basis sets for molecular wavefunctions containing atoms from $Z=2$ to $Z=54$. *J. Chem. Phys.* **60**, 4725–4729.

Rohrer, H. (1992). STM: 10 years after. *Ultramicroscopy,* **42–44**, 1–6.

Rose, J. H., Ferrante, J., and Smith, J. R. (1981). Universal binding energy curves for metals and bimetallic interfaces. *Phys. Rev. Lett.* **47**, 675–678.

Sacks, W., and Noguera, C. (1991). General expression for the tunneling current in scanning tunneling microscopy. *Phys. Rev. B* **43**, 11612–11622.

Sacks, W., and Noguera, C. (1991a). Beyond Tersoff and Hamann: A general formula for the tunneling current. *J. Vac. Sci. Technol. B* **9**, 488–491.

Sakai, A., Cardillo, M. J., and Hamann, D. R. (1985). He–Si(100) potential: Charge superposition and model structures. *Phys. Rev.* **33**, 5774–5781.

Salmink, H. W. M., Meier, H. P., Ellialtioglu, R., Gerritsen, J. W., and Muralt, P. R. M. (1989). Tunneling spectroscopy of the AlGaAs–GaAs heterostructure interface. *Appl. Phys. Lett.* **54**, 1112–1114.

Samsavar, A., Hirschorn, E. S., Leibsle, F. M., and Chiang, T. C. (1989). Scanning-tunneling-microscopy studies of Ag on Si(100)-(2 × 1). *Phys. Rev. Lett.* **63,** 2830–2833.

Samsavar, A., Hirschorn, E. S., Miller, T., Leibsle, F. M., Eades, J. A., and Chiang, T. C. (1990). High-resolution imaging of a dislocation on Cu(111). *Phys. Rev. Lett.* **65,** 1607–1610.

Sarid, D. (1991). *Scanning Force Microscopy,* Oxford University Press, New York.

Sawaguchi, E. (1953). Ferroelectricity versus antiferroelectricity in the solid solutions of $PbZrO_3$, and $PbTiO_3$. *J. Phys. Soc. Japan* **8,** 615–629.

Schottky, W. (1923). Über kalte und warme Electronenentlasungen. *Z. Physik* **14,** 63–75.

Schröder, J., Günter, C., Hwang, R. Q., and Behm, R. J. (1992). A comparative STM study of the growth of thin Au films on clean and oxygen-precovered Ru(0001) surfaces. *Ultramicroscopy* **42–44,** 475-482.

Schuster, R., Barth, J. V., Wintterlin, J., Bohm, R. J., and Ertl, G. (1992). Distance dependence and corrugation in barrier-height measurements on metal surfaces. *Ultramicroscopy* **42–44,** 533–540.

Selloni, A., Carnevali, P., Tosatti, E., and Chen, C. D. (1985). Voltage-dependent scanning-tunneling microscopy of a crystal surface: Graphite. *Phys. Rev. B* **31,** 2602–2605.

Shedd, G. M., and Russell, P. (1990). The scanning tunneling microscope as a tool for nanofabrication *Nanotechnology* **1,** 67–80.

Shih, C. K., Feenstra, R. M., and Martensson, P. (1990). Scanning tunneling microscopy and spectroscopy of thin metal films on the GaAs(110) surface. *J. Vac. Sci. Technol. A* **8,** 3379–3385.

Shih, C. K., Feenstra, R. M., and Chandrashekhar, G. V. (1991). Scanning tunneling microscopy and spectroscopy of BiSrCaCuO 2:2:1:2 high temperature superconductors. *Phys. Rev. B* **43,** 7913–7922.

Shockley, W. (1939). On the surface states associated with a periodic potential. *Phys. Rev.* **56,** 317–323.

Singh, D., and Krakauer, H. (1988). Bonding and reconstruction of the W(001) surface. *Phys. Rev. B* **37,** 3999-4006.

Simmons, J. G. (1963). Generalized formula for the electric tunnel effect between similar electrodes separated by a thin insulating film. *J. Appl. Phys.* **34,** 1793–1803.

Simmons, J. G. (1969). Image force in metal–oxide&ndas.metal tunnel junctions. *Tunneling Phenomena in Solids,* E. Burstein and S. Lundqvist, editors, Plenum, 135–148.

Slater, J. C. (1930). Atomic shielding constants. *Phys. Rev.* **36,** 57–65.

Slater, J. C. (1963). *Quantum Theory of Molecules and Solids, Vol. 1.* , McGraw-Hill, New York.

Slater, J. C., and Koster, G. F. (1954). Simplified LCAO method for the periodic potential problem. *Phys. Rev.* **94,** 1498 –1510.

Sleator, T., and Tycko, R. (1988). Observation of individual organic molecules at a crystal surface with use of a scanning tunneling microscope. *Phys. Rev. Lett.* **60,** 1418–1421.

Smith, D. P. E., and Binnig, G. (1986). Ultrasmall scanning tunneling microscope for use in a liquid-helium storage dewar. *Rev. Sci. Instrum.* **57,** 2630–2631.

Smith, D. P. E., Hörber, H., Gerber, Ch., and Binnig, G. (1989). Smectic liquid crystal monolayers on graphite observed by scanning tunneling microscopy. *Science* **245**, 43–45.

Smith, J. R. (1968). Self-consistent many-electron theory of electron work functions and surface potential characteristics for selected metals. *Phys. Rev.* **181**, 522–529.

Smith, N. V. (1978). Angular dependent photoemission, in *Photoemission in solids,* edited by M. Cardona and L. Ley, Springer-Verlag, Berlin.

Smith, N. V., and Mattheiss. L. F. (1976). Linear combination of atomic orbitals model for the electronic structure of H_2 on the W(001) surface. *Phys. Rev. Lett.* **37**, 1494–1497.

Soler, J. M., Baro, A. M., Garcia, N., and Rohrer, H. (1986). Interatomic forces in scanning tunneling microscopy: Giant corrugations of the graphite surface. *Phys. Rev. Lett.* **57**, 444–447.

Sonnenfeld, R., and Hansma, P. K. (1986). Atomic-resolution microscopy in water. *Science,* **232**, 211–213.

Sonnenfeld, R., Schneir, J., and Hansma, P. K. (1990). Scanning tunneling microscopy: A natural for electrochemistry. *Modern Aspects of Electrochemistry,* **21**, 1–28. Edited by R. E. White and J. O'M. Bockris, Plenum, Oxford.

Spence, J. C. H. (1988). *Experimental High-Resolution Electron Microscopy.* Oxford University Press, New York.

Steele, W. A. (1974). *The Interaction of Gases with Solid Surfaces,* Pergamon, Oxford.

Stoll, E. (1984). Resolution of the scanning tunneling microscope. *Surf. Sci.* **143**, L411–L416.

Stoll, E., Baratoff, A., Selloni, A., and Carnevali, P. (1984). Current distribution in the scanning vacuum tunneling microscope: a free-electron model. *J. Phys. C* **17**, 3073–3086.

Stroscio, J. A., Feenstra, R. M., and Fein, A. P. (1986). Electronic structure of the $Si(100)2 \times 1$ surface by scanning tunneling microscopy. *Phys. Rev. Lett.* **57**, 2579–2582.

Stroscio, J. A., Feenstra, R. M., and Fein, A. P. (1987). Local density and long-range screening of adsorbed oxygen atoms on the GaAs(110) surface. *Phys. Rev. Lett.* **58**, 1668–1671.

Stroscio, J. A., Feenstra, R. M., and Fein, A. P. (1987a). Imaging electronic surface states in real space on the $Si(111)2 \times 1$ surface. *J. Vac. Sci. Technol.* **A5**, 838–845.

Stroscio, J. A., Feenstra, R. M., Newns, D. M., and Fein, A. P. (1988). Voltage dependent scanning tunneling microscope imaging of semiconductor surfaces. *J. Vac. Sci. Technol.* **A6**, 499–507.

Swanson, L. W., and Crouser, L. C. (1966). Anomalous total energy distribution for a tungsten field emitter. *Phys. Rev. Lett.* **16**, 389–392.

Swanson, L. W., and Crouser, L. C. (1967). Total-energy distribution of field-emitted electrons and single-plane work functions for tungsten. *Phys. Rev.* **163**, 622–641.

Takada, Y., and Kohn, W. (1985). Interaction potential between a helium atom and metal surfaces. *Phys. Rev. Lett.* **54**, 470–472.

Takayanagi, K., Tanishiro, Y., Takahashi, S., and Takahashi, M. (1985). Structure analysis of $Si(111)$-7×7 reconstructed surface by transmission electron diffraction. *Surface Sci.* **164**, 367–392.

Tamm, I. (1932) Über eine mögliche Art der Elecktronenbindung an Kristaaoberflächen. *Phys. Z. Sowjetunion* **1**, 733–746.

Tang, K. T., Toennies, J. P., and Yiu, C. L. (1991). The exchange energy of H_2^+ calculated from polarization perturbation theory. *J. Chem. Phys.* **94,** 7266–7277.

Teague, E. C. (1978). Room temperature gold-vacuum-gold tunneling experiments. Thesis at North Texas University. Reprinted in *J. Research of the National Bureau of Standards* **91,** 171–233 (1986).

Tersoff, J., and Hamann, D. R. (1983). Theory and application for the scanning tunneling microscope. *Phys. Rev. Lett.* **50,** 1998–2001.

Tersoff, J., Cardillo, M. J., and Hamann, D. R. (1985). Sensitivity of helium diffusion to surface geometry. *Phys. Rev. B* **32,** 5044–5050.

Tersoff, J., and Hamann, D. R. (1985). Theory of scanning tunneling microscope. *Phys. Rev. B* **31,** 805–814.

Tersoff, J. (1986). Anomalous corrugation in scanning tunneling microscopy: Imaging of individual states. *Phys. Rev. Lett.* **57,** 440–443.

Tersoff, J. (1989). Sample-dependent resolution in scanning tunneling microscopy. *Phys. Rev. B* **39,** 1052–1058.

Tersoff, J. (1990). Role of tip electronic structure in scanning tunneling microscope images. *Phys. Rev. B* **41,** 1235–1238.

Tersoff, J., and Lang, N. D. (1990). Tip-dependent corrugation of graphite in scanning tunneling microscopy. *Phys. Rev. Lett.* **65,** 1132–1135.

Tiedje, T., and Brown, A. (1990). Performance limits for the scanning tunneling microscope. *J. Appl. Phys.* **68,** 649–654.

Timoshenko, S., and Goodier, J.N. (1970). *Theory of Elasticity.* McGraw-Hill, New York.

Timoshenko, S., and Young, D. H. (1962). *Elements of Strength of Materials.* Van Nostrand, Princeton.

Timoshenko, S., Young, D. H., and Weaver, W. (1974). *Vibration Problems in Engineering.* Wiley, New York.

Tinkham, M. (1964). *Group Theory and Quantum Mechanics,* McGraw-Hill, New York.

Tokumoto, H., Miki, K., Murakami, H., Bando, H., Ono, M., and Kajimura, K. (1990). Real-time observation of oxygen and hydrogen adsorption on silicon surfaces by scanning tunneling microscopy. *J. Vac. Sci. Technol.* **A8,** 255–258.

Tosch, S., and Neddermeyer, H. (1988). Initial stage of Ag condensation on Si(111) 7×7. *Phys. Rev. Lett.* **61,** 349–352.

Traum, M. M., Rowe, J. E., and Smith, N. E. (1975). Angular distribution of photoelectrons from (111) silicon surface states. *J. Vac. Sci. Technol.* **12,** 298–230.

Tromp, R. M., Hamers, R. J., and Demuth, J. E. (1985). Si(001) dimer structure observed with scanning tunneling microscopy. *Phys. Rev. Lett.* **55,** 1303–1306.

Tromp, R. M., Hamers, R. J., and Demuth, J. E. (1986). Atomic and electronic contributions to $Si(111)7 \times 7$ scanning tunneling microscopy images. *Phys. Rev. B* **34,** 1388–1391.

Tromp, R. M., Hamers, R. J., and Demuth, J. E. (1986a). Quantum states and atomic structure of Si surface *Science* **234,** 304–309.

Tromp, R. M., Hamers, R. J., Demuth, J. E., and Lang, N. D. (1988). Tip electronic structure in scanning tunneling microscopy. *Phys. Rev. B* **37,** 9042–9045.

Tsong, T. T. (1979). Quantitative atom-probe and field ion microscope studies at atomic resolution. *Direct Imaging of Atoms in Crystals and Molecules: Nobel Symposium 47,* The Royal Swedish Academy of Sciences, 7–15.

Tsong, T. T. (1987). Field ion microscopy and scanning tunneling microscopy: Similarities and differences. *Physica Scripta* **38**, 315–320.

Tsong. T. T. (1990) *Atom-Probe Field Ion Microscopy,* Cambridge University Press, Cambridge.

Tsong, T. T., and Kellogg, G. (1975). Direct observation of the direct walk of single adatoms and the adatom polarizability. *Phys. Rev. B* **12**, 1343–1353.

Tsukada, M., Kobayashi, K., and Isshiki, N. (1991). Effect of tip electronic structure on scanning tunneling microscopy/spectroscopy. *Surf. Sci.* **242**, 12–17.

Tsukada, M., Kobayashi, K., and Ohnishi, S. (1990). First-principle theory of the scanning tunneling microscopy simulation. *J. Vac. Sci. Technol. A* **8**, 160–165.

Udagawa, M., Umetani, Y., Tanaka, H., Itoh, M., Uchiyama, T., Watanabe, Y., Yokotsuka, T., and Sumita, I. (1992). The initial stages of the oxidation of Si(100)-2×1 studies by STM. *Ultramicroscopy,* **42–44,** 946–951.

van Loenen, E. J., Tromp, R. M., and Demuth, J. E. (1987). Local electron states and surface geometry of Si(111)-$\sqrt{3} \times \sqrt{3}$ Ag. *Phys. Rev. Lett.* **58**, 373–376.

van Wees, B. J. et al. (1988). Quantum conductance of point contacts in a two-dimensional electron gas. *Phys. Rev. Lett.* **60**, 848–850.

Varshni, Y. P. (1957). Comparative study of potential energy functions for diatomic molecules. *Rev. Mod. Phys.* **29**, 664–682.

Waclawski, B. J., and Plummer, E. W. (1972). Photoemission observation of a surface state of tungsten. *Phys. Rev. Lett.* **29**, 783–786.

Walmsley, D. G. (1987). Pre-microscope tunneling – Inspiration or Constraint? *Surf. Sci.* **181**, 1–26.

Wang, D. S., Freeman, A. J., Krakauer, H., and Posternak, M. (1981). Self-consistent linearized-argumented-plane-wave-method determination of electronic structure and surface states on Al(111). *Phys. Rev. B* **23**, 1685–1692.

Wang, S. C., and Tsong, T. T. (1982). Field and temperature dependence of the directional walk of single adsorbed W atoms. *Phys. Rev. B* **26**, 6470–6475.

Wang, S. C., and Ehrlich, G. (1988). Adatom diffusion on W(211): Re, W, Mo, Ir, and Rh. *Surf. Sci.* **206**, 451–474.

Weimer, M., Kramer, J., Bai, C., and Baldeschwieler, J. D. (1988). Tunneling microscopy of 2*H*-MoS$_2$: A compound semiconductor surface. *Phys. Rev. B* **37**, 4292–4295.

Weng, S. L. (1977). Surface resonances on the (100) plane of molybdenum. *Phys. Rev. Lett.* **38**, 434–437.

Weng, S. L., Plummer, E. W., and Gustafsson, T. (1978). Experimental and theoretical study of the surface resonances on the (100) faces of W and Mo. *Phys. Rev. B* **18**, 1718–1740.

Wengelink, H., and Neddermeyer, H. (1990). Oxygen-induced sharpening process of W(111) tips for scanning tunneling microscope use. *J. Vac. Sci. Technol. A* **8**, 438–440.

Wiesendanger, R., Güntherodt, H. J., Güntherodt, G., Gambino, R. J., and Ruf, R. (1990). Observation of vacuum tunneling of spin-polarized electrons with the scanning tunneling microscope. *Phys. Rev. Lett.* **65**, 247–250.

Wigner, E. (1930). The operation of time reversal in quantum mechanics. *Göttinger Nachrichten,* 133–146. See *Group Theory and Solid State Physics,* edited by P. H. Meijer, Gordon and Breach Science Publishers, 1964, New York, 265–278.

Willis, R. F., and Feuerbacher, B. (1975). Angular-resolved secondary-electron-emission spectroscopy of clean and adsorbate covered tungsten single-crystal surfaces. *Surf. Sci.* **53,** 144–155.

Wilson, R. J., and Chiang, S. (1987). Structure of the Ag/Si(111) surface by scanning tunneling microscopy. *Phys. Rev. Lett.* **58,** 369–370.

Wilson, R. J., and Chiang, S. (1987a). Registration and nucleation of the Ag/Si(111) surface by scanning tunneling microscopy. *Phys. Rev. Lett.* **59,** 2329–2332.

Wilson, R. J., Chiang, S., and Chambliss, D. D. (1990). Imaging semiconductors, metals and molecules with scanning tunneling microscopy. *Aust. J. Phys.* **43,** 393–400.

Wintterlin, J., Wiechers, J., Brune, H., Gritsch, T., Höfer, H., and Behm, R. J. (1989). Atomic-resolution imaging of close-packed metal surfaces by scanning tunneling microscopy. *Phys. Rev. Lett.* **62,** 59–62.

Wolf, E. L. (1985). *Principles of Electronic Tunneling Spectroscopy,* Oxford University Press, New York.

Wöll, Ch., Chiang, S., Wilson, R. J., and Lippel, P. H. (1989). Determination of atom positions at stacking-fault dislocations on Au(111) by scanning tunneling microscopy. *Phys. Rev. B* **39,** 7988–7991.

Wu, X. L., and Lieber, C. (1990). Direct characterization of charge-density-wave defects in titanium-doped $TaSe_2$ by scanning tunneling microscopy. *Phys. Rev. B* **41,** 1239–1242.

Yamada, H., Fujii, T., and Nakayama, K. (1988). Experimental study of forces between a tunnel tip and the graphite surface. *J. Vac. Sci. Technol. A* **6,** 293–295.

Yang, W. S., Li, Y., Mou, J., Yan, J. (1991). Atom-charge superposition calculation of STM images of glycine and alanine adsorbed on HOPG. *Ultramicroscopy* **42–44,** 1031–1036.

Young, R. D. (1959). Theoretical total-energy distribution of field-emitted electrons. *Phys. Rev.* **113,** 110–114.

Young, R. D. (1971). Surface microtopography. *Physics Today,* November 1971, 42–49.

Young, R. D., and Müller, E. W. (1959a). Experimental measurement of the total-energy distribution of field-emitted electrons. *Phys. Rev.* **113,** 115–120.

Young, R., Ward, J., and Scire, F. (1971). Observation of metal-vacuum-metal tunneling, field emission, and the transition region. *Phys. Rev. Lett.* **27,** 922–925.

Young, R., Ward, J., and Scire, F. (1972). The topografiner: An instrument for measuring surface microtopography. *Rev. Sci. Instrum.* **43,** 999–1011.

Zangwill, A. (1988). *Physics at Surfaces.* Cambridge University Press, Cambridge.

Zaremba, E., and Kohn, W. (1977). Theory of adsorption on simple and noble-metal surfaces. *Phys. Rev. B* **15,** 1769–1781.

Zeglinski, D. M., Ogletree, D. F., Beebe, T. P., Jr., Hwang, R. Q., Somorjai, G. A., and Salmeron, M. B. (1990). An ultrahigh vacuum scanning tunneling microscope for surface science studies. *Rev. Sci. Instrum.* **61,** 3769–3774.

Zener, C. (1930). Analytic atomic wave functions. *Phys. Rev.* **36,** 51–56.

Zeppenfeld, P., Lutz, C. P., and Eigler, D. M. (1992). Manipulating atoms and molecules with a scanning tunneling microscope. *Ultramicroscopy,* **42–44,** 128–133.

Ziman, J. M. (1972). *Principles of the Theory of Solids.* Cambridge University Press, Cambridge.

INDEX